On Integrating Unmanned Aircraft Systems into the National Airspace System

International Series on
INTELLIGENT SYSTEMS, CONTROL, AND AUTOMATION:
SCIENCE AND ENGINEERING

VOLUME 54

For further volumes:
www.springer.com/series/6259

Konstantinos Dalamagkidis • Kimon P. Valavanis •
Les A. Piegl

On Integrating Unmanned Aircraft Systems into the National Airspace System

Issues, Challenges, Operational Restrictions, Certification, and Recommendations

Second Edition

Springer

Konstantinos Dalamagkidis
Technische Universität München
Institut für Informatik VI
Boltzmannstraße 3
85748 Garching bei München
Germany
dalamagkidis@tum.de

Dr. Les A. Piegl
Dept. Computer Science & Engineering
University of South Florida
E. Fowler Ave. 4202
33620 Tampa Florida
USA
piegl@csee.usf.edu

Kimon P. Valavanis
University of Denver
Dept. Electrical & Computer Engineering
S. York Street 2390
Clarence M. Knudson Hall
80208 Denver Colorado
USA
kimon.valavanis@du.edu

ISBN 978-94-007-2478-5 e-ISBN 978-94-007-2479-2
DOI 10.1007/978-94-007-2479-2
Springer Dordrecht Heidelberg London New York

Library of Congress Control Number: 2011940001

Cover design: VTeX UAB, Lithuania

Printed on acid-free paper

Springer is part of Springer Science+Business Media (www.springer.com)

To my family, and the friends who have helped me retain my sanity over the years.
Konstantinos Dalamagkidis

To Stellitsa and Panoulis, my kids.
Kimon P. Valavanis

Dedicated to my loving wife, Kae.
Les A. Piegl

Preface to the Second Edition

Almost two years have passed since the publication of the first edition of the book you are holding. During this time a lot has been accomplished and many things have changed. Nevertheless, there are many that will argue that what has been done is still not enough. It has been a year since the small Unmanned Aircraft Systems (UAS) rulemaking committee submitted its recommendations to the Federal Aviation Administration and although its term ended in the end of 2009, no term extension was given up until the time these lines are written and no policy is expected to be available before 2013. Moreover, certification of civilian UAS in the restricted category is not expected before 2015 and routine access to the NAS for at least five years after that. The challenges identified two years ago, largely remain. Sense and avoid remains an important issue, as is communications latency, security and reliability. It is still not possible to fly in the National Airspace with similar procedures and/or restrictions as those that are currently in effect for manned aviation.

The gap in UAS regulation remains, albeit there are signs from all over the world that this is starting to close. Regardless, one must also consider another gap that exists between the civilian market and universities, research centers and companies developing the UAS of the future. Although there are several civilian applications enumerated in the literature, the civilian market seems mostly oblivious to the benefits of incorporating UAS in its business. There are several reasons for this: UAS are still seen as immature technology; the partners that develop this technology are segregated from the actual civilian market; the lack of regulation lowers interest in UAS, which in turn reduces the pressure for developing regulation faster. In fact, FAA predicts that commercial activities will begin only in 2018 and expects that no extraordinary demand will bring about a quicker introduction of UAS in the NAS.

On the other hand, at any given moment there are over 30 UAS flying over Iraq and Afghanistan, many operated via satellite from the other side of the globe. All together, 135,000 hours were flown by UAS, over these two countries in 2009 alone and the one million cumulative flight hours mark has already been reached. Several UAS are also armed and have deployed weapons on multiple occasions, drawing the criticism of the international community while doing so. The use of UAS in the military domain is expected to result in significant savings in the long-term due to lower acquisition costs, lower fuel consumption and reduced requirements for personnel overseas. We are rapidly reaching the point where military UAS will take over the role of the protagonist from their manned counterparts, as shown by the accumulated flight hours and the announced procurement programs.

In civilian airspace UAS have been extensively used by US Customs and Border Patrol (CBP) where they assisted with over 4,000 arrests and the seizure of approximately 20,000 pounds of illegal drugs between September 2005 and March 2008. Other government agencies have introduced UAS into their activities both in the US and Europe. UAS have already been used successfully for ensuring the security of important events. On the technology front, flight endurance records are broken repeatedly, the Aerovironment Puma, a backpackable UAS, achieved over 9 hours of flight time, the A160T unmanned helicopter edged close to 19 hours, the Ion Tiger accomplished 26 hours in November of 2009 using a fuel cell engine and the solar-powered Zephyr was airborne for over 82 hours in October of 2009 and two weeks in July of 2010! In the meantime DARPA is aiming towards systems that can stay aloft for 5 years! And the advances are not limited to endurance. UAS performance is improving constantly, coupled with increases in capabilities and autonomy.

On a similarly optimistic note, the Teal Group is now forecasting a market worth $62bn worldwide with almost 8% annual compound growth rate over the 2009–2018 period. The capacity for growth is also evident by the number of acquisitions and partnerships between UAS developers that occurred in the last three years. Hopefully the tide is finally turning and although military applications have been the driving force behind this change, civil applications will start to emerge. About 50 countries are currently producing close to 1,000 different UAS models. Some of these models are aimed at non-military applications like earth monitoring or law enforcement. Even with restrictions and problems, UAS are finally taking to the (civil) skies!

For this second edition of the book several things have changed and almost all chapters include major updates and corrections. Nevertheless, the largest revisions are focused in the following three chapters, as well as a new appendix chapter:

Chapter 2: The historical overview has been improved to incorporate information on the early drone designs and how they finally led to the UAS we know today. In addition to that, new high quality photos were used where possible.

Chapter 4: The Chapter on Unmanned Aircraft regulations has been updated to reflect the latest information on the topic. Information on light/small UAS has been moved to a separate section, to highlight the differences of such systems and their importance in the near-term developments.

Chapter 5: This Chapter has been reorganized, rewritten and enhanced with new material. It is now more than twice the length with a new material layout, which the authors believe is both more clear and more comprehensive. To improve the presentation, the Section containing case studies has been moved to a separate chapter.

Appendix A: As a complement to Chapter 5, a chapter on human vulnerability has been added to the appendix that provides an overview of estimating expected injury severity given different trauma mechanisms. This Chapter is accompanied by 6 figures and 3 tables that assist the reader with understanding and comparing the different human vulnerability models available.

The authors are also grateful to all the comments and corrections submitted by readers of the first edition. Without their diligence and willingness to help, this book would be much poorer.

Last, but not least, the authors acknowledge wholeheartedly the continuous support of the Springer group and in particular the support of our Editor, Ms. Nathalie Jacobs.

Munich, Germany
Denver, Colorado
and Tampa, Florida
July 2010

Konstantinos Dalamagkidis
Kimon P. Valavanis
Les A. Piegl

Foreword

When first asked by Kimon to write a foreword to this monograph I felt honored to be selected and began my review of the manuscript with an eye to it's potential contribution to the UAS industry which is currently trying to gain access to the National Airspace System. Due to the nature of the document I soon found myself conflicted on precisely what the contribution might ultimately be. One goal of this publication it appeared was to establish a baseline summary of the state of the industry from airframe/power-plant variability and hence mission complexity perspective. It also organized critically important information related to the United States and International regulatory status and led to suggestions regarding how to overcome remaining hurdles. This, the authors have organized very precisely and place the document in a position for periodic review on regulatory progress. As the results of the current Aviation Rule Committee addressing Small UAS Airworthiness Certification and Operational considerations (sUAS ARC), these results can easily be incorporated into addenda or manuscript revisions.

Another key contribution is the review of UAS flight safety from the perspective of risk (which is defined in early on) one component of which is the level of damage by impact on the ground coupled with the likelihood of an event happening (probability of occurrence), and parsed with respect to flight over areas of varying population density. This was the first comprehensive analysis of risk since the seminal studies by Roland Weibel and Dr. Jim Hansman of MIT. The problem is that this industry has no mechanism yet, is too immature, to have dedicated peer review processes associated with standard journals. What weight or significance then can we apply to the data and interpretation/analysis that the authors have brought forward?

I took it upon myself, on an ad hoc basis, to ask industry colleagues to review the manuscript, and the scientific papers presented at conference by Kimon and Konstantinos, which are used as a basis for the data presented, and to provide feedback addressing the validity of assumptions, models, the data paradigm and workup — in essence a "poor mans peer review".

Several authorities weighed in with comments and suggestions. I am pleased to report that the reviewers report no significant issues with the data models and experimental paradigm used. This is significant because the reported data indicate that the earlier data and analysis by the MIT group was "overly conservative" and UASs of varying weight classes, would pose less risk that what has been reported by earlier studies. (I had intended to name the reviewers in the spirit of full disclosure, however regrettably, publishing deadlines interfered with my ability to get permission to include names and also to acquire permission from their employers to be quoted.)

Further, the data contained within has been submitted to several members in the leadership of the sUAS ARC and it is hoped that it will be given due consideration as the ARC reports in with it's suggested guidance to the FAA as it promulgates sUAS safety regulations beginning with small UASs.

In general I am very pleased that Kimon and his colleagues have put this summary document together for the UAS community and I look forward to semi annual or annual updated to what may become an ongoing, living, compilation of the UAS State of the Industry.

Oyster Bay, NY, September 2008

James E. Jewell
President UAV MarketSpace Inc.
ASTM F38 UAS Executive Committee Membership Secretary

Preface

Unmanned Aircraft Systems (UAS) research and development and procurement, as well as diverse UAS related activities are increasing rapidly, worldwide. UAS have seen unprecedented levels of growth on all fronts over the past ten years; however, the best is yet to come!

What has been an amazing fact is that although in 1997 the total income of the Unmanned Aerial Vehicle (UAV) global market was about $2.27 billion[1], it has been argued that until 2015 the UAV market in the US, as a whole, will reach $16 billion, with Europe as a continent playing the role of the second but distant competitor, spending just about €2 billion[2]. However, a study conducted by the Teal Group[3] claims that UAVs will continue to be the most dynamic growth sector of the world aerospace industry, estimating that UAV spending will more than triple over the next decade, totaling close to $55 billion. An interesting conclusion that the Teal Group reached was that the civil UAV market will slowly emerge over the next decade, starting first with government organizations requiring surveillance systems similar to military UAVs such as coast guards, border patrol organizations and similar national security organizations.

Surprisingly enough, it is this conclusion that, coupled with major initiatives to push for civilian and public use UAS, motivated the authors to write this book. The motivation and rationale becomes more than obvious when one considers that utilization of UAS for civilian applications requires that they fly in civilian, restricted, space, that is, it requires that UAS be integrated in to the National Airspace System (NAS) of the country or continent they fly over.

The challenge is huge because all that is available today, worldwide, reflects manned aviation. Efforts by national and international organizations to produce and develop rules, regulations, procedures and standards for integration of UAS in to

---

[1] "World Markets for Military, Civil and Commercial UAVs: Reconnaissance UAVs and Aerial Targets", Frost & Sullivan, 1998

[2] Dickerson L., "UAVs on the Rise", Aviation Week & Space Technology, Aerospace Source Book 2007, Vol. 166, No. 3, 2007.

[3] http://www.roboticstrends.com/displayarticle880.html, 09/06, Robotics Trends.

the national airspace are on the rise, but, development of a detailed and complete roadmap is far from being complete.

In essence, integration of UAS in the NAS will depend, among other things, on whether UAS achieve, at a minimum, an equivalent level of safety to that of manned aviation. This safety level is defined primarily in terms of the risk posed by UAS operations to human life even though other potential collateral damage may be taken into account.

The authors hope that this project, being far from complete, will serve as a reference text, perhaps a useful tool, which lays the foundation for what needs be done in order to arrive at the desired outcome: integrated manned and unmanned aviation flying in the same airspace.

As such, current manned aviation regulations are reviewed, followed by available unmanned aviation regulations worldwide. UAS safety assessments and functional requirements are presented, which lead to derivation of equivalent levels of safety based on those used for manned aviation. Recommendations for a roadmap that leads to UAS integration in to the national airspace system are also presented.

Last, but not least, it is expected that as the field matures and the roadmap is more concrete, this book will be updated in order to serve as a 'reference manual' or 'handbook'.

Tampa, FL
September 2008

Konstantinos Dalamagkidis
Kimon P. Valavanis
Les A. Piegl

Acknowledgements

Writing this book felt like 'a project different than any other' we have dealt with so far. This was a strange feeling because of the 'unknown outcome' of the book, at least in the very early stages. We felt we were trying to produce 'something' that had no starting and ending point. Later on, we felt like we were involved in a very fast race, trying to catch up with what was going on. This, hopefully, gives the reader an idea of how fast things are moving in the UAS arena, and provides an indication, despite major obstacles that need be overcome, of the number of joint activities and efforts related to creating a solid roadmap that will eventually lead to the integration of UAS in to civilian airspace. To make this point even stronger, one should not be surprised if by the time this book is published, a revised edition will be needed.

Regardless of the challenge, many people encouraged the authors to complete this project and offered their advice, suggestions, recommendations and material to include in the book.

We want to thank wholeheartedly Dr. R. Wallace, Mrs. V. Wallace and Mr. D. Schultz for supporting and funding our research through their organizations and for providing valuable insight on what such a book should include and how it should be presented.

We are grateful to our sponsors from ARL, ARO and SPAWAR, Dr. S. Wilkerson, Dr. R. Zachery, Dr. J. Besser for believing in our research program and giving us the opportunity to explore new research avenues.

Dr. G. Vachtsevanos played a key role throughout this project guiding us and serving as the 'reader' who wants to know who is doing what and how. Along the same lines, Mr. J. Jewell, previously Vice Chair, ASTM International UAS Committee F38 and currently CEO of XUAS LLC, provided very important information related to recent UAS developments, rules, regulations, policies, safety, and at the same time served as a promoter of this project.

This book would not be complete without the major help provided by Peter van Blyenburgh, President, UVS International; Peter forwarded to us a lot of information from his database, including details related to the UAS global perspectives presented in the 2007 and 2008 meetings he organized in Paris, France. In addition, he gave us permission to include in the book, in Appendix B, the most recent list of existing and under development UAS, worldwide.

The authors also wish to thank Dr. Xiaogong Lee and Dr. Bruce Tarbert from the Federal Aviation Administration (FAA) for their encouragement and support in writing this book.

In addition, the authors acknowledge the administrative support provided by Archytas Unmanned Systems LLC.

The authors acknowledge the support of their colleagues at the University of South Florida, Dr. A. Kandel, Dr. E. Stefanakos, Dr. W. Moreno, Dr. C. Ferekides and Dr. A. Yalcin for their encouragement in completing this book. On a personal note, the second author wishes to also acknowledge the Dean of the School of Engineering and Computer Science of the University of Denver, Dr. Rahmat Shoureshi, not only for being enthusiastic about this book, but also for allowing him to complete it, giving him extra time before joining the Department of Electrical and Computer Engineering as Professor and Chair. The first author would also like to acknowledge the invaluable help of Stelios Ioannou for being an excellent sounding board and always there when things got tough or frustrating.

Two of the authors, Dr. Valavanis and Dr. Piegl feel that it is important to state that the first author, Kostas Dalamagkidis is the driving force of this project, which is, on top of and in addition to completing his PhD dissertation research. It is really fun and an honor to have such dedicated and talented students and near future colleagues to work with. They make our lives easier.

Last, but certainly not least, the authors want to thank their Publisher from Springer, Ms. Nathalie Jacobs and her group. Nathalie has been a very strong supporter of our projects; she has gone the extra mile to make everything possible. Nathalie, we thank you, and whenever you have time, we will show you how people live in the Greek Islands — that is a promise.

Contents

Chapter 1
Introduction

Prediction is very difficult, especially about the future.
Niels Bohr

We always overestimate the change that will occur in the next two years and underestimate the change that will occur in the next ten.
Bill Gates

The scope of this Chapter is twofold: It provides a combined overview of the motivation and rationale for writing this book, and also publishing this Second Edition. Since many names have been used for unmanned aircraft, the Chapter begins with an overview of these terms and clarifies on the nomenclature that will be used in the rest of the book. This section is followed by an updated, general and non technical discussion about unmanned aircraft. It supports and justifies the need for such a book, even though the road to fully integrating unmanned aircraft in to the National Airspace System (NAS) remains long and uncertain. Observation of the 2009 value of military Unmanned Aircraft System (UAS) procurement and the worldwide UAV market forecasts for 2010–2018 gives more than sufficient evidence for the exponential growth and utilization of unmanned aircraft. Examples of current use of UAS for non-military applications are demonstrated, justifying the need for civilian operations. Sect. 1.3 presents the main concepts behind safety regulation that also give an indication of the difficulties involved. This first Chapter also provides a summary of what follows in subsequent Chapters and how the material is organized and presented. In the First Edition, some fundamental definitions related to aviation in general and unmanned aircraft in particular were given, which are now presented in a dedicated Glossary section before the Appendices.

1.1 What's in a name ... UAV or UAS?

Several names have already been used to describe unmanned aircraft. UAVs became UAS, the preferred term used by the Federal Aviation Administration (FAA). Other names included Remotely Piloted Vehicles (RPVs), a term that was used in the Vietnam War, while the US military also calls them Remotely Piloted Aircraft (RPA), a term used to include both the aircraft and the pilot [9], while the U.K. has designated them as Remotely Piloted Air System (RPAS), to demonstrate the presence of the man-in-the-loop to control them [7].

K. Dalamagkidis et al., *On Integrating Unmanned Aircraft Systems into the National Airspace System*, Intelligent Systems, Control and Automation: Science and Engineering 54, DOI 10.1007/978-94-007-2479-2_1,

An unmanned aerial vehicle (also known as a drone) refers to a pilotless aircraft, a flying machine without an on-board human pilot or passengers. As such, 'unmanned' implies total absence of a human who directs and actively pilots the aircraft. Control functions for unmanned aircraft may be either on-board or off-board (remote control).

The term UAV or Unmanned Aerial Vehicle has been used for several years to describe unmanned aerial systems. Various definitions have been proposed for this term, like [11]:

> A reusable[1] aircraft designed to operate without an onboard pilot. It does not carry passengers and can be either remotely piloted or preprogrammed to fly autonomously.

A few years ago, the US Department of Defence (DoD), followed by the FAA and the European Aviation Safety Agency (EASA), adopted the term UAS or Unmanned Aircraft System. This was meant to signify that UAS are aircraft and as such airworthiness will need to be demonstrated and they are also systems consisting of ground control stations, communication links and launch and retrieval systems in addition to the aircraft itself.

The FAA has defined an *Unmanned Aircraft* or UA as [6]:

> A device used or intended to be used for flight in the air that has no onboard pilot. This includes all classes of airplanes, helicopters, airships, and translational lift aircraft that have no onboard pilot. Unmanned aircraft are understood to include only those aircraft controllable in three axes and therefore, exclude traditional balloons.

As a comparison, the definition of *Unmanned Vehicle* given in the 2007–2012 Unmanned Systems Roadmap is also provided [15]:

> A powered vehicle that does not carry a human operator, can be operated autonomously or remotely, can be expendable or recoverable, and can carry a lethal or nonlethal payload. Ballistic or semi-ballistic vehicles, cruise missiles, artillery projectiles, torpedoes, mines, satellites, and unattended sensors (with no form of propulsion) are not considered unmanned vehicles. Unmanned vehicles are the primary component of unmanned systems

Similarly, EASA defines UAS as [3]:

> An Unmanned Aircraft System (UAS) comprises individual system elements consisting of an "unmanned aircraft", the "control station" and any other system elements necessary to enable flight, i.e. "command and control link" and "launch and recovery elements". There may be multiple control stations, command & control links and launch and recovery elements within a UAS.

In this book the term UA will be used to refer to an unmanned aircraft, while the term UAS will be used in the instances where other parts of the system like the control station are relevant. The same terms will be used when referring to one or multiple systems.

---

[1] The characterization reusable is used to differentiate unmanned aircraft from guided weapons and other munition delivery systems.

1.2 Motivation and Rationale

Regardless of name, UAS have demonstrated repeatedly major potential for diverse applications in military, civilian and public domains. Unfortunately, with the exception of military applications, this great potential has not been yet exploited and utilized to the maximum, particularly in civil or public domains; this happens mainly because of the limited and/or restrictive 'regulatory framework' currently in place in many countries around the world.

Although UAS for civilian/public domain and other commercial applications are expected to be dominant in the future, one should keep in mind that today UAS have basically revolutionized warfare, and it has been forecasted that 40% of manned military aircraft missions will be undertaken by UAS by 2012 [7]. It is estimated that over 1,200 UAS types have been developed, primarily in the US and Europe, as shown in Fig. 1.1. The same figure shows the value of military UAS procurement in 2009, while Figs. 1.2 to 1.5 show projected numbers of UAS for 2010–2018 worldwide.

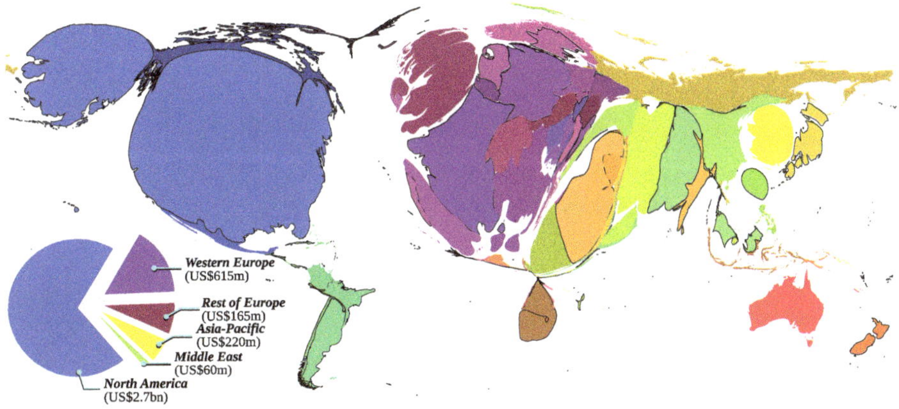

Fig. 1.1 A map of the number of UAS types developed per country. It is evident that the US is leading in number of types (about one third of the total) with Europe following close behind. There are also UAS types that have been developed by international collaborations and as such do not appear on this map. On the lower left corner, the global military UAS procurement in 2009 is provided. Data source: [1, 7]

The unquestionable prerequisite to introducing and/or developing any such (proposed, or) required regulation, is a safety assessment of the design, manufacture, operation and maintenance processes of UAS, which will lead to appropriately defined requirements. It is postulated that what makes this objective feasible, albeit a very difficult one, is that with the exception of UAS operations, the other three processes are essentially the same for both UAS and other manned aircraft. Nevertheless, the need for investigating and evaluating operational safety requirements of UAS is of paramount importance. However, it is essential to understand that when

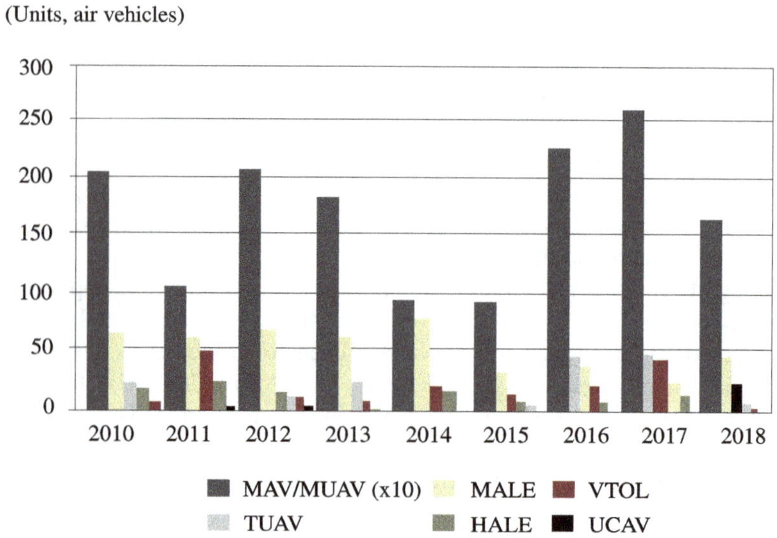

Fig. 1.2 UAS market forecast for North America. Source: [7]

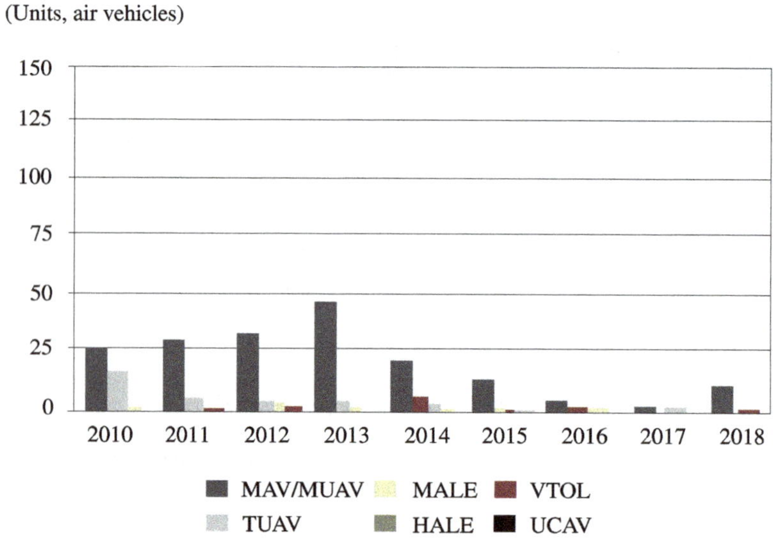

Fig. 1.3 UAS market forecast for South America. Source: [7]

one focuses only on safety levels in general, such levels have so far been considered under certain assumptions that may not apply to UAS. For example, the risk to people on the ground is traditionally evaluated assuming point-to-point operations and that a significant portion of the flight time is spent over less densely populated areas [4]. But this same assumption does not hold for UAS, since several applications require loitering over specific areas. It is reasonable, if not very obvious, to consider

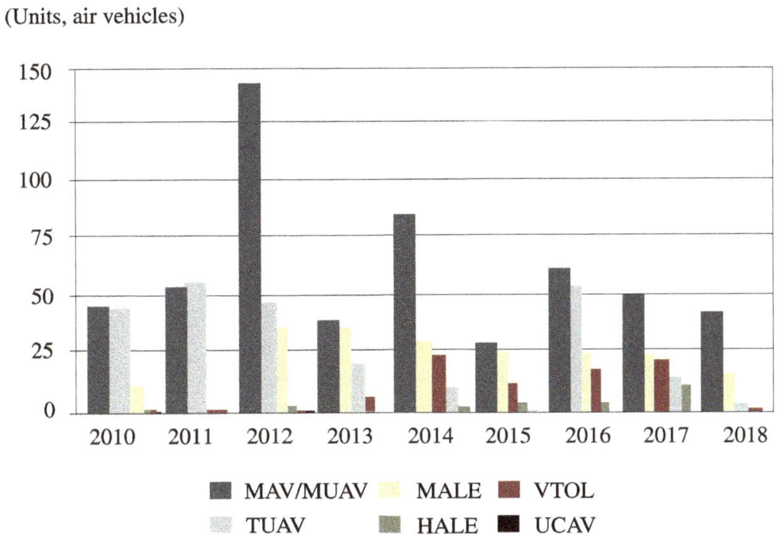

Fig. 1.4 UAS market forecast for Europe. Source: [7]

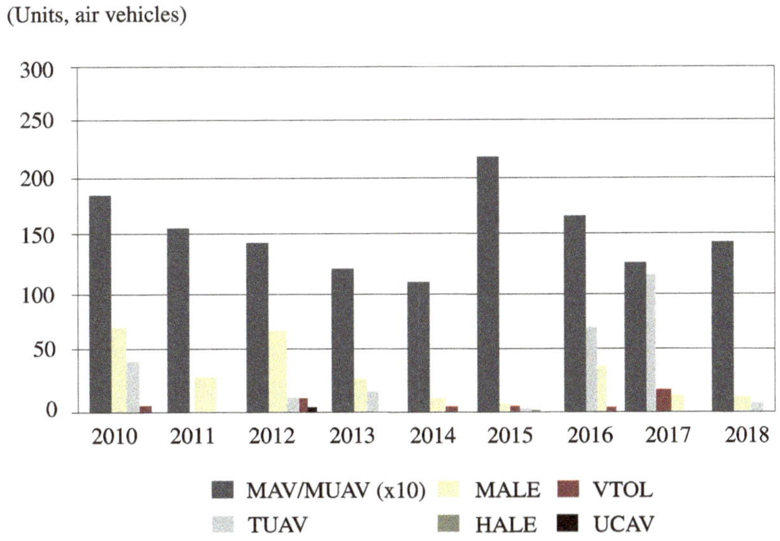

Fig. 1.5 UAS market forecast for Asia and Pacific. Source: [7]

that depending on the population of such areas, current requirements may be over conservative or inadequate. Other differences between UAS and manned aircraft also need be taken into account when drafting regulations. Hence, it is necessary for these differences to be identified and their potential impact quantified, which is part of the rationale behind this book.

Further, in order to define UAS operational reliability requirements, a safety assessment is needed. Such a safety assessment should include drafting a preliminary hazard list, determining failure modes and possible outcomes, defining acceptable risks and deriving target levels of safety (TLS), or equivalent levels of safety (ELOS) to that of manned aviation. The latter has proved a very difficult elusive thing to define, generating a lot of discussion between stakeholders. According to Ed Herlik, lead analyst of the Market Intel Group [9]:

> ... Equivalent Level of Safety: This phrase is deceptive. The FAA, for example, insists that UAVs fly with a level of safety equivalent to the "see and avoid" standard required of pilots. As a practical matter, however, pilots are notoriously bad at seeing other aircraft, especially private pilots. And no commercial aircraft allows pilots any view at all to the rear, below or even very far behind the wingtips to the sides. Which airliner has a rear view mirror? The bottom line? While this standard sounds reasonable, it is actually far more difficult to achieve than is the level of safety actually delivered by today's manned aircraft. UAVs are being held to a higher standard in reality.

Nevertheless, this procedure should ultimately lead to minimum acceptable UAS operational reliability requirements. Deriving and recommending general safety levels and reliability requirements for UAS that are at least as conservative as the ones used for manned aircraft is a rather challenging task.

Currently, most UAS applications and use focus on military domains, with several systems being in service and more under active development. Over the last decade, benefits of UAS use in civil application domains have been noticed by the public sector to the point where several organizations/agencies (including NASA, the US Coast Guard, Customs and Border Protection (CBP), Department of Homeland Security (DHS), Department of Agriculture, as well as local law enforcement agencies) are launching initiatives to introduce UAS in their infrastructure [16]. It is noteworthy that the CBP has employed UAS for several years and has reported that its systems helped with over 4,000 arrests between September 2005 and March 2008 [14]. Encouraging results have also been reported by the US Geological Survey and the US Forest Service after using a UAS to study renewed volcanic activity at Mount St. Helens, Washington in an environment that was particularly hazardous [14].

However, despite significant interest for commercial applications, efforts in that area are limited, mainly because of very strict and prohibitive FAA regulations that do not allow for this kind of operations. Moreover, because of lack of regulations, current UAS operations may be based on the wrong interpretation of FAA policies as admitted by the FAA in [5]. Although the examples above are predominantly about the US and FAA, similar needs and issues are present in all countries involved in UAS activities. As a result, it is essential not only to review the current regulatory status and existing airworthiness certification avenues available, but also evaluate any future possibilities that may arise, allowing UAS operators to fly lawfully as well as safely in the NAS.

1.3 On Regulating Safety

The need to regulate civil aviation ensuring safety and healthy competition dates back to the 1920s, with several relevant conventions addressing such issues and concerns. The most significant such convention took place in Chicago in 1944, right after the end of the Second World War with more than fifty States attending. The accomplishments of that conference set the groundwork for aviation safety and international cooperation on regulations, standards and procedures development, all relevant even to this day. Attending States also founded the International Civil Aviation Organization (ICAO) as a means to secure progress accomplished during the conference, as well as future cooperation [10].

Although UAS operations were very limited before the 1944 Chicago Convention, Article 8 refers specifically to pilot-less aircraft [12] and provisions within still apply to current systems. Some of those provisions are that a UAS cannot fly over another State without special authorization by that State (Article 8); UAS are required to bear registration marks (Article 20) and they must have a certificate of airworthiness (Article 31) [12]. It should be noted that the Chicago Convention applies to civil aircraft and as a result, UAS used in military or law enforcement services may have additional restrictions [12].

The issue of UAS safety, regardless of the specifics it entails and the meaning assigned to it by national or international agencies and organizations, requires special attention, since it will determine the development of UAS airworthiness requirements; in short, it will determine how soon UAS will fly in civilian space. Even though this is studied in detail throughout the book, some fundamentals are also presented in this Section. There are two approaches to defining UAS safety and airworthiness requirements. The first is to determine acceptable levels of risks to third parties. This is usually quantified as the number of fatalities and/or injuries per hour of flight or as an accident rate. It should be noted that the former metric is not an intrinsic characteristic of the platform, since it also depends on the type, frequency and duration of the missions [8]. As a result, application of this method to the commercial sector where UAS roles can change frequently, presents difficulties. On the other hand the use of the accident rate may penalize lighter or smaller vehicles, since after an accident involving such vehicles a lower number of fatalities is usually expected. Regardless of the metric used, this approach has the advantage of allowing UAS to fly without full compliance with a comprehensive code of requirements [12], but at the expense of posing operational restrictions.

The second approach is to produce a code of requirements, usually in the form of standards, for various UAS subsystems and for all stages of its design the final system must adhere to [8]. The advantage of this method is that complete recertification of a system is not required when its mission or one of its subsystems changes. It also allows type certification procedures for UAS similar to manned aircraft instead of a lengthy airworthiness examination of each UAS. This is the primary approach taken by regulatory bodies for drafting requirements for civil, manned aircraft. It should be noted that even in this case, there are provisions that define safety levels used to evaluate new technologies or designs that are not covered by existing code [12]. These requirements can be found in paragraph 1309 of current certification specifications for aircraft and provide a "safety net" by setting a minimum allowed safety performance, the rest of the regulations notwithstanding.

Regardless of the specifics of the approach, the primary intent of current flight regulations has been to reduce the probability of harm to third parties as required by ICAO Annex 8 and the Chicago Convention [8, 12]. Nevertheless new standards are drafted with the safety of the passengers and crew as their goal, under the assumption that it will also reduce the risk to people on the ground [2, 12]. In contrast to their manned counterparts, unmanned systems only pose a risk to people on the ground and a smaller risk to people on board other aircraft from a midair collision. In fact sacrificing the system to avoid fatalities can be an acceptable policy. As a result regulations need reflect this characteristic.

1.4 Book Objectives and Outline

The underlying idea behind writing this book matches the FAA's philosophy that is best stated in presentations by Mr. N. A. Sabatini, Associate Administrator for Aviation Safety, reflecting the "First, do no harm" principle of medicine's Hippocratic Oath, applied to UAS when integrated in to the NAS [13].

Starting from the "First, do no harm" principle, the book objectives focus on:

1. Presenting existing and current manned aviation regulation related to airworthiness certification, pilot certification, operational rules and airspace classes.
2. Discussing UAS regulations and their current status within the US.
3. Presenting and summarizing UAS regulatory efforts at the international level, putting emphasis in the European Union (EU), Canada, Japan and Australia.
4. Providing a detailed safety assessment and functional requirements for UAS that may be used to develop a roadmap for integrating UAS in to civilian airspace.
5. Demonstrating through case studies completed for a wide range of UAS families, where and how UAS can fly worldwide, once safety levels are satisfied.

The book is composed of eight Chapters and three Appendices.

Chapter 2 presents a short review on aviation history and UAS, including a pictorial overview of 50 modern systems of different origins, sizes and designs.

Chapter 3 describes details of current manned aviation regulation, airworthiness certification, special aircraft categories, pilot certification, federal aviation requirements operation rules, airspace classes and regulation development models.

Chapter 4 provides an overview of the history and current status of UAS airworthiness and operational regulation worldwide. Existing regulations have been developed considering the need for a complete regulatory framework for UAS.

As such, national aviation authorities and international organizations are preparing roadmaps, airworthiness and design standards as well as policies in collaboration with academia and industry.

Chapter 5 focuses on UAS safety assessment and functional requirements. This is achieved in terms of defining an "Equivalent Level of Safety", or ELOS, with that of manned aviation, specifying what the ELOS requirement entails for UAS regulations. To accomplish this, the safety performance of manned aviation is first evaluated, followed by models to derive reliability requirements for achieving target levels of safety (TLS) for ground impact and mid-air collision accidents. The chapter also includes thoughts on how to select appropriate models, as well as on design requirements and risk mitigation measures.

Chapter 6 presents some case studies employing the models presented in the previous chapter.

Chapter 7 discusses elements of a viable roadmap leading to UAS integration into the NAS. Key differences between manned and unmanned aviation are presented, followed by information and recommendations provided on key issues like the development of a risk reference system for UAS, classification for regulatory purposes and certification of systems and operators. The Chapter concludes with a discussion of technology-related issues that will need to be adequately resolved before UAS can enjoy unrestricted access to the NAS. When applicable, existing regulations for manned aviation are adapted; however, the need for new rules, procedures and regulations is also essential.

Chapter 8 talks about the road ahead.

Appendix A is a new appendix in this Edition. It presents human vulnerability models that can be taken into account when evaluating risk of UAS operations.

Appendix B provides a sensitivity analysis of the model proposed by the authors in Chapter 5.

Appendix C lists current and older UAS in development, in production and/or in service.

References

1. van Blyenburgh P (2009) UAS Yearbook 2009/2010, UVS International
2. Clothier R, Walker R, Fulton N, Campbell D (2007) A casualty risk analysis for unmanned aerial system (UAS) operations over inhabited areas. In: Proc. 12th Australian International Aerospace Congress and 2nd Australasian Unmanned Air Vehicles Conference, Melbourne, Australia
3. European Aviation Safety Agency (2009) Airworthiness certification of Unmanned Aircraft Systems (UAS). Policy Statement, E.Y013–01

4. European Aviation Safety Agency (EASA) (2005) A–NPA, No. 16/2005, policy for unmanned aerial vehicle (UAV) certification
5. Federal Aviation Administration (2007) Unmanned aircraft operations in the national airspace system. Docket No. FAA-2006-25714
6. Federal Aviation Administration (2008) Unmanned aircraft systems operations in the U. S. national airspace system. Interim Operational Approval Guidance 08-01
7. Fishpool M (2010) International military and civilian Unmanned Aerial Vehicle survey. Tech. rep., Socolofi Research
8. Haddon DR, Whittaker CJ (2002) Aircraft airworthiness certification standards for civil UAVs. UK Civil Aviation Authority
9. Herlik E (2010) Unmanned Aerial Vehicles (UAVs) for commercial applications global market & technologies outlook 2011–2016. Tech. rep., Market Intrel Group LLC
10. International Civil Aviation Organization (ICAO) (2007) Webpage of. [online] http://www.icao.int
11. Joint Capability Group on Unmanned Aerial Vehicles (2007) STANAG 4671 - Unmanned Aerial Vehicle Systems Airworthiness Requirements (USAR). draft, NATO Naval Armaments Group
12. Joint JAA/Eurocontrol Initiative on UAVs (2004) A concept for european regulations for civil unmanned aerial vehicles (UAV). Final Report
13. Sabatini N (2007) Assuring the safe integration of UAS. Unmanned Aircraft Systems, The Global Perspective 2007/2008 p 11
14. United States Government Accountability Office (2008) Unmanned Aircraft Systems: Federal actions needed to ensure safety and expand their potential uses within the national airspace system. Report to Congressional Requesters
15. US Department of Defense Office of the Secretary of Defense (2007) Unmanned systems roadmap 2007-2032. Report
16. Zaloga S (2007) Getting civil with UAVs: How soon? Unmanned Systems 25(3):24–26

Chapter 2
Aviation History and Unmanned Flight

Heavier-than-air flying machines are impossible.
Lord Kelvin, 1895

It is apparent to me that the possibilities of the aeroplane, ... have been exhausted, and that we must turn elsewhere.
Thomas Edison, 1895

Flight by machines heavier than air is unpractical and insignificant, if not utterly impossible.
Simon Newcomb, 1902

This 'pictorial' Chapter presents a historical perspective on unmanned flight starting from the ancient times and reaching the beginning of the 21st Century. This is not aimed to be an exhaustive account of the history of neither aviation nor UAS. It is rather a glimpse of the stages of UAS evolution, complemented by an overview of the broad range of modern UAS sizes, types and capabilities, as well as, the large number of roles they are called upon to play. This will also put into perspective the daunting task of integrating all these different types of unmanned aircraft into an already crowded airspace. We believe that the best way to achieve this goal is through an account of key events and a series of photos.

This Chapter is divided into four sections corresponding to different time periods and — to a degree — to a different concept of what an unmanned aircraft is. The first Section concerns the first flying machines of antiquity and the Renaissance. The second Section is devoted to the first designs of unmanned aircraft that led to target drones and cruise missiles; followed by a section on the developments of the Cold War era, when the focus on research and development turned to unmanned, airborne reconnaissance. Finally a number of modern day systems, including some future designs are presented in the fourth Section.

2.1 Precursors of Flight and Unmanned Aircraft

In modern times, manned aviation appeared in the late 1700s and it took another century for heavier than air machines to take to the skies. Unmanned aircraft followed soon after the advent of the airplane, appearing around the time of the First World War (1916). However, the idea for a 'flying machine' was first conceived close to 2,500 years ago, in ancient Greece and China!

Pythagoras, Archimedes and others studied the use of autonomous mechanisms for a variety of applications. The first known autonomous flying machine has been

K. Dalamagkidis et al., *On Integrating Unmanned Aircraft Systems into the National Airspace System*, Intelligent Systems, Control and Automation: Science and Engineering 54, DOI 10.1007/978-94-007-2479-2_2,

credited to Archytas from the city of Tarantas or Tarentum in South Italy, known as Archytas the Tarantine. Archytas has been referred to as Leonardo Da Vinci of the Ancient World and was also the father of number one in number theory [9] and the solution for doubling the cube. He was also possibly the first engineer, designing and building various mechanisms. In 425 B.C. he built a mechanical bird, which he called "the pigeon", shown in Fig. 2.1. According to Cornelius Gellius in his Noctes Atticae, the bird was made of wood, nicely balanced with weights and flew using air (most likely steam) enclosed in its stomach [1]. It is alleged that Archytas' pigeon flew about 200 meters before falling to the ground, once all energy was used. The pigeon could not fly again, unless the mechanism was reset [2].

Fig. 2.1 An artist's depiction of the flying pigeon, the first documented UAS in history. It is reported that it flew for about 200 meters

During the same era in a different part of the Ancient World — China — at about 400 B.C., the Chinese were the first to document the idea of a vertical flight aircraft. The earliest version of the Chinese top consisted of feathers at the end of a stick. The stick was spun between the hands to generate enough lift before released into free flight.

Over the years, the Chinese experimented with other types of flying machines such as hot air balloons, rockets or kites. It is noteworthy that although some of these machines were used for entertainment, some of the applications were military in nature. In fact there are historical records of a "wooden hawk" that was used for reconnaissance around 450 B.C., as well as a kite in the shape of a crow, which was employed during the Ming Dynasty to bomb enemy positions [11].

Several centuries later, Leonardo Da Vinci, in 1483, designed an aircraft capable of hovering, called aerial screw or air gyroscope, shown in Fig. 2.2. It had a 5 meter diameter and the idea was to make the shaft turn and if enough force were applied, the machine could spun and fly. This machine is considered by some experts as the ancestor of today's helicopter [3]. Da Vinci also devised a mechanical bird in 1508 that would flap its wings by means of a double crank mechanism as it descended along a cable.

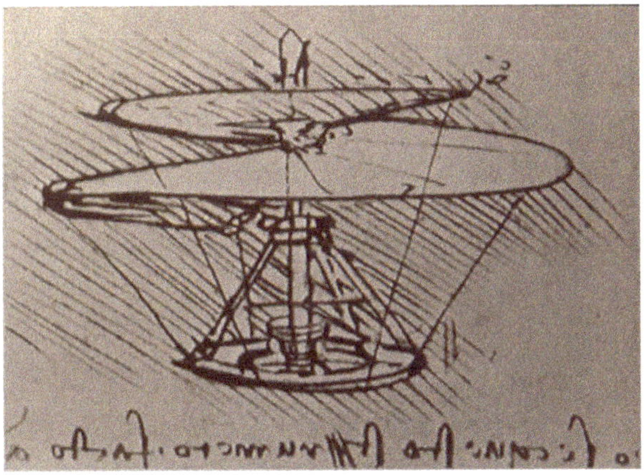

Fig. 2.2 Leonardo Da Vinci's air screw, a forerunner of modern helicopter designs (Public domain photo)

The first widely recognized manned flight took place in 1783 using a hot air balloon designed by the Montgolfier brothers and commemorated in Fig. 2.3. Soon after, similar attempts took place in England and for several years ballooning dominated manned flights, until the first helicopters in the 1860s and later fixed-wing aircraft.

Many flying machines were designed between 1860 and 1909, initially focusing on vertical take-off and landing aircraft because of the limitations of the steam-powered engines that were in use at the time. As the power to weight ratio of engines improved, these early machines were transformed to the helicopter and airplane designs that are in use today.

The main drive behind aircraft development has always been the fast and safe transportation of people and cargo. Nevertheless, the military soon realized the potential benefits of unmanned aircraft and efforts to adapt flying machines to operate without a pilot onboard started. Such systems were initially unmanned ordinance delivery systems, what would now be referred to as 'missiles' or 'smart bombs'. Another use for such systems was to operate as 'target drones' that assisted in the training of anti-aircraft gun operators. Today UAS have been defined as those systems that are designed to be recovered after each mission and although they may carry weapons, the weapon is not by itself an integral component of the airframe [5].

Fig. 2.3 The first manned flight using a hot air balloon took place in 1783 in France (Photo Credit: Bildarchiv Preussuscher Kulturbesitz, Berlin)

Nevertheless, in the early days of unmanned flight, these distinctions were not that important since the technological obstacles were the same and were even shared by manned aircraft as well.

2.2 1916–1944

In 1916, less than 15 years after the Wright brothers historical flight, the first modern unmanned aircraft was demonstrated. It was the Hewitt-Sperry Automatic Airplane, named after the two inventors that designed it. This aircraft could not have become a reality without the previous work of Sperry on gyroscopic devices that were needed to provide flight stabilization. Sperry managed to attract the interest of the US Navy resulting in the development of the Curtiss-Sperry Aerial Torpedo, while at the same time the US Army Air Force sponsored the Liberty Eagle Aerial Torpedo of Charles Kettering shown in Fig. 2.4 [12]. Due to technical problems and lack of accuracy, interest on "automatic" planes was lost but the potential for use of remotely operated drones for target practice was soon realized.

Fig. 2.4 The USAF Liberty Eagle Aerial Torpedo, also known as the Kettering Bug after its creator Charles Kettering (Photo Credit: National Museum of the USAF)

In Britain, experiments with unmanned aircraft took place throughout the 1920s with the RAE 1921 Target. In 1933, the Royal Navy used the Queen Bee target drone (Fig. 2.5) for the first time [5]. It was a modified version of the DeHavilland Tiger Moth biplane and was successfully employed for gunnery practice.

Fig. 2.5 The DH.82B Queen Bee drone (Photo Credit: Adrian Pingstone)

Remote operation required the perfection of radio control, a concept proposed and demonstrated in 1895 and 1898, respectively, by Tesla [5]. Yet again private industry spurred developments, when actor Reginald Denny opened his "Reginald Denny Hobby Shops" in 1934 and started selling radio controlled airplanes. A few years later he demonstrated his work for the US Army that lead to the development of a very successful target drone used extensively during WWII.

2.3 The Machines of the Cold War

Soon after the end of WWII, interest in reconnaissance missions increased. The descendants of Reginald Denny's target drones became the basis of the first reconnaissance drone, the SD-1 [5]. Also known as the MQM-57 Falconer, it was developed in the mid 1950s and by the end of its career close to 1,500 had been built [4]. The SD-1 (Fig. 2.6) was remotely operated, carried a camera and after a 30 minute flight returned to base and was recovered with parachute [12].

Fig. 2.6 The SD-1, also known as the MQM-57 Falconer, was the first reconnaissance drone of the US Army and remained in service until the 1970s (Photo Credit: National Museum of the USAF)

The loss of the U-2 spy plane over the Soviet Union in 1960 gave a new push towards unmanned reconnaissance drones and the loss of a second U-2 over Cuba two years later helped circumvent any doubts and funding problems [12]. The USAF supported the Ryan Model 147 drone that evolved into a series of models with different capabilities. Two variations of the more than two dozen available, are shown in Figs. 2.7 and 2.8. They were also based on a drone design and were used for reconnaissance missions by the US over China, Vietnam and other countries in the 1960s and 70s [5, 12]. During this time, close to 3,500 Lightning Bugs were launched and close to 84% returned [10]. It is noteworthy that these drones were credited with kills that occurred when fighters tried to shoot them down and one drone was given ace status after being responsible for the loss of five North Vietnamese MIGs [12].

The Ryan Model 147, which became known as the Lightning Bug was probably the first unmanned aircraft that can withstand today's definition of a UA. In fact, modernized versions of this drone are still being built and used to carry out missions to this day.

Fig. 2.7 The AQM-34Q, one of several variations of the Ryan Model 147 unmanned reconnaissance drone, used in the 1960's and 1970's (Photo Credit: National Museum of the USAF)

Fig. 2.8 The BGM-34C was another member of the Ryan Model 147 family. It was a multi-role drone capable of performing reconnaissance, electronic counter-measure and strike missions (Photo Credit: US Air Force)

Meanwhile, the US Navy acquired a helicopter drone from the Gyrodine Company called the QH-50 DASH (Fig. 2.9) [12]. This design was preferred because it could be launched from smaller vessels. Its main mission was to launch antisubmarine torpedoes, nevertheless, it was also used for surveillance, cargo transport and other applications. This was despite reliability issues with its electrical system that led to large number of peacetime losses [12].

Fig. 2.9 A QH-50 DASH-drone armed with torpedoes over the flight deck of the USS Hazelwood (Photo Credit: US Navy)

In the late 1960s, the CIA was involved in its own hypersonic, long-range reconnaissance drone, designed to be launched from another "mother" aircraft. The D-21 Tagboard (Fig. 2.10) program was mired with technical problems, accidents and failed missions that led to its ultimate cancellation in the early 1970s [12].

Fig. 2.10 The Lockheed D-21B Tagboard (Photo Credit: National Museum of the USAF)

In parallel with US efforts, the Soviet Air Force developed its own reconnaissance drones. The first system was the TBR-1 based on a popular target drone and it was soon followed by the DBR-1 (Fig. 2.11) that allowed for higher range and capabilities [12]. The DBR-1 was not designed to be fully recovered, instead when it reached the recovery area it dumped fuel, ejected the nose containing the sensor package and the rest of the airframe crashed. As a result, the DBR-1 involved high operational costs which led to its replacement in the mid 1970s by the Tu-141/143 (Fig. 2.12), a mid- and short-range, respectively, reconnaissance drone that were fitted with parachutes for recovery.

Fig. 2.11 The DBR-1 also known as the Tupolev Tu-123 Yastreb (Photo Credit: Tupolev)

Fig. 2.12 The Tupolev Tu-143 Reys reconnaissance drone with SPU-143 launcher at the Ukrainian Air Force Museum (Photo Credit: George Chernilevsky)

In Europe the unmanned system of choice of the time was funded by Canada and the UK and was developed by Canadair [12]. This resulted in the CL-89 Midge (Fig. 2.13) that was also acquired by the French and German armies. It was designed to follow a pre-programmed course, take photographs (day or night) and return to be recovered by parachute [12]. A more sophisticated version, the CL-289, that also featured better range was developed in the late 1970s with major funding coming now from Germany [12].

Fig. 2.13 German reconnaissance drone CL-289 (Photo Credit: Karsten Franke)

Another major player in the area of unmanned aircraft was the Israeli Air Force, that acquired and operated a squadron of American drones for reconnaissance purposes during the Yom Kippur War [12]. Later the Israeli Aircraft Industries and Tadiran developed their own aircraft, the Scout (Fig. 2.14) and Mastiff, respectively [12]. The Mastiff was the basis of the very popular Pioneer system and the Israeli designs have also influenced the construction of the Predator and Shadow UAS [5].

Fig. 2.14 The Israeli Aircraft Industries Scout drone. The similarity with the Pioneer UAS that was later bought by the US Armed Forces are evident (Photo Credit: Wikimedia Commons User Bukvoed)

2.4 Modern Systems

This Section is devoted to modern systems. Rather than following a chronological order, the systems are presented in groups of similar design or functionality. Furthermore, this Section only presents a small percentage of the vast number of operational UAS that are currently available.

The first group concerns the direct descendants of the reconnaissance drones. It includes the very popular Pioneer (Fig. 2.15) of the 1980s, which had over 1,500 flight hours in the Desert Storm operation [12]. Another very successful system is the Predator (Fig. 2.16), the improved version of the Gnat-750.

Fig. 2.15 An RQ-2 Pioneer ready for launch during operation Desert Shield (Photo Credit: US Marine Corps)

Fig. 2.16 An armed MQ-1 Predator flying a training mission. The MQ-1 is used for armed reconnaissance, airborne surveillance and target acquisition, can carry two laser-guided missiles and employs a crew of two [6] (Photo Credit: US Air Force)

Other systems include the RQ-4 Global Hawk (Fig. 2.17) a large, high altitude, long endurance system, the MQ-9 Reaper (Fig. 2.18), which besides reconnaissance can also be used as a hunter-killer, the Neptune (Fig. 2.19) that is used for water operations and finally the RQ-7 Shadow (Fig. 2.20), the Scan Eagle (Fig. 2.21) and Silver Fox (Fig. 2.22) systems.

Fig. 2.17 The RQ-4 Global Hawk is a high altitude, long endurance UAS . It was the first UAS to do a trans-pacific crossing [5] (Photo Credit: US Air Force, Master Sgt. Jason Tudor)

Fig. 2.18 The MQ-9 Reaper is an updated version of the Predator UAS. It is primarily used as a persistent hunter-killer UAS for critical time-sensitive targets and secondarily for intelligence gathering [8] (Photo Credit: US Air Force, Staff Sgt. Brian Ferguson)

Fig. 2.19 The Neptune, a reconnaissance UAS capable of water landings (Photo Credit: US Navy)

Fig. 2.20 The RQ-7 Shadow is developed by AAI for the US Army and has been deployed in Iraq (Photo Credit: US Army)

Fig. 2.21 Scan Eagle is a low-cost, long endurance UAS used by the US Marine Corps [7] (Photo Credit: US Air Force)

Fig. 2.22 Silver Fox is a reconnaissance UAS used by the US Marine Corps. (Photo Credit: US Navy, Mate 2nd Class Daniel J. McLain)

Of course besides the US, other countries have developed their own similar systems. Specifically, more than 50 countries are currently involved in the development of over 1,000 different UAS. Israel has been updating its own fleet with systems such as the IAI Heron (Fig. 2.23). In Britain the BAE Phoenix (Fig. 2.24) has been used for combat surveillance while the french-built SPERWER (Fig. 2.25) has been acquired by a number of other European armed forces.

Fig. 2.23 The IAI Heron is medium altitude long endurance UAS. Specifically it can fly for close to two full days. (Photo Credit: US Air Force)

Fig. 2.24 The BAE Systems Phoenix was a British surveillance drone used in Kosovo and Iraq until its retirement (Photo Credit: Wikimedia Commons User BilborneOne)

Fig. 2.25 The french SPERWER. It is used by the military forces of several European nations (Photo Credit: David Monniaux)

In the Russian Federation there are several companies involved with UAS development. Yakovlev has fielded several systems, an example of which is the Pchela (Fig. 2.26). Similarly, the Irkut Corporation is developing a range of UAS systems, including the Irkut 2M, 3, 10, 200 (Fig. 2.27) and 850 systems. The Dozor-600 (Fig. 2.28) is another recent, Russian UAS. Argentina, Australia, Belgium, Brazil, Canada, Chile, Greece, Italy, Iran, South Africa, Spain, Sweden Switzerland and Turkey are only some of the other countries around the world developing their own systems, sometimes in collaboration.

Fig. 2.26 The Yakovlev Pchela surveillance and observation UAS (Public domain photo)

Fig. 2.27 The Irkut-200 Aerial Remote Sensing System with up to 12 hours of flight endurance (Photo Credit: Wikimedia Commons User Allocer)

Fig. 2.28 The Dozor-600 is a long endurance surveillance platform similar to MQ-1 Predator or the Hermes 450 (Public domain photo)

Although some of the aforementioned UAS like the MQ-9 Reaper can be armed, there is an entire class of systems being developed that their primary mission is combat operations. Although many of these systems known as Unmanned Combat Aircraft Systems (UCAS) are still in experimental stages, there are several that are operational. Examples of UCAS include the Neuron (Fig. 2.29), the Barracuda, the Italian Sky-X (Fig. 2.30), the MiG Skat, the General Atomics Avenger, the BAE Mantis (Fig. 2.31) and the Northrop Grumman X-47 system (Fig. 2.32).

Fig. 2.29 The Neuron is an experimental UCAS being developed by a consortium of European companies (Photo Credit: Wikimedia Commons User Tangopaso)

Fig. 2.30 The Sky-X is an Italian UCAS built by the Finmeccanica group. It first flew in 2005 (Photo Credit: Wikimedia Commons User Duch.seb)

Fig. 2.31 The BAE Mantis is a demonstrator platform for an advanced UCAS that first took flight towards the end of 2009 (Public domain photo)

Fig. 2.32 The X-47 started from the DARPA sponsored J-UCAS and is now part of a US Navy program to develop a carrier-based UCAS (Photo Credit: DARPA)

Almost all of the aforementioned systems are of the fixed-wing design. However, there is a number of helicopter UAS available, several of which are currently operational in military and civil applications. Some examples of helicopter UAS include the A-160 Hummingbird (Fig. 2.33), the APID55 (Fig. 2.34), the Schiebel S-100 (Fig. 2.35) and the MQ-8 Firescout (Fig. 2.36).

Fig. 2.33 The A-160 Hummingbird built by Boeing/Frontier. It is a demonstrator for improvements in range endurance and controllability [7] (Photo Credit: US DoD)

Fig. 2.34 The APID55 is a multi-purpose, fully autonomous helicopter platform with an endurance of over three hours and payload capacity of over 50 kg. It is built by Cybaero (Photo Credit: Cybaero)

Fig. 2.35 The Schiebel S-100 is a helicopter UAS used by a number of nations particularly for navy operations (Photo Credit: Matthias Kabel)

Fig. 2.36 The MQ-8 Fire Scout is being developed by Northrop Grumman Corporation to support US Army and Navy requirements (Public domain photo)

In recent years there is an increasing interest for long endurance UAS, systems that can fly for several days. Besides the RQ-4 Global Hawk previously presented, which was the first UAS to do a trans-pacific crossing, the Aerosonde Laima (Fig. 2.37) was the first to do the trans-atlantic crossing. A large number of long endurance systems are used for civilian applications. For example NASA employs the Helios, Altair and Ikhana UAS (Figs. 2.38 to 2.40) primarily for earth science missions. Finally the Zephyr (Fig. 2.41) managed to remain for almost four days airborne, taking into advantage its lightweight design and solar power.

Fig. 2.37 The Aerosonde Laima was the first UAS to make a trans-atlantic crossing [5] (Public domain photo)

Fig. 2.38 The Helios UAS developed by NASA and AeroVironment. During its second high-altitude flight, it reached 96,863 ft, shattering the existing world altitude record for sustained level flight for both propeller and jet-powered aircraft (Photo Credit: NASA)

Fig. 2.39 The Altair is a Predator variant developed for NASA for high altitude, earth science missions (Photo Credit: NASA)

Fig. 2.40 The Ikhana is a Predator B UAS acquired by NASA and retrofitted for earth science missions (Photo Credit: NASA)

Fig. 2.41 The Zephyr is a solar-powered, high endurance UAS developed by Qinetiq. In July of 2008 it performed an 82 hour long flight (Photo Credit: Qinetiq)

When the size of a UAS is reduced, its payload capacity will need to be reduced as well. Nevertheless, there is a number of small and miniature UAS in operation or active development. This is because they are versatile, portable, and easy to maintain; they can be employed for the same applications as larger UAS on a smaller scale and at a lower cost. In fact several of these systems are considered expendable and no recovery is attempted especially if there is any risk perceived.

Fig. 2.42 The Skylark is a hand-launched system for tactical surveillance and reconnaissance developed by Elbit (Photo Credit: MORH)

Examples of small UAS include the Skylark (Fig. 2.42), the Evolution (Fig. 2.43) the Puma (Fig. 2.44) and the Aerocon Inspector (Fig. 2.45). In the miniature category, there exist systems that weigh less than 1–3 kg and are easily back-packable (some are even foldable). Such systems include the Cyberbug, Raven, WASP, BATCAM, Nighthawk and the Dragon Eye (Figs. 2.46 to 2.51).

Fig. 2.43 The L-3 BAI Aerospace Evolution XTS is a portable UAS with EO/IR payload. It was employed in the rescue efforts after hurricane Katrina (Photo Credit: NIST)

Fig. 2.44 The Puma is another evolution from the Pointer UAS. There are two versions available, the Aqua Puma and the Terra Puma for marine and land environments respectively (Photo Credit: US DoD)

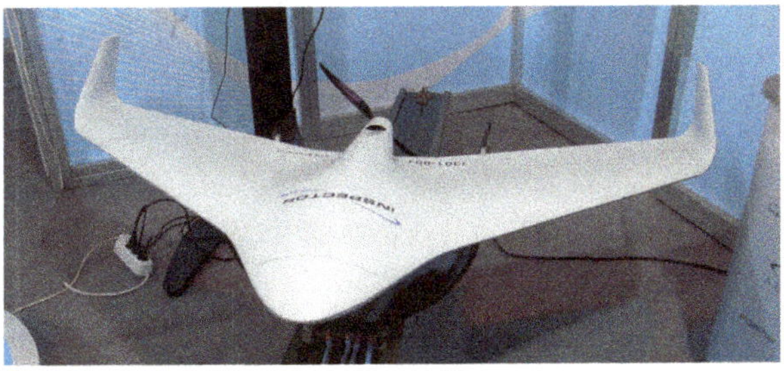

Fig. 2.45 The Aerocon Inspector weighs just 7 kg and has an operational range of 25 km (Photo Credit: Wikimedia Commons User Allocer)

Fig. 2.46 The Cyberbug is a light-weight UAS for urban surveillance applications (Photo Credit: NIST)

Fig. 2.47 The Raven is a back-packable UAS for "over the hill" and route reconnaissance that evolved from the Pointer UAS. In the photo paratroopers are training with an RQ-11 Raven (Photo Credit: US Army, Sgt. Amanda Jackson)

Fig. 2.48 The WASP is a MAV that can carry an EO payload and has an endurance of over one hour (Photo Credit: NIST)

Fig. 2.49 The Battlefield Air Targeting Camera Micro Air Vehicle (BATCAM) is a miniature, short-range UAS (Photo Credit: National Museum of the USAF)

Fig. 2.50 The Nighthawk is the evolution of the BATCAM foldable UAS (Photo Credit: NIST)

Fig. 2.51 The Dragon Eye is a small UAS used by the US Marine Corps for short-range Surveillance and Reconnaissance (Photo Credit: US DoD)

Another UAS design concerns the lighter-than-air systems or airships, a distinct UAS category with significantly higher endurance than other systems that makes them ideal for a variety of operations. The main advantage of such systems is that they provide an airborne sensor platform that can be used for persistent area surveillance. Such systems include the Joint Land Attack Elevated Netted Sensor (Fig. 2.52), the Persistent Threat Detection System (Fig. 2.53) and the Rapid Aerostat Initial Deployment system (Fig. 2.54). The High Altitude Airship (Fig. 2.55) is an untethered platform concept that will be able of providing satellite-like services to ground forces.

Fig. 2.52 Joint Land Attack Elevated Netted Sensor (JLENS) by Raytheon/TCOM capable of providing over-the-horizon surveillance (Photo Credit: US Army)

Fig. 2.53 The Persistent Threat Detection System (PTDS) is equipped with a high resolution EO/IR payload used for surveillance (Photo Credit: US Army)

Fig. 2.54 The Rapid Aerostat Initial Deployment (RAID) is a smaller version of the JLENS, used in missions of area surveillance and force protection against small arms, mortar and rocket attacks. (Photo Credit: US Army)

Fig. 2.55 High Altitude Airship (HAA) developed by Lockheed Martin. It is a solar powered, untethered, long endurance, high altitude demonstrator (Photo Credit: US Missile Defense Agency)

In addition to the popular fixed-wing and helicopter systems, other designs are also used for UAS. Figures 2.56 to 2.58 show the iStar MAV, the Sikorsky Cypher and the SELEX Galileo Spyball, all employing duct-fan designs. The IT-180 (Fig. 2.59) and the KOAX X-240 (Fig. 2.60) use two counter-rotating rotors, thus eliminating the need for a tail-rotor. The Eagle Eye (Fig. 2.61) and X-50 (Fig. 2.62) are systems that combine some of the advantages of fixed-wing and helicopter designs. The CyberQuad (Fig. 2.63) and the AirRobot AR 100-B (Fig. 2.64) are both examples of the quad-rotor design that is especially popular in academic environments.

Fig. 2.56 The iSTAR MAV duct-fan aircraft built by Allied Aerospace (Photo Credit: US Navy)

Fig. 2.57 The Sikorksy Cypher uses a shrouded twin-rotor design (Photo Credit: US Navy)

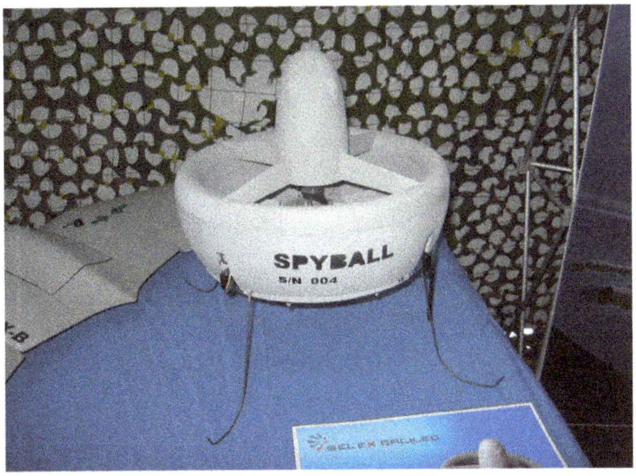

Fig. 2.58 The SELEX Galileo Spyball is an ultra-light UAS that can fit inside a rucksack (Photo Credit: Wikimedia Commons User Causa83)

Fig. 2.59 The IT-180 is a French-built light-weight UAS that uses a twin, counter-rotating rotor system (Photo Credit: Infotron)

Fig. 2.60 The KOAX X-240 is a Swiss coaxial unmanned helicopter with 90 min endurance (Photo Credit: Wikimedia Commons User ILA-boy)

Fig. 2.61 The Eagle Eye is a tilt-rotor UAS developed by Bell Helicopter. It was selected by the US Coast Guard for its Deepwater program but was put on hold in 2007 (Photo Credit: US Coast Guard)

Fig. 2.62 The X-50 aircraft built by Boeing Corp. It is a technology demonstrator for the Canard Rotor Wing (CRW) configuration which combines hovering capabilities with high cruise speeds [7] (Photo Credit: US DoD)

Fig. 2.63 The CyberQuad uses a shrouded quad-rotor design that offers simple dynamics, stability and agility (Photo Credit: CyberTech)

Fig. 2.64 The AirRobot AR 100-B features autonomous landing when it is out-of-range or when the battery is low (Public domain photo)

2.5 Remarks

Although this Chapter was not meant to be a comprehensive presentation of all UAS in development or service, it does indicate the range of designs and operational characteristics available. It is noteworthy that unique applications of UAS in environments traditionally inaccessible to aircraft, as in the case of low altitude urban operations, have led to the development of equally unique solutions.

A comprehensive listing of UAS developed and/or in operation around the globe is provided in App. C.

References

1. Gellius C (1927) Attic nights. Book 10, Translated by Rolfe, J. C., [online] http://www.penelope.uchicago.edu/Thayer/E/Roman/Texts/Gellius/10*.html
2. Guedj D (1998) Le Theoreme du Perroquet. Editions du Seuil

3. Hiller Aviation Museum (2004) History of helicopters. Online, [online] http://www.hiller.org/
4. National Museum of the USAF (2009) Radioplane/northrop MQM-57 falconer factsheet. [online] http://www.nationalmuseum.af.mil/factsheets/factsheet.asp?id=7684, retrieved Oct 2009
5. Newcome L (2004) Unmanned Aviation: A brief history of UAV's. American Institute of Aeronautics and Astronautics
6. US Air Force (2009) MQ-1 predator unmanned aircraft system. [online] http://www.af.mil/information/factsheets/factsheet.asp?fsID=122, retrieved Oct 2009
7. US Department of Defense Office of the Secretary of Defense (2005) Unmanned aircraft systems roadmap 2005-2030. Report
8. US Department of Defense Office of the Secretary of Defense (2007) Unmanned systems roadmap 2007-2032. Report
9. Valavanis KP, Vachtsevanos GJ, Antsaklis PJ (2007) Technology and autonomous mechanisms in the mediterranean: From ancient greece to byzantium. In: Proc. European Control Conference (ECC2007), Kos, Greece, pp 263–270
10. Wagner W (1982) Lightning Bugs and Other Reconnaissance Drones. Aero Publishers
11. Yinke D (2005) Ancient Chinese Inventions. China Intercontinental Press
12. Zaloga SJ (2008) Unmanned Aerial Vehicles: Robotic Air Warfare 1917–2007. No. 144 in New Vanguard, Osprey Publishing Ltd.

Chapter 3
Current Manned Aviation Regulation

. . . φύσις ἐστὶν ἄτακτον καὶ κατ' ἄνδρ' ἴδιον τοῦ ἔχοντος, οἱ δὲ νόμοι κοινὸν καὶ τεταγμένον καὶ ταὐτὸ πᾶσιν . . .
Δημοσθένης, κατ' Αριστογείτονος Α (25.15)

. . . nature is something irregular and incalculable, and peculiar to each individual; but the laws are something universal, definite, and the same for all . . .
Demosthenes, Against Aristogeiton A (25.15)

3.1 Introduction

United States federal law gives the Secretary of Transportation and the Administrator of the Federal Aviation Agency (FAA) the responsibility of the economic and safety regulation of the aviation industry. To fulfill this obligation, they are given the authority to *conduct investigations, prescribe regulations, standards, and procedures, and issue orders* [49 USC §40113(a)]. Federal law assigns great importance to safety. The paragraph on safety considerations in public interest [49 USC §40101(d)] reads:

> . . . the Administrator shall consider the following matters, among others, as being in the public interest:
>
> 1. assigning, maintaining, and enhancing safety and security as the highest priorities in air commerce.
> 2. regulating air commerce in a way that best promotes safety and fulfills national defense requirements.
> 3. encouraging and developing civil aeronautics, including new aviation technology.
> 4. controlling the use of the navigable airspace and regulating civil and military operations in that airspace in the interest of the safety and efficiency of both of those operations.
> 5. consolidating research and development for air navigation facilities and the installation and operation of those facilities.
> 6. developing and operating a common system of air traffic control and navigation for military and civil aircraft.
> 7. providing assistance to law enforcement agencies in the enforcement of laws related to regulation of controlled substances, to the extent consistent with aviation safety.

The statutory mandate of the FAA also includes regarding safety:

> . . . before authorizing new air transportation services, evaluating the safety implications of those services; and preventing deterioration in established safety procedures, recognizing the clear intent, encouragement, and dedication of Congress to further the highest degree of safety in air transportation and air commerce, and to maintain the safety vigilance that has evolved in air transportation and air commerce and has come to be expected by the traveling and shipping public. [49 USC §40101(a)]

K. Dalamagkidis et al., *On Integrating Unmanned Aircraft Systems into the National Airspace System*, Intelligent Systems, Control and Automation: Science and Engineering 54, DOI 10.1007/978-94-007-2479-2_3,

Aviation regulations are collected and codified in the Code of Federal Regulations (CFR), Title 14, Chapter I, also known as Federal Aviation Regulation (FAR). Similarly, in Europe the Joint Aviation Authorities (JAA) has issued the Joint Aviation Requirements (JAR), while other countries may have other similar regulatory documents. Due to an ongoing effort for harmonization between the aviation regulations, part and section numbers between the JAR and the FAR are largely the same.

This Chapter presents an overview of key parts of current manned aviation regulations as defined in the FAR, with the understanding that the provisions of other aviation regulations will be similar if not the same. Where appropriate, the relative section in the FAR will be given. In any case, the reader is advised to consult with the civil aviation authority of his/her country and the current version of the regulations for the latest and most accurate information. The FAR is publicly available both online and in print from the Government Printing Office.

The FAR is comprised of several parts related to airworthiness certification (21–39), maintenance (43), aircraft registration and marking (45–49), pilot certification (61–67), airspace classes (71–77), operating rules (91–99) and special classes of vehicles (101–105). Responsible to oversee access to the NAS is the FAA, a federal agency belonging to the Department of Transportation.

The provisions of the CFR notwithstanding, the FAA issues supplementary material like handbooks, orders, Advisory Circulars (AC) and Technical Standard Orders (TSO) that clearly define appropriate procedures, standards and practices required to comply with current regulations. This material helps ensure that aircraft manufacturers and operators are able to establish the minimum level of safety and reliability required for civil operations [3]. Several of these documents adopt established standards prepared by government agencies like the US Department of Defence, standards development organizations as well as other organizations, national or international. A list of organizations that have been involved with the development of aerospace-related standards is provided below:

- Aeronautical Radio, Incorporated (ARINC)
- American Institute of Aeronautics and Astronautics (AIAA)
- American National Standards Institute (ANSI)
- American Society of Testing & Materials (ASTM)
- American Welding Society (AWS)
- Electronic Industries Alliance (EIA)
- Electrostatic Discharge Association (ESDA)
- European Organisation for Civil Aviation Equipment (EUROCAE)
- Institute of Electrical and Electronics Engineers (IEEE)
- Institute of Environmental Sciences and Technology (IEST)
- International Civil Aviation Organization (ICAO)
- International Electrotechnical Commission (IEC)
- International Organization for Standardization (ISO)
- National Aeronautics and Space Administration (NASA)
- National Institute of Standards and Technology (NIST)

- The North Atlantic Treaty Organization Standards Agency (NSA)
- Occupational Safety & Health Administration (OSHA)
- Radio Technical Commission for Aeronautics (RTCA)
- Society of Automotive Engineers (SAE)

3.2 Airworthiness Certification

In order for any aircraft to fly legally in the US, it must carry an airworthiness certificate issued by the FAA [FAR §91.203]. Airworthiness certification covers a wide spectrum of areas related to aspects of the aircraft design, construction and operation. Presented below are some of these areas along with a selection of the various aspects investigated during certification:

Flight:	Performance, flight characteristics, controllability, maneuverability and stability.
Structure:	Loads, control surfaces, stabilizing and balancing surfaces and fatigue evaluation.
Design & Construction:	Wings, control surfaces, control systems, landing gear and pressurization.
Powerplant:	Fuel system, oil system, cooling system, induction system, exhaust and control.
Equipment:	Instruments' installation, electrical systems, lights and safety equipment.

In addition to aircraft, airworthiness directives exist for aircraft engines and propellers.

According to the FAA, there are two conditions that need be met in order for an aircraft to be considered airworthy; it must conform to its type certificate including any supplemental certificates, and it must be in a condition that ensures safe operation [4]. For aircraft that are not type certified, compliance with the second condition is adequate. Besides standard certification, special airworthiness certificates are also available, usually for experimental or special purpose aircraft.

It should be noted that the FAR allows the FAA administrator to prescribe additional requirements and special conditions for aircraft, aircraft engines or propellers when due to a novel or unusual feature, current airworthiness regulations are inadequate or inappropriate [FAR §21.16].

3.2.1 Type Certificate

A type certificate is a collection of documents, drawings, specifications, datasheets and any related information needed to demonstrate compliance with the applicable paragraphs of the FAR [FAR §21.41]. These may also include inspection and preventive maintenance programs and instructions for continued airworthiness [FAR §21.31]. During the application for type certificate, the FAA administrator may require an inspection and test of the aircraft [FAR §21.33], which may also include flight tests [FAR §21.35].

Once a type certificate has been issued, it is in effect until surrendered, suspended or revoked [FAR §21.51]. Nevertheless, after modifications to an aircraft a new certificate may be required. When the extend of the changes is not significant, the type certificate can be amended [FAR §21.91] or a supplemental certificate will be issued [FAR §21.113].

3.2.2 Standard Certificates

Standard airworthiness certificates are given to aircraft that are type certificated in any of the categories defined in [FAR §21.175], including:

- Normal, utility, acrobatic and commuter aircraft (FAR Part 23).
- Transport aircraft (FAR Part 25).
- Normal rotorcraft (FAR Part 27).
- Transport rotorcraft (FAR Part 29).
- Manned free balloons (FAR Part 31).

In addition to the above categories, type certification is available for primary [FAR §21.24], restricted [FAR §21.25], US Army surplus [FAR §21.27] and imported [FAR §21.29] aircraft, as well. An overview of the applicability requirements for each of the aforementioned categories is given in Table 3.1.

3.2.3 Special Certificates

For aircraft that do not meet requirements for a standard certificate but are still capable of safe flight, special airworthiness certificates are available [4]. More specifically special certificates can be given in the primary [FAR §21.184], restricted [FAR §21.185] and limited [FAR §21.189] categories, for aircraft type certificated under these categories. In addition to that, special airworthiness certificates are available for aircraft belonging to the light-sport category and for experimental aircraft. Finally special flight permits are also available.

3.2.3.1 Light-sport (LSA)

This category is for aircraft other than helicopters that do not exceed 600–650 kg, have a maximum speed of not more than 120 knots and a capacity of not more than

Table 3.1 Aircraft types with standard airworthiness certificates along with occupancy, weight and other restrictions (compiled from information in the FAR).

Category	Max. seats	MTOW (kg)	Notes
Normal	$\leq 9^a$	$\leq 5{,}670$	Non-acrobatic operations
Utility	$\leq 9^a$	$\leq 5{,}670$	Limited acrobatic operations
Acrobatic	$\leq 9^a$	$\leq 5{,}670$	No restrictions
Commuter	$\leq 19^a$	$\leq 8{,}600$	Non-acrobatic operations
Transport	N/A	N/A	
Primary	$\leq 4^b$	$\leq 1{,}225$	Limited power/unpressurized cabin
Restricted	N/A	N/A	Special purpose operations[d]
Normal rotorcraft	$\leq 9^a$	$\leq 3{,}175$	
Transport rotorcraft[c]	$\leq 9^a$	$\leq 9{,}070$	
Manned free balloons	N/A	N/A	

[a] Excluding pilot seats
[b] Includes the pilot
[c] Transport rotorcraft are type-certificated in two categories (A and B). Rotorcraft that meet the above restriction may be certificated in the B category, while those with higher seating capacity must be certificated in the A category.
[d] Includes agricultural, forest and wildlife conservation, aerial surveying, patrolling, weather control and aerial advertising operations.

two persons. Additional requirements are made based on the presence of certain equipment [4].

A special certificate of airworthiness is issued for aircraft of this type after successful inspection of the aircraft and its documentation. The latter includes operating instructions and maintenance procedures and a statement from the manufacturer that the aircraft complies with the provisions of the appropriate consensus standards [FAR §21.190]. Upon successful completion of the inspection, the FAA may amend the certificate with operational restrictions, if deemed necessary [4].

3.2.3.2 Experimental

Experimental certificates are given for a variety of purposes [FAR §21.191]:

- Research and development of equipment, operating techniques or new aircraft designs.
- Showing aircraft compliance with a type certificate or a supplemental certificate after major changes.
- Crew training.
- Exhibitions at air shows or movies. This includes required pilot training and flight from and to the exhibition area.
- Air racing, including practicing and flight from/to the area.
- Market surveys, sales training and customer flight crew training.
- Operating of amateur-built aircraft.

- Operating of primary kit-built aircraft that have not been assembled under the supervision and control of a production certificated entity.
- Operation of certain types of light-sport aircraft.

Before a special certificate in this category is issued, the applicant must submit appropriate documentation. In the case of aircraft used for research and development purposes, this documentation includes the purpose of the experiment along with the number of flights, the location and drawings/photographs of the aircraft [FAR §21.193]. Several operational requirements exist for experimental aircraft depending on their characteristics [4]. The duration of experimental certificates is one year or less except for kit-built aircraft, that typically do not expire [FAR 21.181].

3.2.3.3 Special flight permits

These permits are given to aircraft that would not qualify for other airworthiness certificates, but are capable of safe flight [FAR §21.197]. The purpose of these permits is to allow the aircraft to fly to a different location for storage, repairs, maintenance or to avoid areas of impending danger. The permit is issued after an application where the purpose and characteristics of the flight are detailed, and it may include limitations or special instructions from the FAA [FAA §21.199].

Special flight permits may also be given for airworthy aircraft, to allow them to fly with excess fuel weight, beyond their certificated capacity, when flying over areas where refueling is not possible [FAR §21.197].

3.3 Special Aircraft Categories

Although normally all aircraft need either a standard or a special airworthiness certificate to fly, there is a category of aircraft (classified as vehicles in the FAR) for which this requirement is waived. The other special category concerns remote-controlled (R/C) model aircraft, that also operate under few restrictions. Although not mentioned in the FAR, R/C aircraft are of interest since they present the basis of many UAS designs. It should be stressed however that R/C models are allowed to operate only for recreational purposes and that the FAA has made clear that UAS operations can not be based on R/C model procedures [6].

3.3.1 Vehicles

This category of aircraft includes moored balloons, unmanned balloons, unmanned rockets defined in FAR Part 101 and ultralights defined in FAR Part 103. Ultralights are single-occupant, manned aircraft used for recreation or sport purposes only, with a maximum empty weight of 70 kg for unpowered and 115 kg for powered vehicles [FAR §103.1].

Many of the requirements regarding pilot certification, operating and flight rules, vehicle registration and marking, maintenance certification, including the requirement to carry an airworthiness certificate that are normally applicable to aircraft, are waived for this category [9]. Nevertheless, operational restrictions may be in place.

For example, the following pertain to the operation of ultralight vehicles:

- Daylight operations only [FAR §103.11].
- Yield the right-of-way to all aircraft [FAR §103.13].
- No operations allowed over congested areas [FAR §103.15].
- No operations allowed in Class A, B, C and D airspace. For operations in Class E near airports, ATC authorization is required first [FAR §103.17].
- Pilot must operate by visual reference with the surface [FAR §103.21].

3.3.2 R/C Models

Model airplanes are regulated on a voluntary basis, based on AC91-57 with few operational restrictions. In addition to that an independent organization, the Academy of Model Aeronautics (AMA) issues normal or restricted flight permits after inspection of the model, provides insurance for its members and organizes areas to safely practice aeromodeling. It is noteworthy that the AMA poses additional restrictions to the ones in FAA AC91-57, both in design (e.g. the weight of the models and their propulsion methods) as well as in operation [1].

3.4 Pilot Certification

FAR Part 61 is involved with the requirements for issuing pilot, flight instructor and ground instructor certificates, ratings and authorizations [FAR §61.1]. An appropriate pilot certificate is required for a person to assume the role of pilot in command or of required crew member [FAR §61.3]. Some operators are also required to possess a current medical certificate issued based on procedures described in FAR Part 67.

There are several types of pilot certificates with different training and certification requirements and with different privileges for their holders [FAR §61.5]:

1. Student pilot.
2. Sport pilot.
3. Recreational pilot.
4. Private pilot.
5. Commercial pilot, and,
6. Airline transport pilot certificate.

Each pilot certificate (with the exception of a student certificate), comes with ratings for aircraft categories, classes and types the holder may operate as well as the instrument rating for private and commercial pilots. Table 3.2 summarizes the aircraft category and class ratings. There are also instrument ratings for airplanes, helicopters and powered lifts [FAR §61.5].

Similar ratings are placed on flight instructor and ground instructor certificates when all the training and certification requirements are met.

Table 3.2 Pilot certificates summarized from [FAR §61.5].

Category	Class
Airplane	Single-engine land Multiengine land Single-engine sea Multiengine sea
Rotorcraft	Helicopter Gyroplane
Lighter-than-air	Airship Balloon
Weight-shift-control aircraft	Weight-shift-control aircraft land Weight-shift-control aircraft sea
Powered lift	N/A
Powered parachute	Powered parachute land Powered parachute sea

FAR Part 61 also includes the level of knowledge, training, operations proficiency and experience a pilot must possess before being issued a certificate. This includes training and testing procedures. FAR Part 63 is involved with certification of crew members other than pilots and FAR Part 65 with airmen certification.

3.5 FAR Operation Rules

Operational rules for manned aircraft operating in the US NAS are prescribed in FAR Part 91, which applies to all aircraft with the exception of moored balloons, kites, unmanned rockets, unmanned free balloons and ultralights [FAR §91.1]. Part 91 also establishes the responsibility for aircraft operators to support the continued airworthiness of each airplane [FAR §91.1].

The person ultimately responsible for the operation of the aircraft is the pilot in command [FAR §91.3]. The pilot is also responsible for evaluating the airworthiness of the aircraft and determining if it is in a condition safe to fly [FAR §91.7]. After the aircraft has been deemed safe to fly and before take-off, the pilot needs to be familiar with any information concerning the flight, such as weather reports, fuel requirements, airport characteristics and aircraft performance characteristics [FAR §91.103].

To minimize the risk of collisions, no person is allowed to operate an aircraft in close proximity to another [FAR §91.111] and when the weather conditions permit, the pilot should be alert in order to *see and avoid* other aircraft [FAR §91.113]. Additionally right-of-way rules are established [FAR §91.113]. With the exception of water operations, typically the aircraft with less maneuverability has the right-of-way. This rule is superseded when an aircraft is in distress, at which time it has the right-of-way with respect to all other air traffic. In general during emergencies pilots are allowed to deviate from the requirements of Part 91, even contrary to ATC instruction, provided that ATC is notified of this deviation as soon as possible [FAR §91.3,§91.123]. In any other situation, no one is allowed to deviate from ATC clearance and instructions [FAR §91.123].

Additional safety regulations do not permit pilots to fly below 10,000 ft or in proximity of Class B, C and D airspace at speeds exceeding 250 and 200 knots respectively [FAR §91.117]. Similarly, minimum safe altitudes are established so that upon catastrophic failures, an emergency landing can take place without undue risk to people or property [FAR §91.119].

3.5.1 Flight Rules

FAR Part 91 defines two types of flight rules; visual flight rules (VFR) and instrument flight rules (IFR). In addition to the normal operations, FAR Part 91 includes guidelines for emergencies as well as special operations like aerobatics, towing and parachuting.

3.5.1.1 Visual Flight Rules

Under VFR the pilot is expected to control the aircraft's trajectory and avoid other aircraft based on visual cues, although separation instruction may be provided by ATC when flying in certain classes of controlled airspace.

A prerequisite to flying under VFR is the presence of enough fuel onboard, so that the aircraft can reach its first landing destination and fly for 30 minutes or 45 minutes after that during the day or night, respectively [FAR §91.151]. Similar requirements exist on the flight altitude and weather conditions [FAR §91.155]. The minimum weather conditions for VFR operations are summarized in Table 3.3.

The information required when filing a VFR flight plan includes besides the aircraft and pilot identification data, the point and time of departure, the route, altitude and airspeed, the amount of fuel onboard and the point and time of arrival [FAR §91.153].

Table 3.3 Weather minimums for VFR operations. Source: [FAR §91.155]

Airspace	Visibility (statute miles)	Distance from clouds Above	Below	Horizontal
Class A	N/A	N/A		
Class B	3	Clear of clouds		
Class C	3	1,000 ft	500 ft	2,000 ft
Class D	3	1,000 ft	500 ft	2,000 ft
Class E ($< 10,000$ft)	3	1,000 ft	500 ft	2,000 ft
Class E ($\geq 10,000$ft)	5	1,000 ft	1,000 ft	1 stat. mile
Class G (day, $\leq 1,200$ft)	1	Clear of clouds		
Class G (day, $> 1,200$ft and $< 10,000$ft)	1	1,000 ft	500 ft	2,000 ft
Class G (night, $< 10,000$ft)	3	1,000 ft	500 ft	2,000 ft
Class G ($\geq 10,000$ft)	5	1,000 ft	1,000 ft	1 stat. mile

3.5.1.2 Instrument Flight Rules

Flight under IFR is subject to similar restrictions in terms of fuel availability [FAR §91.167], but in addition to that the presence of an operational and properly maintained VHF omnidirectional range (VOR) radio navigation system is also required [FAR §91.171].

In order to operate IFR in controlled airspace, the pilot must first submit an IFR flight plan and receive appropriate ATC clearance [FAR §91.173]. Once in controlled airspace, the pilot is required to monitor the appropriate communication frequency and report to ATC when she/he reaches predetermined points, encounters unforecast weather conditions or other problems that may affect flight safety [FAR §91.183].

3.5.2 Emergency Rules

In the case of disasters the FAA administrator can issue a notice to airmen (NOTAM) designating an area within which temporary flight restrictions apply [FAR §91.137]. In this case no aircraft can enter the designated area except for aircraft participating in hazard relief, carrying law enforcement officials or any other category specified in [FAR §91.137].

The FAA can also use the NOTAM system to issue new emergency traffic rules or regulations, whenever conditions for safe operations under normal rules are or will not be sufficient [FAR §91.139].

3.5.3 Maintenance Requirements

The person responsible for the maintenance of the aircraft is either the owner or the operator who may not operate the aircraft unless the inspection and replacement intervals as well as any other prescribed maintenance procedures have been complied with [FAR §91.403]. More specifically an annual inspection is required along with an inspection for the issuance of an airworthiness certificate [FAR §91.409]. Additional inspections may be required for other types of aircraft.

After any maintenance procedure, the aircraft must be approved to return to service by an authorized person and the maintenance record has been updated [FAR §91.407]. In some cases an operational inspection by a pilot with adequate rating may be required [FAR §91.407].

In addition to that, FAR Part 91 requires operators to "support the continued airworthiness" of each airplane, by revising the inspection schedule, incorporating any design changes and revisions to the Instructions for Continued airworthiness [FAR §91.1501].

3.6 Airspace Classes

Depending on the altitude and proximity to airports, the NAS is segregated into several classes as shown in Fig. 3.1. For each airspace class, different operating rules may be in effect, based on the stipulations of FAR Part 91.

Classes A through E, ordered from most restrictive to less restrictive, correspond to controlled airspace. Wherever different airspace classes overlap, the most restrictive designation applies [FAR §71.9].

Airspace between 18,000 ft above mean sea level (MSL) to about 60,000 ft, belongs to the Class A airspace [FAR §71.33]. Class A airspace is reserved for IFR traffic and an aircraft needs to receive ATC clearance before entering [FAR §91.133]. There are also requirements for communication and transponder equipment.

Classes B [FAR §71.41], C [FAR §71.51] and D [FAR §71.61] include the airspace above and around airports of different sizes. They are designed to include traffic from/to the airport and ensure appropriate separation. In order to operate an aircraft in Class B airspace, ATC clearance is required from the facility controlling that area and the pilot must possess at least a private pilot certificate although exceptions exist for other pilot certificates when specific requirements are met [FAR §91.131]. In addition to the above, the aircraft needs to be equipped with appropriate communications and navigation equipment as well as an operating transponder [FAR §91.131].

Requirements for operations in Class C and D airspace are similar, although more relaxed. In general an aircraft entering these classes of airspace must establish and maintain communications with ATC while within that airspace, regardless of whether it is arriving or being only through traffic [FAR §91.129, §91.130]

Finally, Class E corresponds to the rest of the controlled airspace. It includes the airspace from 14,500 ft up to the Class A boundary, as well as the airspace above 60,000 ft [FAR §71.71]. In the vicinity of airports, class E airspace may extend down to the surface. Similarly in the proximity of federal airways, class E airspace extends upwards from 700 or 1,200 ft above ground level [5]. In the vicinity of an airport in Class E airspace, the pilot must establish two-way radio communication with ATC, although in the case of radio failure landing is still allowed, provided that VFR conditions exist, visual contact with the tower is maintained and a clearance to land has been received [FAR §91.127].

The last class is Class G airspace, which normally includes the space up to 1.200 ft above the ground but can be up to 14.500 ft. Although Class G airspace is also known as uncontrolled airspace, operating rules do apply. Helicopters and aircraft flying below 1.200 ft should typically operate clear of clouds and at speeds that allow the pilot to see and avoid other traffic as well as any obstructions on the ground [FAR §91.155]. In the vicinity of airports two-way radio communication with ATC must be established and in addition to that all turns must be made to the left, unless the airport indicates the opposite [FAR §91.126]. Other restrictions may also be in effect depending on the type of the aircraft, such as avoiding crowded areas, noise limits, etc.

Every year the FAA publishes a revised Order 7400.9, which includes the current airspace designations for the US NAS [5].

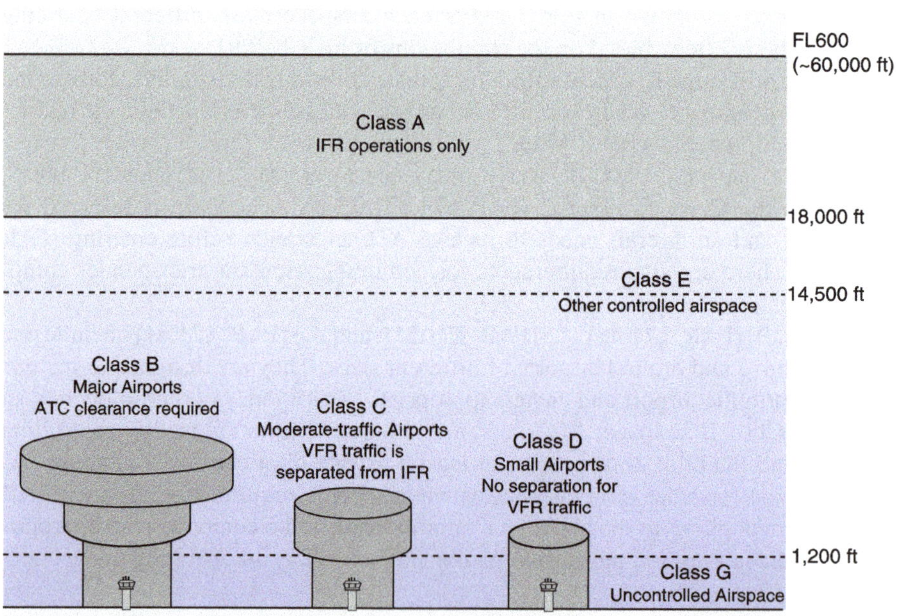

Fig. 3.1 Overview of the NAS classes. This figure depicts only a general view of the airspace classes. For the accurate extends of each airspace class the reader is referred to the latest FAA Order 7400.9.

3.7 Regulation Development Models

The traditional model of regulation development has been based on sufficiently mature technologies for which standards had been developed and possibly implemented. The regulatory body, in this case the FAA, undertakes the task of assessing the technology and standards available and develops appropriate regulations. Because of the aforementioned requirements this process is slow, costly and in some cases counter-productive since developed technology and standards are not necessarily adopted.

In March of 1996, the US Congress recognizing the need to advance cooperation between industry and the Federal Government, signed Public Law 104-113, also known as the National Technology Transfer and Advancement Act of 1995. Section 12(d)(1) reads:

> Except as provided in paragraph (3) of this subsection, all Federal agencies and departments shall use technical standards that are developed or adopted by voluntary consensus standards bodies, using such technical standards as a means to carry out policy objectives or activities determined by the agencies and departments.

This "industry consensus" model was recently used for the regulation of the LSA category. In this case, the FAA participated actively in the development of standards and as a result these standards were immediately incorporated into the regulatory framework upon publication. This approach is faster and more cost-effective, since the burden of drafting the standards is mostly with the industry. In addition to that conformance with the standard is self-regulated and FAA involvement is limited to oversight and penalizing non-conformance thus further reducing the cost to the federal government [2].

Regardless of the regulation model, another key characteristic of regulation development is the basis of the airworthiness certification. Military systems have been traditionally evaluated under a safety target approach [7]. In this case the aircraft is designed and operated with a particular role and operating environment in mind and the airworthiness certification includes appropriate operational restrictions to ensure that an adequate level of safety is achieved [7].

According to manned aviation regulations airworthiness is based on compliance with a code of requirements [7]. This approach has the advantage of no interlinking between airworthiness and operation, facilitating a variety of applications as well as interoperability. Nevertheless special safety targets are also included to evaluate new technologies or designs, not covered by existing code [8]. These can be found in *paragraph 1309* of current Certification Specifications (CS) or the corresponding Acceptable Means of Compliance (AMC) sections.

References

1. Academy of Model Aeronautics (2007) 2008 Official national model aircraft safety code. Effective January 1, 2008

2. Anand S (2007) Domestic use of unmanned aircraft systems: Evaluation of policy constraints and the role of industry consensus standards. Journal of Engineering and Public Policy 11
3. ASTM International (2006) Standard practices for unmanned aircraft system airworthiness. Standard F 2501 - 06
4. Federal Aviation Administration (2004) Airworthiness certification of aircraft and related products. Order 8130.2F
5. Federal Aviation Administration (2007) Airspace designations and reporting points. Order 7400.9R
6. Federal Aviation Administration (2007) Unmanned aircraft operations in the national airspace system. Docket No. FAA-2006-25714
7. Haddon DR, Whittaker CJ (2002) Aircraft airworthiness certification standards for civil UAVs. UK Civil Aviation Authority
8. Joint JAA/Eurocontrol Initiative on UAVs (2004) A concept for european regulations for civil unmanned aerial vehicles (UAV). Final Report
9. Schultz R (2006) Ultralights, LSAs and kit airplanes – what's the difference? Florida Aviation and Business Journal [online] http://www.airportjournals.com/Display.cfm?varID=0609005, accessed April 2008

Chapter 4
Unmanned Aircraft Systems Regulation

Nulla lex satis commoda omnibus est: id modo quaeritur, si majori parti et in summam prodest.

No law can possibly meet the convenience of every one: we must be satisfied if it be beneficial on the whole and to the majority.
Livy Ab Urbe Condita ca. 29 B.C.

This Chapter provides an overview of the history and current status of UAS airworthiness and operational regulation worldwide. Due to the different nature of operations and the regulations enforced, civilian and military systems are addressed separately. In addition, the new Sect. 4.9 is devoted to light UAS regulation, since such systems are expected to see a large growth in the civilian sector over the coming years.

4.1 Introduction

Over the last decade the interest for civil, as well as public UAS operations has steadily increased. Stakeholders are requesting NAS access with rules similar to those for manned aviation. In the US alone, about 100 companies, universities and public organizations are developing over 300 different UAS designs [45].

On the other hand, safety concerns are working against a quick integration of UAS in the NAS. The following excerpt from a talk of Mr. N. A. Sabatini, Associate Administrator for Aviation Safety before the House aviation subcommittee [32] is indicative of the concerns of all aviation authorities:

> ...there is a missing link in terms of technology today that prevents these aircraft from getting unrestricted access to the NAS ...

This concern is also shared by other NAS stakeholders. The Aircraft Owners and Pilots Association (AOPA) has repeatedly stated that UAS (even small ones) should be governed by the same rules that apply to manned aircraft [41]. AOPA is also against the use of restricted airspace to segregate UAS operations, since pilots would need to circumnavigate that airspace raising their expenses [119].

Despite these problems, many countries have established preliminary operational guidelines that allow limited operations in their respective NAS. For safety reasons UAS flight is currently segregated from the rest of the air traffic, mainly with the use of NOTAMs [47]. Current regulations are interim measures, drafted in prepara-

K. Dalamagkidis et al., *On Integrating Unmanned Aircraft Systems into the National Airspace System*, Intelligent Systems, Control and Automation: Science and Engineering 54, DOI 10.1007/978-94-007-2479-2_4,

tion for the development of a complete regulatory framework. In cooperation with industry and academia, national aviation authorities and international organizations are preparing roadmaps, airworthiness and design standards, as well as policies.

4.2 International Civil Aviation Organization

The large number of stakeholders in many different countries and the need for international operations and interoperability led to the involvement of many different organizations in the regulatory efforts. In many cases the progress has been shared among different groups and there have been also joined efforts like the first EUROCAE/RTCA joined meeting in Florida, in January of 2007. This Section provides a brief overview of related efforts undertaken by the International Civil Aviation Organization (ICAO).

Several states like Australia, Canada, Finland, Italy, Malaysia, Sweden, South Africa, the UK and the US, are currently implementing procedures to issue special operating authorizations for UAS [65]. Furthermore, many states foresee international civil UAS operations in the near future [65], a fact that has motivated the ICAO to explore UAS regulations. ICAO is an international organization with the primary goal of promoting collaboration between civil aviation authorities [48]. Although ICAO does not regulate military aviation, it has been active in activities aimed to improve civil/military cooperation [48].

ICAO involvement with UAS dates back to April 2005, when it decided to consult some of its member states regarding current and future UAS activities in their NAS, and the need for ICAO guidance material [65]. An informal, exploratory meeting followed in May 2006 in Montreal, Canada, where attending delegates of fifteen states and seven international organizations agreed that the ICAO was not the appropriate body to lead the regulatory effort, which should be based on the work of RTCA, EUROCAE and other bodies [65]. Nevertheless, several states proposed that ICAO assume the role of the coordinator, to ensure harmonization of regulations between member states [63].

In a second ICAO meeting during January 2007 in Florida, a UAS study group (UASSG) was established with the goal of supporting the regulation and guidance development within the ICAO [67]. The UASSG has members from 15 member states and 8 organizations like EASA, EUROCAE, IAOPA and others [12]. The interest of ICAO in UAS was also shown by a working paper presented by the US in the 36th ICAO Assembly in September of 2007. In that paper, the need to amend the accident definition with occurrences involving UAS and appropriate investigation of such accidents was put forth [66].

The first meeting of the UASSG took place in Montréal, Canada in April of 2008 [68]. This meeting was preparatory in nature and involved deciding on the role and objectives of the group, electing officials, reviewing other regulatory activities and agreeing on a work program [68]. Since then, additional meetings have taken place on a semiannual basis. Currently, UASSG has drafted a circular on UAS and on how they are going to be addressed within ICAO, a document that is going to be published in early 2011 [12]. The focus of the circular will be on international UAS operations in non-segregated airspace [12, 69]. The longterm goal of the study group is the development of ICAO standards and recommended practices, procedures and guidance material for civil unmanned aircraft systems [12]. Towards that goal, a study of current ICAO regulations has been performed to identify gaps and to organize future work [12]. It should be reiterated though that ICAO focuses on requirements and regulations that enable cross-border operations [9].

With respect to the radio frequency spectrum allocation for UAS, ICAO has drafted an initial position for the 2011 World Radiocommunication Conference [69, 108]. This position recognizes the importance of allocating and protecting spectrum for UAS operations [108]. Dedicated spectrum is vital for UAS safety since it is a prerequisite for command and control, communications with ATC, as well as sense and avoid. Payload communications requirements were also recognized, but they are deemed not critical and as such, ICAO recommends that it is not addressed in the same agenda item [108]. The ICAO position was finalized with the assistance of workgroup F of the ICAO Aeronautical Communications Panel and was sent to the ICAO member states in June of 2009 [13].

4.3 United States

The first efforts towards UAS regulation took place as early as 1991, when, at the time the FAA issued a notice for proposed rule making and formed an industry support group [90]. Over the following years, work progressed mostly with development of ACs regarding design, maintenance, pilot qualification and equipment requirements, among other topics.

In 2001, the New Mexico State University published the first version of a document titled "High Altitude Long Endurance, HALE, UAV Certification and Regulatory Roadmap" [90]. That work was sponsored by the NASA Erast Project and since then, newer versions have been published with feedback from other stakeholders. The goal of that document was to be a basis of discussion between the FAA, the industry and other stakeholders for establishing regulation for aircraft airworthiness, flight standards and air traffic that will allow safe operation of HALE UAS in the NAS. This effort was continued with the Access 5/UNITE program also sponsored and funded primarily by NASA with participation of the UNITE consortium, the FAA, the DoD and other stakeholders. The aim of this project was to integrate HALE UAS in the NAS [1]. Although the program was terminated early in February of 2006 due to budgetary reasons [1], it has been claimed that it was successful in forging solid relationships between industry partners and public agencies, as well as producing valuable recommendations for the integration of UAS into the NAS [112]. The UNITE organization with members from US-based UAS manufacturers was initially disbanded after the completion of the Access 5 program, but has since been reanimated [75]. Its new goals include the acceleration of approval of UAS for certain applications and the streamlining of the request and approval processes of current certification paths [75].

Although the LSA model (see Sect. 3.7) for regulation development was successful in enabling accelerated NAS access and reducing costs without compromising public safety [2], FAA has not used it for UAS regulation. Instead, the FAA seems to adopt a more cautious approach, mainly because UAS technology is still under development and cannot be considered mature. This is further indicated by the FAA's administration belief that the process of UAS airworthiness certification is riddled with technical challenges such as the sense and avoid system and the communication issues between ground station, ATC and aircraft [97].

In October of 2003, the FAA published Order N8700.25 [33], as a response to inquiries it had received regarding UAS operations. Until that time the use of Certificates of Authorization (COA) for UAS operations had been limited for military operations based on an agreement between the FAA and the DoD. With this order, the FAA opened the door for non-military operations, mainly for proof-of-concept demonstrations. It was also made clear that it was not applicable to model aircraft that can only be used in non-commercial, recreational applications.

Two years later, in September of 2005, the FAA issued "AFS-400 UAS Policy 05–01" [35], which provides guidance for issuing COA. With this policy it was clarified that COA applications would not be available for civil UAS. This policy required the presence of observers to ensure that the UAS complies with right-of-way rules and the see-and-avoid requirement of FAR Part 91. The policy also provided operational guidelines, as well as minimum pilot and observer qualifications. Regarding the risk to other aviation, the policy requires a safety analysis indicating that mid-air collisions are extremely improbable. Similarly for operations over populated areas, the estimated risk of injury to people on the ground must be highly unlikely. Nevertheless an exception is made allowing the approval of COA applications from the DHS or DoD, even for non-conforming UAS, when the operations are a matter of national security.

In March of 2006 the FAA established the Unmanned Aircraft Program Office to facilitate the UAS regulation process [97]. This office, in cooperation with the FAA Air Traffic Organization, is responsible for evaluating COA applications [115]. A few months later, in September, it contracted the Lockheed Martin to begin development of a five year roadmap for integration of UAS in the NAS [106, 118]. Although the first version of the roadmap was supposed to be published in March of 2007 [106], publication is still delayed pending review and approval.

In February of 2007, the FAA issued a notice on UAS policy, seeking feedback from stakeholders [37]. The notice retains the pilot-in-command and UA observer requirement. Furthermore a new definition for UA was given, as follows:

> ... Unmanned Aircraft is a device that is used, or is intended to be used, for flight in the air with no onboard pilot ...

According to the FAA this definition includes everything from small R/C model aircraft to large full scale aircraft and from remotely operated to fully autonomous systems [37]. The current FAA policy is also stipulated as follows:

No person may operate a UAS in the national airspace system without specific authority.

The authority is the COA for public UAS, a special airworthiness certificate in the experimental category for civil UAS and the AC91-57 for models. The latter is available only to modelers and only for non-commercial, recreational purposes. In addition to that, the FAA is currently pursuing a Special Federal Aviation Regulation (SFAR) for small UAS [23] with an instrument similar to AC91-57 [37, 105]. Towards that end, a Small UAS Aviation Rulemaking Committee was established in 2008 [41]. This committee submitted its recommendations a year later [103]. The FAA is currently drafting a proposed rule, aimed to be published in 2011, while the final rule will be available a year later [44]. More information on the recommendations of the aforementioned committee are provided in Sect. 4.9.

In March of 2008, the FAA issued an updated guidance document, titled "Interim Operational Approval Guidance 08–01", that replaced the AFS-400 policy [42] and is applicable to both civil and public UAS operations. With this document, the FAA first makes note that all UAS operators must at a minimum comply with FAR Parts 61 and 91. Other important changes include the incorporation of a continued airworthiness requirement and the provision for alternate means of compliance to the observer requirement previously maintained. Although the issue of sense and avoid is still considered important, the presence of observers is no longer required, provided that other risk mitigation measures are proposed and then found through appropriate safety studies to be adequate.

Later that month, the FAA issued "Order 8130.34 Airworthiness Certification of Unmanned Aircraft Systems" that defines the requirements for issuance of special airworthiness certificates in the experimental category for UAS.

For 2008, the FAA had declared an initiative to "Develop policies, procedures, and approval processes to enable operation of unmanned aircraft systems (UAS)" [36], a statement that was repeated in its Flight Plan for 2009–2013 [40]. The issues to be resolved include frequency spectrum in densely populated areas and at low altitudes as well as the completion of an appropriate safety analysis that will determine the size and speed restrictions [23]. The business plan for 2010 also includes the introduction of UAS in the FAA aviation forecasts [43]. In the currently published forecast, 15,000 UAS are expected to be operated by civilian/commercial entities by 2020, with that number doubling by 2030 [45].

At the moment, the FAA is active in reviewing current regulations and performing a gap analysis with respect to UAS [49]. The areas of interest unmanned rotorcraft, small airplanes and transport aircraft, as well as engine and propellers [49]. In parallel with UAS integration, the FAA is also working on the next generation of the ATM system, called NextGen. Part of the research for NextGen will involve solutions for integration of UAS, as well as interim measures for improving performance in areas such as see and avoid [6]. Regarding NextGen and UAS, the FAA has already implemented the initial operational capability for UAS simulation into a tool that is used to assess NextGen technologies and their integration [91]. This was followed by a demonstration of a 4D trajectory flight management system. For the following years, the demonstrations will be enhanced to include collision avoidance technologies such as ADS-B, while work is also planned on joint network enabled operations that include UAS [91].

A major issue for current UAS developers is the ability to test and evaluate the performance of their UAS, which at the moment is limited by FAA restrictions. To alleviate this problem, a cooperative agreement currently coordinated between the FAA and the New Mexico State University led to the creation of the first UAS flight test center in the US [23, 98]. The test center opened in the spring of 2008 [112] and operates with procedures approved by the FAA and collects important information regarding component as well as system performance [62]. Another FAA/DoD UAS joint research lab is going to be founded in the FAA Technical Center [23]. At the same time, the FAA is collecting operational data from UAS operators authorized to flight either by a COA or an experimental certificate [112]. Such data will be useful for the development of policies, but the Government Accountability Office (GAO) reported in 2008 that due to lack of resources the FAA had been thus far merely archiving the data collected [112]. Lately, the FAA launched a new laboratory that scientists can use to simulate UAS in the future ATM environment, as well as made new cooperative research agreements with Insitu and GE [18]. In the second half of 2009, the state of Ohio released funding for the "Center for Unmanned Aerial Vehicles Exploitation", a collaboration between industry, academia and the military.

The FAA currently believes that UAS technology is still not mature for integration into the NAS. The following excerpts from a speech given to aerospace executives in late 2009 by the FAA administrator, Mr. J. Babbitt are characteristic of the stance of the agency [6]:

> ...the level of [UAS] technical maturity is not where it needs to be for full operation in the NAS ... [UAS present] an exponential leap in capability, and that leap needs a contemporaneous jump in technology and procedures to do so safely ...
> ...complexity [of the UAS see-and-avoid system] is what makes it difficult to meld the UAS safely into a mature system like the NAS ...
> ...most UAS have a single point of failure for hydraulics, electrical, flight control and satellite link. That is a concern. When there is a single point of failure for something that runs into trouble every thousand hours, that is a problem ...

Especially regarding the issue of sense and avoid, FAA officials reported to the GAO in 2008 that a reliable system could cost as much as $2 billion to develop and is in any case many years away [112].

4.3.1 RTCA

To assist with UAS-related technical issues the FAA contacted the RTCA which, in October of 2004, formed committee SC-203 with participation of government and industry representatives from several countries. The first task was to develop "Guidance Material and Considerations for UAS", a document that was issued in March of 2007. In addition to that, the committee has been working on Minimum Aviation System Performance Standards (MASPS) for:

- UAS;
- Command, Control and Communication Systems for UAS;
- Sense and Avoid Systems for UAS.

These standards were initially expected to be completed by 2013 [44]. Nevertheless, RTCA quickly realized that because of the large scope of the project, as well as the unique characteristics of UAS and the way they operate, a thorough safety evaluation is required that may take as long as 2017 or 2019 to complete [112, 116]. This in turn means that UAS may not receive routine access to the US NAS before 2020, that is before the standards developed by the RTCA can be incorporated into FAA regulations [112].

4.3.2 ASTM

Another organization that has been particularly active with the development of standards for UAS is the ASTM that has also formed a specialized committee for this reason. The goal of the ASTM F38 committee is to build technical standards that will support the MASPS developed by the RTCA [72]. The ASTM created four subcommittees, each tackling different subjects:

1. Airworthiness;
2. Flight Operations;
3. Personnel Training, Qualification and Certification;
4. Executive.

The ASTM performed a gap analysis to identify points were current manned aircraft regulation had to be adapted or amended to apply to UAS. This analysis included airworthiness requirements, as well as UAS operations and crew qualifications. For these identified gaps, the ASTM proceeded with planning for and developing appropriate standards. So far the ASTM has produced more than 10 such standards, one of the most known being the F2411-07 Standard Specification for Design and Performance of an Airborne Sense-and-Avoid System, which according to the ASTM has been adopted by the DoD [3]. Others include "Standard Practices for Unmanned Aircraft System Airworthiness", "Standard Practice for Quality Assurance in the Manufacture of Light Unmanned Aircraft System" and "Standard Practice for Unmanned Aircraft System (UAS) Visual Range Flight Operations". There are also several standards on UAS propulsion systems, launch systems and maintenance manuals. Moreover, there are several standards currently in development that relate to small UAS, including "Selecting sUAS Takoff and Landing Sites" and "Design and Construction of Group III sUAS".

The ASTM through its standard practice document [4], has proposed two certification pathways; type certification leading to a standard airworthiness certificate for large UAS and a "Light UAS" special airworthiness certificate similar to that for LSA. The special airworthiness certificate for the LSA category is issued by the FAA if the aircraft complies with all eligibility requirements in [34] and after the manufacturer of the aircraft provides all the necessary documents that certify compliance with industry consensus standards [101]. The only requirement mentioned by the ASTM for eligibility in the "Light UAS" category is a maximum take-off weight (MTOW) of at most 600 kg. In addition to that, the ASTM is currently working on a standard guide document for mini UAS airworthiness, as well as a review of requirements for unmanned rotorcraft.

4.3.3 SAE

The Society of Automotive Engineers (SAE) has been active for several years in the area of aerospace standards, with close to 6,700 already published, many of which apply to UAS, as well [79]. To better address the unique requirements of UAS, SAE has created the specialized AS-4 committee which is comprised of the following three subcommittees [79]:

AS-4A Architecture Framework
AS-4B Network Environment
AS-4C Information Modeling and Definition

The AS-4 in collaboration with other relevant SAE committees has already been working on a number of new and revised standards, especially with respect to the Joint Architecture for Unmanned Systems (JUAS) that was migrated from the US DoD to SAE [79].

4.3.4 Current Certification Paths and Operational Guidelines

Currently, flight of public UAS in the US is authorized on a per-case basis and after a COA application is filed at least 60 days prior to commencement of operations. The COA is issued after submission of required documentation and an analysis performed by the FAA Air Traffic Division to determine that an ELOS with that of manned aviation is achieved. COA applications for public UAS are approved based on compliance with MIL-HDBK-516 "Airworthiness certification criteria" or other approved policies listed in [42] and are normally effective for up to one year. It should be noted that the certification basis is the responsibility of the public agency operating the UAS [105]. It is noteworthy that a COA is typically issued for a specific region of operations, UAS and operation type. As a result an agency operating different UAS types or under different operating scenarios would need multiple COAs, increasing FAA workload and possibly delaying operations [112]. Nevertheless an exception to that rule was made with a national COA that was issued to the United States Air Force for operating the Global Hawk in the NAS, primarily for training purposes [114].

As mentioned in the previous Section and according to current FAA policy, COA applications are accepted only for public UAS. Civil UAS can get a special certificate under the experimental category, a policy prototyped in 2007 [105]. Currently experimental certificates are available for research and development, crew training and market survey purposes and are issued based on Order 8130.34 [39]. As a result, no instrument is currently available for commercial (for-hire) operations. According to that order, the application must be accompanied by a program letter, a safety checklist, charts of the area of operations, training manuals, pilot and medical certificates. The program letter details the characteristics of the UAS, the purpose and type of operations, the area of operations, safety measures taken, etc. Before issuance of the certificate, FAA personnel will conduct a safety evaluation of the documentation provided, followed by an on-site inspection.

Operations under an experimental certificate are possible for up to one year and are subject to the same restrictions imposed for that category in FAR Part 21 [38] and possibly additional provisions set by the FAA, specifying other operational requirements [59]. Additionally, market survey operations are not allowed unless the manufacturer has accumulated 50 flight hours under a COA or experimental certificate. It should be noted that the FAA considers both the COA and the special airworthiness certificate processes as interim measures [97].

Despite the regulatory problems, a significant interest for the use of UAS was demonstrated with the number of COA applications. In 2005 the FAA issued 50 COA and more than 100 were issued in 2006 [21, 121]. By the end of 2009, close to 500 COA had been issued (146 that year alone) and almost 200 were still pending [44]. The interest for special airworthiness certificates was also significant, since by the same time, close to 60 of those had already been issued [44] and several more were pending [23]. FAA had announced a reduction of the number of special airworthiness certificates issued, to four per year due to high work load [105]. Nevertheless, the current actual number is quite higher, with about 20 being issued in 2009 alone [40, 44]. It is also noteworthy, that through the use of the aforementioned instruments, in 2009 alone there were 20,000 UAS flights totaling 2,500 flight hours in civilian airspace [6].

UAS operations are also possible without a COA or an experimental airworthiness certificate for operators that have access to restricted airspace. These kind of operations can take place in coordination with the authority responsible for controlling the airspace and under any restrictions deemed necessary.

All UAS operations are subject to the guidelines established in the "Interim Operational Approval Guidance 08-01" [42]. As mentioned earlier compliance with the see-and-avoid requirement is of particular importance to the FAA. Current guidance presents three alternatives; segregation of operations, the presence of qualified UA observers unless operating in IFR conditions or adequate onboard see and avoid (S&A) capability. Observers can be either on the ground or onboard a chase aircraft, but must maintain constant communication with the UA operator to assure collision avoidance. In addition to that, radio communication with the ATC should be available to the UA pilot. It should be noted that operations over trafficked roads and open-air assemblies are to be avoided while flight over populated areas is allowed only in disaster relief or other emergency situations. To enhance safety, guidelines require the presence of a facility allowing the pilot to take over control, sufficient system redundancy or when not possible a flight termination system, as well as of provisions to recover the UA in the case of loss of the communications link.

4.4 Europe

The activities regarding UAS operation and regulation in Europe are slightly more complicated due to the fact that the individual states are responsible for their own airspace. As a result, certain countries may appear to be ahead or lagging behind in UAS regulation. On the other hand there have been initiatives aimed at harmonizing the regulations of the different nations. Originally, the Joint Aviation Authorities (JAA) developed such harmonized codes, until a few years ago when the European Aviation Safety Agency (EASA) was founded and took over some of the responsibilities of the JAA. In contrast with the JAA, the EASA is also capable of passing rules into law through the European Commission. In addition to the EASA, there is a separate body (EUROCONTROL) that is responsible for ATM over most of Europe. EUROCONTROL has 38 member states and is an integral part of the "Single European Sky" program that is set to redefine the airspace management over Europe. As a result, this Section will start with an overview of the activities of the EASA and EUROCONTROL. This will be followed by EUROCAE, an aviation standardization body that produces test specifications, system performance standards and operational guidance, similar in scope with RTCA. The section on Europe closes with a brief overview of the activities of other EU and national organizations.

4.4.1 EASA

In 1999 and after a joint EUROCONTROL/NATO workshop, the JAA was requested to prepare a document on UAS certification [26]. Five years later the JAA/EUROCONTROL UAV Task force issued its report [74] on civil UAS regulation. A year later, EASA issued an Advance Notice for Proposed Amendment (A-NPA) based on that report, titled "Policy for Unmanned Aerial Vehicle (UAV) certification" [30]. The scope of the A-NPA was limited to airworthiness of UAS with MTOW over 150 kg and contained provisions regarding environmental protection as well [89]. Nevertheless, EASA believes in the necessity of complete UAS regulations that cover airworthiness, environmental protection, operations, crew licensing, ATM and airport [88]. As a result, although S&A operational requirements are not part of the A-NPA, EASA has contacted EUROCAE to develop an appropriate specification [89]. The purpose of the A-NPA was to provide a view of future UAS policy in Europe and solicit comments from stakeholders [89].

The Comment Response Document (CRD) was published online in June of 2007 and was open for feedback for one year [89]. Several comments in the CRD critique the lack of a comprehensive regulatory framework for UAS. EASA has acknowledged that this is indeed an issue but believes that this is a longterm goal and has chosen to defer it until the policy as prescribed in the A-NPA is available. Another important item in the CRD is the regulatory approach chosen. In the A-NPA, EASA had presented two approaches; the "conventional" approach that is based on system specifications and the "safety target" approach that is based on a specific UAS and a specific operational environment. In the CRD, EASA has clarified that a "safety target" approach is inadequate and was only included to solicit comments. It should be noted that, although in the initial A-NPA text the term UAV was used, in the CRD EASA has decided to adopt the term UAS and as a result align itself with other international partners.

In the summer of 2009, EASA published a policy statement regarding UAS certification based on the A-NPA and the CRD. This document does not apply to military or public UAS, systems with weight less than 150 kg, as well as research, experimental or UAS used for scientific purposes [28]. Moreover, the policy only covers procedures for type certification and does not include operational guidelines [28]. According to EASA, for certification of civil UAS, the procedures of Part 21 apply, leading to the issuance of an airworthiness certificate [28]. Furthermore, the option of restricted type certificate and a corresponding restricted airworthiness certificate is given to systems that cannot comply with the regular type certificate requirements [28]. In the latter case, it is possible to employ a safety target approach to demonstrate that operation of the UAS does not violate the safety target defined by EASA [28]. In the paragraph that follows, some of the key elements of the aforementioned policy statement are provided.

Since by definition a UAS also includes the ground control station, launcher and other components, EASA has determined that any *"remote equipment performing functions that can prejudice take-off, continued flight, landing or environmental protection, shall be considered as part of the aircraft and included in the type-certification basis"*. Since UAS-specific Certification Specifications (CS) have not been developed, certification will be based on equivalent CS for manned aircraft. The determination of which airworthiness standards are applicable in each case will be made based on the kinetic energy at impact after two different failure scenarios. When the latter is determined, the manned aircraft category that best corresponds to these impact characteristics will be used as a basis for certification. In addition to that, special conditions that concern among others emergency recovery, human-machine interface and the link between the ground control station and the UA, will also need to be included in the certification basis. Under certain conditions, EASA will also accept the airworthiness requirements defined in STANAG 4671 (*see* Sect. 4.10.3) as the certification basis. Finally, although EASA considers the issue of S&A of major importance, it believes that the responsibility lies with the authorities tasked with regulation of air navigation devices. These authorities should determine the performance requirements depending on airspace class and other aircraft operating together with the UAS. When these requirements have been determined, the S&A system will be approved similarly to other avionic systems.

Although EASA delegates responsibility for Light and other classes of UAS to national authorities, it is involved with the latter to provide guidance and assistance as needed [50]. For the future, EASA has programmed the development of additional UAS related regulation, including acceptable means of compliance (AMC), with a final decision for the latter currently scheduled for mid-2015 [29].

4.4.2 EUROCONTROL

EUROCONTROL is an organization tasked with the planning and development of a pan-European ATM. In a parallel effort to EUROCAE and EASA, it is currently investigating ATM requirements for UAS integration in the NAS. In December of 2007, "Specifications for the use of Military UAVs as Operational Air Traffic" were published [83]. This work provides the basis of specifications for ATM for the operation of the Global Hawk UAS in European airspace, after a request from NATO to the EUROCONTROL. The next step is to develop similar specifications for UAS as General Air Traffic, that will include civil UA as well, but with the requirement of equivalent performance with that of manned aviation [83].

The goal of EUROCONTROL is to avoid segregation and satisfy the requirements of all legitimate airspace users [82]. It should be noted that EUROCONTROL considers UAS integration only under the condition that no disruption will occur to current civil aviation that is expected to grow significantly in Europe (240% from 2005 to 2025) [82]. As a result, it is the view of EUROCONTROL that UAS should adapt to ATM requirements, rather than the reverse [85]. To better assess the interaction between UAS and ATC, EUROCONTROL proceeded with real-time simulation in September 2008, that provided invaluable insight on ATM issues with respect to UAS [60]. To better coordinate Agency activities, the "EUROCONTROL UAS ATM Integration Activity" was established in 2008, that will investigate all aspects of UAS technology and operations from ATM and safety, to human factors and communications [84]. The activity takes place into two concurrent streams, one dealing with short integration and one dealing with integration of UAS into SESAR (the Single European Sky ATM Research Programme) [84] which is currently in development and planned to be deployed in 2014–2020. The SESAR program involves key technology areas that are shared with UAS, such as Sense and Avoid, 4D trajectory management and advanced communications [85].

4.4.3 EUROCAE

EUROCAE is a standardization body, traditionally responsible for developing minimum performance requirements as a basis for EASA TSOs [57]. In April of 2006 workgroup 73 (WG-73) was launched to provide expertise on UAS and assist EASA in the development of appropriate airworthiness criteria that will supplement the policy that will follow from the A-NPA [26, 57]. Currently EUROCAE has taken the lead to develop UAS standards and guidance based on the recommendations of the JAA/EUROCONTROL and EASA reports [54]. To that end four subgroups have been formed, each focusing on a different area [57]:

1. UAS operations and S&A;
2. Airworthiness and continued airworthiness;
3. Command and control, communications, spectrum and security;
4. Small UAS (<150 kg).

EUROCAE is currently considering six classes of UAS operations, for which tailored operational standards need be developed [81]. These standards will include all aspects of UAS operations, from flight in non-segregated airspace, to crew qualifications and UAS operator organization [81]. On the issue of S&A, discussions have commenced with RTCA to achieve a common technical standard [81]. In general, the two organizations have decided on tighter collaboration to better take advantage of available resources and to expedite the development of UAS technical specifications [56].

The goal of EUROCAE is the development of airworthiness requirements compatible with current ATM infrastructure and procedures without affecting current operations [57]. The next major deliverable is a concept document on airworthiness certification and operations of UAS in non-segregated airspace that was initially slated to be released by the end of 2008 [57], but the final version will not be ready before 2010 [56].

Communications and security is another major issue, for which EUROCAE is planning to produce a special document on UAS Command Control and Communication Systems [56]. Towards this goal, EUROCAE is currently in cooperation with several stakeholders in Europe and internationally to bid for the allocation of UAS spectrum [55]. Regarding security, WG-73 is in cooperation with WG-72 which is involved with aeronautical systems security [55].

4.4.4 Other European Union Activities

The EU has sponsored other concurrent activities by means of the Fifth European Community Framework Programme that covered research, as well as technology development and demonstration. These activities included the UAVNET network that begun its activities during October of 2001. The goal of this network has been to coordinate research and development activities in the area of civil UAS. Under the UAVNET two "critical technology" programmes have been funded by the EU; USICO and CAPECON. The goal of the first was to investigate regulation, procedures and technology related to improving operational safety and was completed in April of 2004. CAPECON's goal was to identify civil UAS applications and configurations to comply with safety criteria in a cost-effective way and was completed in April of 2005. UAVNET itself issued in March of 2005 a "European Civil Unmanned Air Vehicle Roadmap" [93] that identifies strategies to enhance the European UAS activities and bridge the gap with the US. Another program funded this time from the Sixth Framework Programme was IFATS, which investigated the concept of an autonomous air transportation system. This latter project was concluded in March of 2007.

In 2007, the European Commission launched the INnovative Operational UAS Integration (INOUI) program, also funded by the Sixth Framework Programme. The goal of the INOUI is to complement SESAR activities by developing requirements for insertion of UAS into non-segregated civilian airspace [107]. The work of INOUI is divided into several work packages. The first one is concerned with UAS procedures and requirements for integration into the SESAR in 2020 [64]. For this purpose INOUI gathered data from previous flight tests and simulations and identified key procedures (e.g. those relating to loss of data link or transfer between control stations), as well as the main technological enablers required. Other work packages focused on technology issues like ATM architecture and the System Wide Information Management (SWIM). The latter is a key element of ATM in SESAR and refers to the timely and accurate sharing of important data between airspace users and other stakeholders [64]. The INOUI program also investigated the regulatory roadmap in Europe and issues of operational safety, proposing its own acceptable accident frequency limits [64].

Several projects undertaken within the EU were also successful in demonstrating safety in key aspects of UAS operations. Interaction with ATC was demonstrated using the conclusions of the USICO project for operations over a busy German airport [93]. Similar flights over inhabited areas and in some cases in the vicinity of major airports were performed in Netherlands, Sweden and Belgium [93]. The latter has certified the B-Hunter UAS for civil operations over most of its airspace. Netherlands is currently in the process of certifying an unmanned helicopter, allowing other states to observe and/or participate [22].

For the future, the European Commission is looking to improve public acceptance of UAS and provide a step-wise integration of UAS into the European Airspace [107].

The European Defence Agency (EDA) announced in January of 2008 that it had awarded a contract to the Air4All consortium, consisting of various defense and aerospace companies. The contract called for the development of a "UAV Air Traffic Insertion Road Map" by 2015 [112] and was complemented by a follow-up study. The Air4All consortium provided a number of recommendations regarding the certification of ground control stations, UAS crew, UAS technology such as obstacle avoidance, weather avoidance and automatic take-off and landing, as well as requirements for cross-border operations in non-segregated airspace [11].

4.4.5 National Activities

Many UAS regulation-related activities have been done on a national level or with participation of few nations. In the UK, the Civil Aviation Authority (CAA) — referred to as CAA-UK in this book to avoid confusion with other civil aviation authorities— has been particularly active with civil UAS regulation development, issuing guidance on civil as well as military UAS operations in May 2002 [25]. That guidance document, titled "CAP 722, Unmanned Aerial Vehicle Operations in UK Airspace — Guidance", is currently in its third edition, dated April 2008. Further promoting UAS operations in the UK is ParcAberporth, a technology park with a dedicated UAS operations center [7] that hosts an annual UAS conference. In addition to that, a significant portion of the airspace over the airport has been designated restricted, thus, allowing unencumbered UAS operations.

Also in the UK, the Autonomous Systems Technology Related Airborne Evaluation & Assessment (ASTRAEA), a consortium of UAS stakeholders, has emerged with the objective of developing enabling technologies for communications, S&A and emergency flight termination [27]. The consortium is funded by public agencies and its goal is to improve UK positioning in the area of UAS technology [86]. Since its inception in 2006, ASTRAEA has performed technology demonstrations in ParcAberporth showcasing collision avoidance technology, object identification and communications [85, 86].

In central Europe, the "UAV DACH" working group has been working on developing standards and regulations covering UAS operations, as well as technical solutions for issues like sense and avoid [112]. Although established for the German-speaking countries in central Europe, its activities can be regarded as international.

The German Aerospace Center (DLR) tested UAS operations in controlled airspace using existing technologies and infrastructure. The project was completed in 2005 after a successful flight demonstration. The German Aviation Safety Agency (DFS) started the "Validation of Unmanned Aircraft Systems Integration into the Airspace" (VUSIL) project, sponsored by the German government in 2007 [109]. A UAS manufacturer as well as the German military were also involved in this project, evaluating S&A from the perspective of the UAS operator and the manned aircraft pilot, using both simulated and real flights [109]. Currently the trend in the area is to try to match the see and avoid capability of the pilot to on-board, using technology-based sense and avoid. In the VUSIL project, radar information was transferred back to the pilot, to re-enable the see and avoid function [110]. A number of technologies that assisted with traffic monitoring and collision avoidance were tested and the feedback is going to be used in the second phase of the project that started in 2009 [109].

Other countries with active participation in UAS regulation in Europe include Austria, Belgium, the Czech Republic, Finland, France, Italy, the Netherlands, Norway, Spain and Switzerland among others.

4.5 Australia

The Australian Civil Aviation and Safety Authority (CASA) was the first agency to issue civil UAS regulations in 2002 [24, 93]. The regulatory approach taken by CASA, is based on risk management. This means that regulation is analogous to the perceived risk stemming from UAS operations [17]. Furthermore, regulation is based on operational requirements and not on design standards. In general, operations are not allowed over populous areas or at altitudes not sufficient to clear the area after component failures [17]. To that end CASA issued CASR Part 101 [92] that includes the regulations for UA and rockets [10]. It should be noted that the distinction between model aircraft and UAS is based solely on the purpose under which they are operated. Specifically, CASR Part 101 and AC101-3(0) state that any model aircraft, which is used for purposes other than sport or recreation, is considered a UAV and thus, Section F of CASR 101 applies [15, 92]. Moreover, UAS are divided into three groups, micro UAVs (MTOW less than 0.1 kg), small UAVs (that have MTOW up to 150 kg) and large UAVs [92]. It should be noted that in the case of rotorcraft the limit for the large category is 100 kg, whereas for an airship the limit is given as a function of the total volume it envelopes ($100\,m^3$) [92].

Similarly to other countries, CASR 101 explicitly forbids the operation of a UAS in any way that may compromise the safety of another aircraft, person or property. In general, operational restrictions typically preclude flight over populated areas (within 30 m of people not directly associated with the operation of the UAS) or at altitudes over 400 ft. These requirements can be partially or completely waived when the operation is to be restricted in an area designated for UAS operations or CASA has been notified and given permission [92]. On the other hand, the CASA may include additional operational restrictions like allowing operations only during daylight and/or within line of sight of the operator when deemed necessary. Although UAS operations for hire are allowed, they require a UAV operator certificate [92].

In the case of large UAS, an operating certificate must be issued before operations can take place [92]. AC101-1(0) provides guidance material on how to obtain an approval for UAS operations [17]. In addition to an operating certificate, an approved maintenance program and an airworthiness certificate in the experimental or restricted category are required to operate UAS [10]. Airworthiness certificates are issued based on AC21.43. It should be noted, that operations in controlled airspace are allowed under certain preconditions mainly relating to equipment and operator training [92]. For light UAS (less than 150 kg MTOW), the airworthiness requirements are waived, but an operating certificate may still be required [10]. Exempted from any restrictions are micro UAS (less than 0.1 kg) [47].

Although the guidelines provided by CASA are general enough to afford adequate flexibility to UAS developers and operators, it is possible that they contribute to increased cost and reduced functionality and as such they are considered only a temporary measure [17]. Nevertheless, by March of 2009 there were eight certified UAS operators in Australia, with the first having been certified for nearly 7 years [80]. In an interview, some of these operators expressed satisfaction with current regulation, requiring only minor changes [80]. It is noteworthy, that their main concern was with individuals that operated UAS without certification, a concern that was brought to the forefront in April of 2009 when an individual flew a model plane in a collision course with a passenger jet, albeit without success [80].

In addition to the activities of the CASA, the Australian Research Centre for Aerospace Automation (ARCAA) has also been active in supporting regulatory efforts. ARCAA, a joint venture of Queensland University of Technology and CSIRO, was founded with a mission to facilitate UAS research and certification by providing simulation, development and testing facilities for all aspects of UAS operations [117]. Recent work involved collision avoidance systems with encouraging results, when a collision between a real UAS and a virtual aircraft was successfully avoided [80]. Additionally, CASA in its 2007–2008 annual review has declared an initiative to establish a national UAS office so that UAS activities in Australia can expand without compromising safety [16].

4.6 Canada

Currently and since 2005, UAS operations in Canada are possible after an application for a Special Flight Operation Certificate (SFOC) that describes the specifics of the operation as well as safety and contingency planning to ensure the safety of people in the area. UAS operations beyond visual range are possible, but normally the applicant must demonstrate a capacity for safe operations, appropriate risk mitigation measures and sufficient S&A capability (either onboard or through the operator).

In December of 2006 and due to the increasing interest for UAS operations, the "Unmanned Air Vehicle Working Group" was convened by the Canadian aviation authority, Transportation Canada, with participation of industry stakeholders. The goal of this working group was to develop a regulatory framework for all aspects of UAS operations. The final report [113] was issued a little less than a year later. The report identifies key issues with current Canadian Aviation Regulations (CAR) that lack comprehensive safety requirements relating to S&A, flight termination systems and command, control and communication systems, among others. As a result specific amendments are proposed to the CAR, although for systems larger than 150 kg harmonization with international and NATO regulations is preferred.

The final report also proposed a roadmap to fully integrate UAS operations in the Canadian NAS by the year 2012. To achieve this goal several parallel efforts are proposed:

- Revision of current SFOC procedures by the end of 2008, to better facilitate UAS operations in the short term.
- Implement a UAS registration program starting the summer of 2008.
- Creation of working groups to adopt and/or develop UAS airworthiness and design standards to be completed by the end of 2010.

- Develop NPAs for operating and flight rules, pilot qualifications and maintenance procedures during 2009.
- Develop NPAs for airworthiness certification and operating certificates during 2010.

Other Canadian organizations that have activities in the UAS sector include Unmanned Vehicle Systems Canada (UVS) and Association for Unmanned Vehicle Systems International, Canada (AUVSI-Canada) as well as the Canadian Centre for Unmanned Vehicle Systems (CCUVS). The latter is a not for profit company governed by a panel drawn from the industry, academia and government, which provides expertise on UAS research and development, testing, evaluation and training.

4.7 Japan

Japan started using unmanned helicopters for agricultural applications (mostly pesticide spraying) almost 20 years ago, complementing a fleet of manned helicopters that had been used since 1958 for the same purpose [58, 71]. More than 2,000 Yamaha Rmax models were in service by 2002 and several are added each year [120]. Currently, over 12,000 operators are licenced to fly unmanned helicopters [99] and the fleet of unmanned helicopters has surpassed the number of manned ones used for agriculture [71].

Unmanned aircraft operations have been under the scope of the Japanese Ministry of Agriculture, Forest and Fisheries and its affiliated association, the Japanese Agriculture Aviation Association [114]. The latter has been active in developing safety standards, as well as operator certification programs.

During the last decade, unmanned systems in Japan are expanding into new applications beyond agriculture [71]. Realizing this need, four major manufacturers established a consortium, which in September of 2004 lead to the founding of the Japan UAV Association (JUAV). This organization that has since been reinforced with several new members from the industry as well as the cooperation of the research and academic community, aims to establish safety standards and guidance for unmanned systems, including fixed-wing, employed in a variety of applications.

Initially safety standards were available for remotely operated unmanned helicopters with weights up to 50kg, to fly over unpopulated areas [99]. Two such systems have been already certified to those standards; the Yamaha Rmax and the Fuji RPH2 [71]. Since then, the JUAV has revised those safety standards to include autonomous helicopters recently certifying the Yamaha Rmax G1 and has established new standards for fixed-wing UAS [100].

In 2005, the Japan Aerospace Exploration Agency (JAXA) in cooperation with other aviation agencies produced its own safety standard to guide its experimental UAS activities [70]. Currently JUAV and JAXA are collaborating to discuss standards for commercial UAS operations in non-segregated airspace. It should be noted though, that neither JAXA nor JUAV are regulatory authorities.

4.8 International UAS Associations

The Association for Unmanned Vehicle Systems International (AUVSI) is one of the oldest UAS associations, funded as early as 1973 [24]. It is also one of the largest, with more than 2,000 member organizations and 24 international chapters.

Unmanned Aerial Vehicle Systems (UAVS) Association was established in 1998 in the UK and has since been participating in different UAS-related working groups with the goal of paving the way for a vibrant UAS market [24].

In 1997 the EuroUVS association was founded in France, but changed its name to UVS International in 2004 to account for its now global scope. By 2007 the association had reached several hundred members from 35 countries encompassing industry and academia. UVS International has participated in various UAS-related studies including the JAA/EUROCONTROL task force.

The Euro UAV Industry Consultative Body was formed in early 2004 by seven EU states with the goal of paving the way for routine access to national airspace [24]. The initiative is now taken over by the European Aerospace & Defence Association (ASD).

4.9 Light and Small UAS

Before moving on with this Section, it is necessary to underscore the difference in using the term 'light' in the airworthiness certification literature of manned aircraft versus the one used for UAS. In the former category, light aircraft are those that do not exceed an MTOW of 600–650 kg depending on their use. On the other hand, EASA and before that JAA and NATO have defined light UAS as those not exceeding 150 kg. The aforementioned weight requirements for light UAS correspond better to the ultra light category as defined in the FAR Part 103. The categorization is further complicated by the use of the small UAS term by national aviation agencies to refer to systems that are significantly lighter than 150 kg and closer to the weight range of R/C systems.

4.9.1 United States

In April of 2008, FAA established the Small Unmanned Aircraft System Aviation Rulemaking Committee (sUAS ARC) in anticipation of the high, short-term interest for small, low-cost systems for aerial photography and law enforcement applications [41]. Another factor that influenced the decision of creating this committee was the fact that the FAA received criticism both from stakeholders in the UAS industry for not allowing "for hire" operations, as well as other NAS stakeholders like the Aircraft Owners and Pilots Association (AOPA) because of the risk posed by UAS to general aviation operations [41]. In parallel to the efforts of the sUAS ARC, the ASTM F38 committee has been selected by the FAA to develop the necessary standards for sUAS [5]. Despite the fact that the sUAS ARC has already published its recommendations, the FAA proposed rule for light UAS is not expected to be available before the spring of 2011 and its final version will require approximately a year more [44, 46]. Even so, FAA believes that even sUAS will require secure spectrum for command and control, an issue that is not expected to be resolved before then [112]. In the following paragraphs the key points of the proposed regulations will be provided. Since the regulations are currently fluid and the committee itself did not reach unanimous decisions on all matters, a detailed account would be of limited value.

The key principles behind the sUAS ARC recommendations involve the mitigation of risks stemming from sUAS operations by visual see-and-avoid, yielding right-of-way to all other aircraft and operating limitations [103]. Moreover, it was decided to follow existing manned aviation procedures, standards and regulations when possible [103]. The proposed regulations also include exemptions for universities and UAS manufacturers that allow them to operate such systems with higher flexibility for educational and developmental reasons respectively [103]. The proposed rules involve systems up to 25 kg in weight, divided into four classes, each with increasing requirements as the perceived risk increases [103]. A fifth class is reserved for lighter-than-air systems while additional requirements are proposed for sUAS test centers [103]. Significant emphasis is given to the issue of see-and-avoid, with the pilot-in-command playing a major role, although there are cases where the presence of visual observers is also mandated [103]. Nevertheless, provisions are mentioned to avoid encounters and if not possible maintain appropriate separation with other traffic, to avoid the need for collision avoidance. In the presence of collision risk, the sUAS should give the right-of-way, preferably descending and approaching the position of the pilot-in-command [103]. Other recommendations include the provision for fly-away protection mechanisms, the use of high-contrast painting to increase the visibility of the sUAS and the availability of telemetry data to the operator [103].

4.9.2 Europe

In Europe, the responsibility for developing airworthiness certification procedures and policies for lighter vehicles, as well as public UAS remains with national authorities [30]. Although national authorities retain control for certification of vehicles lighter than 150 kg, there is currently little or no information available on general certification requirements for this category of UAS; the only exception being a recommendation from the CAA-UK [53], which was later adopted by the JAA/EUROCONTROL UAS task force [74]. In the CRD, EASA acknowledges the significant interest for light UAS, but maintains that it is not competent to regulate this class of aircraft. Nevertheless, EASA recommends coordination between member states through EUROCAE, so that the individual state regulations are harmonized [89].

4.9.2.1 CAA-UK

The UK is leading in the development of R/C model regulation, allowing even commercial applications under certain conditions. In fact, the UK was one of the first countries to publish UAS guidance in May of 2002, when it published "CAP 722 Unmanned Aerial Vehicle Operations in UK Airspace — Guidance". CAA-UK believes that light UAS, equivalent in terms of kinetic energy to the R/C models described, can be regulated on the same basis and similar requirements. As a result, realizing the potential of light UAS for early applications and the issues of requiring compliance of such systems with normal UAS regulations, a "Policy for light UAS systems" was published in 2004 [53]. This policy divides light UAS into three categories, small UAS weighing less than 7 kg, small UAS up to 20 kg and light UAS 20–150 kg. Eligible UAS under the latter category are those that do not exhibit a maximum kinetic energy on impact over 95 kJ and an operational velocity of more than 70 kts [53]. UAS also need be operated within 500 m of the pilot and at altitudes not exceeding 400 ft [53]. In order for such vehicles to be certified, a positive recommendation is required from an accredited organization that has inspected the design and manufacture of the vehicle followed by successful completion of a reliability flight test program [53]. Additional restrictions may be imposed by the CAA-UK, if and when required. For systems lighter than 20 kg permission by the CAA-UK is still required, but airworthiness requirements are waived. Operation is restricted within a specified safety distance from airports, congested areas, third party vehicles, structures, etc. These systems are also required to provide a "fail-safe" mechanism and may fly for commercial purposes [53]. Finally the CAA-UK classifies vehicles weighing less than 7 kg as small aircraft and most of the requirements are waived. According to the CAA-UK, there is currently no system available that can perform adequately for collision avoidance. As a result, Visual Line-of-Sight (VLOS) operation is normally required. Moreover, to facilitate UAS development and testing, the CAA-UK has established several areas of segregated airspace where such systems can be operated even Beyond Line-of-Sight (BLOS) [14].

In April of 2008, CAP 722 was revised and renamed "Unmanned Aircraft System Operations in UK Airspace — Guidance". At the same time the Light Unmanned Aircraft System Scheme (LUASS) was launched and is currently run by EuroUSC. This scheme was created to ease the process of evaluating light UAS airworthiness and obtaining permission to operate by the CAA-UK [31].

4.9.2.2 EUROCAE

EUROCAE, also realizing the importance of lighter systems in the initial UAS market, created a "Small UAS" group in its WG-73 committee. This group is involved with lighter systems operating within Visual Line-of-Sight (VLOS) and uses the guidance developed by the CAA-UK as its starting base [77]. The guidance document is scheduled to be presented soon to the EUROCAE council for adoption and release [78].

4.9.2.3 Joint Authorities for Rulemaking on Unmanned Systems (JARUS)

In 2007, the CAA of Netherlands (CAA-NL) received a UAS certification request for an unmanned rotorcraft weighing 80 kg [61, 78], a system that according to EASA rules in under the purview of national authorities. The agency has since been working with the applicant on the development of appropriate certification procedures and operational requirements [61]. At the same time and to better satisfy the request, CAA-NL contacted other national aviation authorities. Soon after the JARUS group was formed to assist with harmonization of regulation of Light UAS across Europe, with participation from many countries of the world as well as EASA and EUROCONTROL [78].

JARUS was initially involved with drafting certification specifications based on those for very light rotorcraft (CS-VLR) due to the nature of the first certification request [78]. They are also oriented towards VFR operations using well established manned aviation standards wherever possible. For the future though, work will expand with very light aircraft (CS-VLA), as well as the drafting of operational requirements [78].

4.10 Military Regulations

4.10.1 United States

The US Congress, identifying the advantages of UAS technology, has mandated an increase in the use of such systems by the military. The goal as published in Section 220 of the Floyd D. Spence National Defense Authorization Act for Fiscal Year (FY) 2001 (Public Law 106-398), is that one third of the aircraft in the operational deep strike force should be unmanned by 2010. To facilitate UAS planning and coordinate between military departments, the UAS Planning Task Force (UAS PTF) was established in 2001 [115]. The UAS PTF promotes interoperability via common standards and interfaces, contributes in the transition of UAS technologies and provides long term guidance with the publication of technology roadmaps [115].

The importance of military UAS has been also demonstrated by the increase in federal expenditure in that sector. It is noteworthy that for 2007 the DoD budget for UAS research and development was $760.8M while $878.4M were allocated for procurement and $590.0M for operations and maintenance [115]. For 2008 the total budget amounts to $2.5bn, an increase of approximately 14% compared to 2007 and almost equal to the entire expenditure of a five year period between 1999 and 2003 [20]. The year 2010 is expected to be the first year where the Pentagon will procure more unmanned aircraft than manned [87].

The OSD initially envisioned unencumbered access to the NAS for adequately equipped UAS by 2010 and published an "Airspace Integration Plan for Unmanned Aviation" [114]. This document provides three guiding principles:

- *Do no harm* by avoiding initiatives that may impact current practices, procedures and operations.
- *Conform rather than create* by avoiding the development of regulations exclusive to UAS.
- *Establish the precedent* that will lead to adoption of regulations by other parties and nations.

Currently, the US DoD aims to gain "file and fly" access in the NAS by 2015 [8]. Nevertheless, military UAS operators also have access to a large number of restricted and warning airspace areas, encompassing $0.15 \cdot 10^6\,km^2$ and over $10^6\,km^2$ respectively [94].

Over the last few years the US DoD has initiated a number of programs that are involved with UAS systems. The US GAO in a 2008 report presents 7 such programs that started between September 2006 and June 2008 alone [111]. These programs involved a study of current capabilities to identify gaps and maximize effectiveness, an evaluation of future operational needs and requirements, interoperability improvement initiatives, UAS training activities coordination, UAS logistics, as well as coordination between different departments and agencies to reduce duplication and acquisition costs. In fact, the US DoD was criticized by GAO for not designating a single entity for coordinating and integrating all US DoD efforts in the area of UAS [111].

4.10.1.1 Airworthiness and operations regulations

Authorization of military UAS operations has been based on COA, in accordance with Order 7610.4J "Special Military Operations" [33]. For UAS operations in the US NAS, the FAA has signed a memorandum of agreement with the DoD, allowing access in class G airspace over DoD owned property for UAS under 9kg [23]. Additionally UAS operations where enabled in class D airspace over military bases, adding $100{,}000\,km^2$ of available airspace [94]. This memorandum was finalized in September of 2007 and allows increased military UAS access to the NAS without COA [98].

Military, as well as most public aircraft, are exempt from several FAA regulations regarding airworthiness and pilot certification but are still required to comply with FAA mandated traffic rules as well as airworthiness standards set by the operating agency [114]. As a result, military UAS will also typically undergo an airworthiness certification program, although it may be more flexible than that for civil aircraft. It is also conceivable that operations may take place even when the UAS is not fully compliant, provided that the risks have been identified and deemed acceptable. This is demonstrated by the case of the Global Hawk UAS, for which Special tailored airworthiness certification criteria have been developed [19]. Nevertheless, it has been issued only a restricted certificate by the USAF, since it doesn't fully comply [19].

Military aircraft are required to follow an airworthiness certification process, similar to that of their civil counterparts. For USAF aircraft this requirement is established in the Air Force Policy Directive 62–6, which also defines the certifying official and an airworthiness certification criteria control board to enforce the policy [114]. Similar to the CFR, the USAF follows relevant policy directives and military handbooks that contain certification criteria as well as maintenance and operational requirements [114]. The other Services follow similar policies and procedures for their aircraft.

The DoD has several groups/organizations that are actively participating in the development of UAS-related regulations. The DoD Policy Board on Federal Aviation (PBFA) provides policy and guidance on ATC/ATM, NAS access and similar matters, in cooperation with the FAA [114]. The DoD-FAA NAS Integration Subgroup is a joint project that aims to investigate issues that might impact DoD operations from the NAS modernization efforts of the FAA [114]. The Air Force Flight Standards Agency (AFFSA) develops regulations as well as standards for procedures, aircraft equipment, navigational facilities and other ATC/ATM matters [114]. The USAF Air Combat Command has sponsored a study on S&A system requirements that led to the publication of a white paper in June of 2004, titled "See and Avoid Requirement for Remotely Operated Aircraft" [115].

4.10.1.2 Standardization and interoperability

Similar to the requirements of other Federal Agencies, and according to the provisions of DoD 4120.24-M12, the DoD is also required to adopt, support and develop industry standards whenever possible [115]. Towards that end, the DoD has established the Defense Standardization Program (DSP) [114], which is actively participating in the developments within related Standards Development Organizations (SDO) and ensures that generated standards are compliant with DoD goals [115]. An example of such collaboration involves the Joint Architecture for Unmanned Systems (JAUS), a program that specifies data formats and methods of communication among computing nodes. Although initially a project within the Joint Ground Robotics Enterprise (JGRE), currently all documentation is transitioned to SAE Aerospace Standards.

Other efforts to improve interoperability and standards development include the Joint Unmanned Aircraft System Center of Excellence (JUAS CoE) as well as the UAS Airspace Integration Joint Integrated Product Team (JIPT) [115]. The latter will also focus on the development of enabling technology to achieve non-segregated access to the NAS, as well as a new COA process designed to better suit DoD's requirements [115].

4.10.1.3 Operator and crew training

Military pilots and crew are exempt from FAA mandated qualification requirements and are required to undergo training and certification programs designed by the military. Nevertheless according to a memorandum of agreement between the DoD and the FAA, the former agrees to meet or exceed civil standards and the FAA agrees to offer DoD pilots access to the NAS [114].

Minimum qualifications for non-military UAS pilots (e.g. pilots hired by a military contractor for flight testing) are provided in the DCMA Instruction 8210.1§3.6. This instruction requires pilots flying outside restricted airspace, to hold a private pilot certificate, instrument rating and have a total of 300 flight hours as pilot-in-command. Pilots operating within restricted airspace are required to hold qualifications consistent with the contract or Service requirements. Currently the USAF has started training UAS pilots, instead of reassigning experienced pilots [87].

4.10.2 Europe

In Europe there is no "umbrella" regulatory body for military aviation like the EASA is for civil. As a result, although military aviation agencies do take into consideration international law and NATO requirements for member states, significant differences may exist between regulations of different states.

One of the first European initiatives in military UAS regulation was undertaken by the Royal Netherlands Army that through a lengthy, complex process certified the Sperwer UAS.

The French Flight Test Center (CEV) that belongs to the Test Directorate (DE) of the French Military Defence Procurement Agency (DGA) has been active for several years performing system tests for various UAS systems. In January of 2005 it was also the first organization to develop UAS airworthiness requirements; developed within 18 months in cooperation with the industry and specialists from nine states. The code was adopted from certification specifications for normal, utility, aerobatic and commuter manned airplanes [30]. This 163 page document has since been adopted as the basis of NATO Standardization Agreement (STANAG) 4671, which is described in the next Section.

To bring harmonization and improve interoperability between European military systems, the EDA proposed the founding of a coordinating agency, a counterpart of EASA for military UAS [96]. The long-term goal is to create a joint aviation authority for both civil and military aviation [96]. EDA has nevertheless suggested that EASA should in the mean time lead the regulatory effort, involving in the process military authorities [96].

4.10.3 NATO

NATO established the Joint Capability Group UAV (JCGUAV) under the Conference of National Armaments Directors to develop guidance and regulations for military UAS operations. The JCGUAV aims to expand UAS and UAS capabilities to meet emerging requirements, to improve autonomy, reliability and interoperability and to assist with airspace integration [76].

Under the JCGUAV, the Flight In Non-Segregated Air Space (FINAS) working group has been established to produce guidelines for cross-border UAS operations in non-segregated airspace [104]. One of the most important works of this group was the approval of the first draft of STANAG 4671 on Unmanned Aerial Vehicles Systems Airworthiness Requirements (USAR) in March of 2007 [73] and was later ratified by NATO member states. Also planned is the review of the STANAG leading to amendments and the publication of a second edition [51]. FINAS has also initiated work on USAR for light UAS (less than 150 kg) and for rotary-wing aircraft [76].

STANAG 4671 only applies to fixed-wing UA with MTOW between 150 kg and 20,000 kg, although certifying authorities can apply it to other UA as well [73]. It is compiled into two books; the first book contains the UAS airworthiness code equivalent to CS-23 [104], while the second book presents possible means of demonstrating compliance [73]. The second book also contains a risk reference system, presented in Fig. 4.1, that defines acceptable failure probabilities to ensure an acceptable level of safety for UAS operations. It should be noted that although it provides guidance for minimum performance requirements of UAS, many issues are not addressed, including S&A, security, environmental impact and operator training among others [102]. Nevertheless any equipment used to address these issues will fall under the directives of the STANAG [73]. The purpose of STANAG 4671 is to allow compliant UAS to overfly other NATO member's airspace, something that is not currently allowed by the ICAO without permission from the countries whose airspace the UAS will enter [52, 102]. Although STANAG 4671 does not consider S&A operational requirements, FINAS, has suggested minimum separation and collision avoidance limits [102]. The proposed TLS is 5×10^{-9} collisions per flight hour [104].

Before that, the FINAS group developed STANAG 4670, that provided guidance for the training of UAS operators [102], tailored to different UAS types and roles [104]. This document is currently undergoing ratification and will remain in the public domain, just like STANAG 4671 [104]. FINAS is planning to produce airworthiness requirements for rotary wing UAS based on CS-27 as well as light

UAS (<150 kg) [104]. Finally investigation has started on a study of human factors influencing UAS safety [104].

Besides FINAS, JCGUAV teams include the Control Systems Specialist team involved in communications and interoperability that has produced STANAG 4586 for a standard interface to UAS control systems, STANAG 4660 for interoperable command and control links (low-bandwidth and STANAG 7085 for interoperable data links for imaging systems (high-bandwidth) [76]. The RTCA and EUROCAE are also using these standards to evaluate the bandwidth requirements for UAS command and control links, in preparation for the 2011 World Radiocommunication Conference [95]. In addition to the aforementioned standards, the JCGUAV has carried out a study on medical standards and published a working paper on ATM requirements [104].

		Catastrophic[a]	Hazardous[b]	Major [c]	Minor [d]	No safety effect
Frequent	$> 10^{-3}/hr$					
Probable	$< 10^{-3}/hr$					
Remote	$< 10^{-4}/hr$					
Extremely remote	$< 10^{-5}/hr$					
Extremely Improbable	$< 10^{-6}/hr$					

[a] Uncontrolled flight and/or uncontrolled crash, which can potentially result in a fatality. Potential fatality to UAV crew or ground staff.
[b] Controlled trajectory termination or forced landing potentially leading to the loss of the UAV where it can be reasonably expected that a fatality will not occur. Potential serious injury to UAV crew or ground staff.
[c] Emergency landing of the UAV on a predefined site where it can be reasonably expected that a serious injury will not occur. Potential injury to UAV crew or ground staff.
[d] Slight reduction in safety margins or functional capabilities and slight increase in UAV crew workload.

Fig. 4.1 UAV operations risk reference system (the grayed areas signify unacceptable risk). Source: [73]

References

1. Access 5 (2006) Home page. No longer available, [online] http://www.access5.aero/
2. Anand S (2007) Domestic use of unmanned aircraft systems: Evaluation of policy constraints and the role of industry consensus standards. Journal of Engineering and Public Policy 11
3. ASTM International (2005) ASTM International sense-and-avoid standard, for use with UAV certifications, adopted by U.S. department of defense. ASTM Tech News, [online] http://www.astm.org/SNEWS/AUGUST_2005/uav_aug05.html, retrieved April 2008

4. ASTM International (2007) Standard practice for application of federal aviation administration (FAA) federal regulations part 21 requirements to unmanned aircraft systems (UAS). Standard F 2505 - 07
5. ASTM International (2010) ASTM committee to aid federal aviation administration by developing standards for small unmanned air vehicle systems. ASTM Newsroom, retrieved March 2010
6. Babbitt JR (2009) Safety must come first. Speech given at Scottsdale, AZ, [online] http://www.faa.gov/news/speeches/news_story.cfm?newsId=10964, accessed February 2010
7. Bennett M (2007) When will UAVs take to the european skies? Unmanned Systems 25(5):27–28
8. Burns J, Capriglione C, Duma KM, Hamilton BA (2010) Unmanned aircraft systems standards the role of standards and standardization in unmanned aircraft systems airspace integration. Bulletin of the Survivability/Vulnerability Information Analysis Center, Vol. 27, No. 1
9. Carey L (2009) ICAO UAS study group. UAS Yearbook 2009/2010, UVS International pp 47–50
10. Carr G (2007) Unmanned aircraft CASA regulations. [online] http://www.uatar.com/CASA%20Presentation%20-%20Unmanned%20Aircraft%20CASA%20Regulations.pdf, retrieved April 15, 2008
11. Cartmell D (2009) ASD — Air4All Consortium. UAS Yearbook 2009/2010, UVS International p 19
12. Cary L (2009) ICAO activities toward integration of UAS into the global ATM system. Presented at the ICAO Global Air Traffic Management Forum on Civil/Military Cooperation
13. Chérif T (2009) ICAO position on issues of critical concern to international civil aviation to be discussed at the international telecommunication union (ITU) world radiocommunication conference (2011) (WRC-11). Letter to ICAO member states
14. Civil Aviation Administration (2010) Airspace and avoiding collisions. Online, [online] http://www.caa.co.uk/, accessed Mar 2010
15. Civil Aviation Safety Authority Australia (2002) Unmanned aircraft and rockets — model aircraft. AC101-3(0)
16. Civil Aviation Safety Authority Australia (2008) Annual review 2007–2008
17. Clothier R, Walker R, Fulton N, Campbell D (2007) A casualty risk analysis for unmanned aerial system (UAS) operations over inhabited areas. In: Proc. 12th Australian International Aerospace Congress and 2nd Australasian Unmanned Air Vehicles Conference, Melbourne, Australia
18. Cooney M (2010) Faa adds brainpower to mix unmanned aircraft with commercial aviation. Online, [online] http://www.networkworld.com/community/node/62271, accessed June 2010
19. Cripps D (2007) Airworthiness and the UAS. Presentation to the AUVSI Pathfinder Chapter Annual Symposium
20. Curtin NP, Francis PL (2004) Unmanned aerial vehicles, major management issues facing DOD's development and fielding efforts. Testimony before the

Subcommittee on Tactical Air and Land Forces, Committee on Armed Services, House of Representatives
21. Davis KD (2007) Federal aviation administration: UAS program office. Unmanned Aircraft Systems, The Global Perspective 2007/2008 p 51
22. Davis KD (2008) Federal aviation administration. UAS Yearbook 2008/2009, UVS International
23. Davis KD (2008) Unmanned aircraft in the national airspace system - the certification path. Presented at the EASA Workshop on UAV
24. DeGarmo MT (2004) Issues concerning integration of unmanned aerial vehicles in civil airspace. Tech. rep., MITRE, Center for Advanced Aviation System Development, McLean, Virginia
25. Directorate of Airspace Policy, Civil Aviation Administration (2004) CAP 722, unmanned aerial vehicle operations in UK airspace — guidance
26. Donnithorne-Tait D (2007) Responding to standardization challenges of the future air transport system. Presented at the ICAS Workshop on UAV Airworthiness, certification and access to the airspace
27. Dopping-Hepenstal L (2007) Autonomous systems technology related airborne evaluation and assessment (ASTRAEA - UK access to airspace programme. Presented at the ICAS Workshop on UAV Airworthiness, certification and access to the airspace)
28. European Aviation Safety Agency (2009) Airworthiness certification of Unmanned Aircraft Systems (UAS). Policy Statement, E.Y013-01
29. European Aviation Safety Agency (2009) Rulemaking programme 2010–2013
30. European Aviation Safety Agency (EASA) (2005) A-NPA, No. 16/2005, policy for unmanned aerial vehicle (UAV) certification
31. EuroUSC (2010) Light unmanned aircraft system scheme. Brochure
32. Ewing L (2007) The quest for the missing link, sense-and-avoid technology edges forward. Unmanned Systems 25(5):18–21
33. Federal Aviation Administration (2003) Inquiries related to unmanned aerospace vehicle operations. Order N8700.25
34. Federal Aviation Administration (2004) Airworthiness certification of aircraft and related products. Order 8130.2F
35. Federal Aviation Administration (2005) AFS-400 UAS policy 05-01, unmanned aircraft systems operations in the U. S. national airspace system. Interim Operational Approval Guidance
36. Federal Aviation Administration (2007) FAA flight plan 2008–2012: Charting the path for the next generation
37. Federal Aviation Administration (2007) Unmanned aircraft operations in the national airspace system. Docket No. FAA-2006-25714
38. Federal Aviation Administration (2007) Unmanned aircraft systems (UAS) questions and answers. [online] http://www.faa.gov/aircraft/air_cert/design_approvals/uas/uas_faq, accessed April 2008
39. Federal Aviation Administration (2008) Airworthiness certification of unmanned aircraft systems. Order 8130.34
40. Federal Aviation Administration (2008) FAA flight plan 2009–2013

41. Federal Aviation Administration (2008) Small unmanned aircraft system aviation rulemaking committee. Order 1110.150
42. Federal Aviation Administration (2008) Unmanned aircraft systems operations in the U. S. national airspace system. Interim Operational Approval Guidance 08-01
43. Federal Aviation Administration (2009) Aviation policy, planning and environment fiscal year 2010 business plan
44. Federal Aviation Administration (2009) Unmanned aircraft systems fact sheet. [online] http://www.faa.gov/news/fact_sheets/news_story.cfm?newsId=6287, accessed February 2010
45. Federal Aviation Administration (2010) FAA Aerospace Forecast Fiscal Years 2010–2030
46. Federal Aviation Administration (FAA) (2010) FAA Civil/Public UAS Roadmap
47. FSF editorial staff (2005) See what's sharing your airspace. Flight Safety Digest 24(5):1–26
48. Galotti V (2009) The way forward: Common understandings, outcomes and way forward. Global Air Traffic Management Forum on Civil/Military Cooperation
49. Gilligan P (2009) Foreword by the FAA. UAS Yearbook 2009/2010, UVS International p 10
50. Goudou P (2009) Foreword by EASA. UAS Yearbook 2009/2010, UVS International p 7
51. Grainger R (2008) NATO - UAS airworthiness requirements group. UAS Yearbook 2008/2009, UVS International p 56
52. Haddon DR, Whittaker CJ (2002) Aircraft airworthiness certification standards for civil UAVs. UK Civil Aviation Authority
53. Haddon DR, Whittaker CJ (2004) UK-CAA policy for light UAV systems. UK Civil Aviation Authority
54. Hawkes D (2007) EUROCAE WG-73: Unmanned aircraft systems. Unmanned Aircraft Systems, The Global Perspective 2007/2008 pp 36–37
55. Hawkes D (2008) Unmanned aircraft systems EUROCAE activities - communications & security. Presented at the EASA Workshop on UAV
56. Hawkes D (2009) EUROCAE WG-73 on unmanned aircraft systems. UAS Yearbook 2009/2010, UVS International pp 35–37
57. Hawkes D, Mardine G, Allouche M (2008) Unmanned aircraft systems EUROCAE activities. Presented at the EASA Workshop on UAV
58. Hayhurst KJ, Maddalon JM, Miner PS, Dewalt MP, Mccormick GF (2006) Unmanned aircraft hazards and their implications for regulation. In: Proc. 25th IEEE/AIAA Digital Avionics Systems Conference, Portland, OR, USA, pp 1–12
59. Hempe D (2006) Unmanned aircraft systems in the United States. Presented to the US/Europe International Safety Conference
60. Hendriks A (2009) EUROCONTROL. UAS Yearbook 2009/2010, UVS International p 39

61. Hermans R (2009) Certification of a light rotorcraft UAS. UAS Yearbook 2009/2010, UVS International p 115
62. Hottman SB (2009) NMSU-PSL-TAAC. UAS Yearbook 2009/2010, UVS International pp 61–62
63. ICAO/IMO Joint Working Group on harmonization of aeronautical and maritime search and rescue (ICAO/IMO JWG-SAR), International Civil Aviation Organization (ICAO) (2008) SAR operational principles, procedures and techniques — development of procedural strategies for the practical provision of SAR services. Information Paper
64. INnovative Operational UAS Integration (INOUI) (2009) Proposal for the integration of UAS into non-segregated airspace. Booklet
65. International Civil Aviation Organization (ICAO) (2006) ICAO exploratory metting on UAVs. Working Paper 2, The Sixteenth Meeting of the APANPIRG ATM/AIS/SAR Sub-Group (ATM/AIS/SAR/SG/16)
66. International Civil Aviation Organization (ICAO) (2007) Addressing unmanned aircraft system (UAS) accident investigation and prevention by ICAO member states. Working Paper 217, ICAO 36th Assembly
67. International Civil Aviation Organization (ICAO) (2007) Progress report on unmanned aerial vehicle (UAV) work. AFI Planning and implementation regional group sixteenth meeting (APIRG/16)
68. International Civil Aviation Organization (ICAO) (2008) ICAO UAS study group. UAS Yearbook 2008/2009, UVS International pp 43–44
69. International Civil Aviation Organization (ICAO) (2009) 2008 annual report of the council. Doc 9916
70. Japan Aerospace Exploration Agency (2008) UAS activities of JAXA's aviation programme group. UAS Yearbook 2008/2009, UVS International pp 50–51
71. Japan UAV Association (2007) Update of UAS-related activities. Unmanned Aircraft Systems, The Global Perspective 2007/2008 pp 59–60
72. Jewell J (2007) ASTM international committee F38 on unmanned aircraft systems. Unmanned Aircraft Systems, The Global Perspective 2007/2008 pp 21–22
73. Joint Capability Group on Unmanned Aerial Vehicles (2007) STANAG 4671 - Unmanned Aerial Vehicle Systems Airworthiness Requirements (USAR). draft, NATO Naval Armaments Group
74. Joint JAA/Eurocontrol Initiative on UAVs (2004) A concept for european regulations for civil unmanned aerial vehicles (UAV). Final Report
75. Karolenko R (2009) UNITE — UAV National Industry Team. UAS Yearbook 2009/2010, UVS International p 82
76. Kessler G (2009) NATO Joint Capability Group UAV. UAS Yearbook 2009/2010, UVS International pp 60–62
77. van de Leijgraaf R (2009) EUROCAE WG-73 SG4 on light UAS. UAS Yearbook 2009/2010, UVS International p 113
78. van de Leijgraaf R (2009) JARUS. UAS Yearbook 2009/2010, UVS International pp 53–54

79. Manns EM (2009) SAE International. UAS Yearbook 2009/2010, UVS International pp 71–72
80. Marchbank M (2009) UAS — the story continues. Flight Safety Australia 68:20–23
81. Mardine G (2008) Unmanned aircraft systems EUROCAE activities - UAS operations. Presented at the EASA Workshop on UAV
82. Matthiesen H (2007) Unmanned aircraft systems pan-european ATM network integration. Presented at the ICAS Workshop on UAV Airworthiness, certification and access to the airspace
83. Matthiesen H (2008) EUROCONTROL ATM integration. Presented at the EASA Workshop on UAV
84. Matthiesen H (2009) EUROCONTROL. UAS Yearbook 2009/2010, UVS International pp 40–41
85. McMillan D (2009) Foreword by EUROCONTROL. UAS Yearbook 2009/2010, UVS International p 6
86. Miller N (2009) ASTRAEA — mapping the future. UAS Yearbook 2009/2010, UVS International pp 25–26
87. Mockenhaupt B (2009) We've seen the future, and it's unmanned. Esquire Magazine, [online] http://www.esquire.com/features/unmanned-aircraft-1109, accessed December 2009
88. Morier Y (2007) EASA update on activities relative to UAS. Unmanned Aircraft Systems, The Global Perspective 2007/2008 pp 31–33
89. Morier Y (2008) EASA and ICAO activities. Presented at the EASA Workshop on UAV
90. New Mexico State University, Unmanned Aerial Vehicle Technical Analysis and Applications Center (2002) High altitude long endurance unmanned aerial vehicle - certification and regulatory roadmap. Version 1.3, [online] http://www.psl.nmsu.edu/uav/roadmap/, accessed April 2008
91. NextGen Integration and Implementation Office, Federal Aviation Administration (2010) NextGen implementation plan
92. Office of Legal Counsel, Civil Aviation Safety Authority (CASA) (2003) CASR part 101, unmanned aircraft and rocket operations
93. Okrent M (2005) 25 nations for an aerospace breakthrough; european civil unmanned air vehicle roadmap. Submitted on behalf of the European Civil UAV FP5 R&D Program members, volume 1 - Overview
94. Pease F (2008) Four toolsets to achieve integration of UAS into the national airspace. In: UVS International Conference, Paris, France
95. Pelmoine C, Neale M, Tränapp N, Bittner JR, Klaeyle A (2008) Uas airspace integration – the communications challenges. In: UVS International Conference, Paris, France
96. Possel H (2008) Military airworthiness and UAS - a european perspective. Presented at the EASA Workshop on UAV
97. Sabatini N (2007) Assuring the safe integration of UAS. Unmanned Aircraft Systems, The Global Perspective 2007/2008 p 11

98. Sabatini N (2008) Progress on the safe integration of unmanned aircraft systems. UAS Yearbook 2008/2009, UVS International p 9
99. Sato A (2008) Japan UAS association. UAS Yearbook 2008/2009, UVS International p 53
100. Sato A (2009) Japan UAV association. UAS Yearbook 2009/2010, UVS International p 57
101. Schultz R (2006) Ultralights, LSAs and kit airplanes – what's the difference? Florida Aviation and Business Journal [online] http://www.airportjournals.com/Display.cfm?varID=0609005, accessed April 2008
102. Seagle D (2007) NATO developments in UAS airworthiness and sense/avoid functional requirements. Presented at the ICAS Workshop on UAV Airworthiness, certification and access to the airspace
103. Small Unmanned Aircraft System Aviation Rulemaking Committee (2009) Comprehensive set of recommendations for sUAS regulatory development
104. Snow M (2008) NATO FINAS. Presented at the EASA Workshop on UAV
105. Tarbert B (2007) UAS airworthiness, certification and access to the airspace. Presented at the ICAS Workshop on UAV Airworthiness, certification and access to the airspace
106. Tarbert B (2007) Unmanned aircraft systems: Charing the path to the future. Presented at New Technologies Workshop III Flying into the Future
107. Tytgat L (2009) Foreword by the European Commission. UAS Yearbook 2009/2010, UVS International p 8
108. UAS Study Group, International Civil Aviation Organization (ICAO) (2008) Preparatory work on the agenda item on unmanned aircraft systems for the ITU world radiocommunication conference 2011. Information Paper, UASSG/2-IP No.1
109. Udovic A, Vielhauer J (2009) VUSIL project: Results and further development. UAS Yearbook 2009/2010, UVS International pp 116–117
110. Udovic A, Vogt J, Vielhauer J (2008) Validation of unmanned aircraft systems' (UAVs) integration into the airspace — the VUSIL project. In: UVS International Conference, Paris, France
111. United States Government Accountability Office (2008) Unmanned Aircraft Systems: Additional actions needed to improve management and integration of DOD efforts to support warfighter needs. Report to the Subcommittee on Air and Land Forces, Committee on Armed Services, House of Representatives
112. United States Government Accountability Office (2008) Unmanned Aircraft Systems: Federal actions needed to ensure safety and expand their potential uses within the national airspace system. Report to Congressional Requesters
113. Unmanned Air Vehicle (UAV) Working Group (2007) Final report
114. US Department of Defense Office of the Secretary of Defense (2004) Airspace integration plan for unmanned aviation
115. US Department of Defense Office of the Secretary of Defense (2007) Unmanned systems roadmap 2007-2032. Report

116. Walker J, Geiselhart K (2008) RTCA Special Committee 203. UAS Yearbook 2008/2009, UVS International pp 66–67
117. Walker R, Gonzalez LF (2007) Australian research centre for aerospace automation. Unmanned Aircraft Systems, The Global Perspective 2007/2008 pp 17–18
118. Washington Watch (2007) FAA roadmap for UAS. Unmanned Systems 25(1):51
119. Williams HJ (2010) Draft environmental impact study for the BRAC beddown and flight operations of remotely piloted aircraft, Grand Forks Air Force Base. Letter of the Aircraft Owners and Pilots Association to Mr. Allbright, Scott AFB
120. Zaloga S (2007) Getting civil with UAVs: How soon? Unmanned Systems 25(3):24–26
121. Zinser TJ (2006) Observations on FAA's oversight of aviation safety. Statement of the Acting Inspector General, US DOT before the Commitee on Transportation and Infrastructure, Subcommittee on Aviation, US House of Representatives

Chapter 5
UAS Safety Assessment and Functional Requirements

> *If one took no chances, one would not fly at all. Safety lies in the judgment of the chances one takes. That judgment, in turn, must rest upon one's outlook on life.*
>
> Charles Lindbergh
> (Journal entry, August 26, 1938)

The primary goal of regulating UAS operations is to assure an appropriate level of safety. This goal is quantified by most national aviation agencies as an "Equivalent Level of Safety" with that of manned aviation, also known as the ELOS principle.

Since many UAS are based on military or general aviation aircraft, they share aspects of the design and construction of the airframe and mechanical components. This of course does not mean that unique designs have not evolved as well. The key difference between manned and unmanned aviation lies in the separation of the pilot from the cockpit and the level of automation introduced. This difference introduces new failure modes and, as a result, an increased perceived risk that needs to be evaluated and mitigated. Nevertheless, it is noteworthy that manned aviation has also benefited from increased automation. A considerable percentage of modern commercial aviation operations —including landing— takes place autonomously with the pilots responsible only for monitoring the computers [11].

This Chapter specifies what the ELOS requirement entails for UAS regulations. To accomplish this, the safety performance of manned aviation is first evaluated. Then, an analysis is carried out to derive a model for estimating UAS risk as the number of fatalities following an accident. Methods to derive estimates for the parameters that affect risk and the rationale for choosing between the alternatives are also presented. This analysis targets two major accidents types, ground impact and mid-air collisions. The final Section of this Chapter gives a brief overview on how to use the risk analysis to derive functional and operational requirements.

It should be noted that this Chapter cites material from two publications of the US Range Commanders Council. The first is the RCC 321-07 titled "Common Risk Criteria Standards for National Test Ranges" [20, 21], while the second is RCC 323-99 titled "Range Safety Criteria for Unmanned Air Vehicles" [18, 19]. Both are to be used by range commanders to ensure that operations within their ranges adhere to safety standards that are acceptable by the military. The following is quoted from the first chapter of [20]:

> The policies and criteria in this document are intended for use by members of the DoD national ranges and Major Range and Test Facility Base (MRTFB). These policies and criteria apply to launch and reentry hazards generated by endoatmospheric and exoatmospheric

K. Dalamagkidis et al., *On Integrating Unmanned Aircraft Systems into the National Airspace System*, Intelligent Systems, Control and Automation: Science and Engineering 54, DOI 10.1007/978-94-007-2479-2_5,

> range activities including both guided and unguided missiles and missile intercepts, space launches, and reentry vehicles. This does not include aviation operations or UAV operations. The RCC Document 323-99 (Range Safety Criteria for Unmanned Air Vehicles) provides criteria for unmanned air vehicles.

Although this distinction is made, it was felt that in certain cases, the reader would benefit from a presentation of certain, selected information from RCC 321-07. This is because either this information is general and applies to most risk/reliability assessments, or valuable insight is to be gained from contrasting it with information specific to UAS.

Two key terms used in this Chapter are "hazard" and "accident"; the first is defined as the necessary conditions that may lead to the second, and the latter as an unwanted outcome with associated damages. As a result, the expected rate of occurrence of an accident can be calculated from the expected rate of hazards, which in turn can be traced back to the rate of causal factors (e.g. equipment failure). Equivalently, given an accident rate limit, a UAS can be designed so that its components have sufficient reliability to ensure that the set limit is not violated.

5.1 Equivalent Level of Safety

According to the JAA/EUROCONTROL UAS Task Force as well as the EASA, one of the guiding principles for UAS regulation should be *equivalence*, and based on that, they assert the following [5, 15]:

> Regulatory airworthiness standards should be set to be no less demanding than those currently applied to comparable manned aircraft nor should they penalize UAS systems by requiring compliance with higher standards simply because technology permits.

This principle has been widely adopted by most national aviation agencies worldwide and is known as the ELOS requirement. For example, the Range Commanders Council in its guidance on UAS operations it states [18]:

> Any UAV operation or test must show a level of risk to human life no greater than that for an operation or test of a piloted aircraft.

To define the ELOS, requirements of current regulations for manned aviation need to be investigated first. Nevertheless, it should be noted that there is also some criticism aimed at the usefulness of this principle. This is because of the difficulty inherent in quantifying what exactly the ELOS requirement entails.

5.1.1 Manned Aviation Requirements

As mentioned in Chap. 3, manned aviation is regulated through a code of requirements. These requirements usually take the form of standards for various aircraft subsystems the final system must adhere to, as well as for all stages of design, manufacture and operation [11]. Use of standards ensures that the components of the system are reliable enough so that the whole system is compliant with a set Target Level of Safety (TLS).

As mentioned in Sect. 3.7, regulations contain safety targets found in *paragraph 1309* of current CS or the corresponding AMC sections. These targets are typically presented as a risk reference system. Such a system categorizes events based on their severity and assigns a maximum rate of occurrence for each event category.

Figure 5.1 presents the risk reference system proposed in the *1309 AMC* Section of EASA CS 25. There, a failure condition that includes injuries and/or fatalities is categorized as hazardous and as such it should be extremely remote ($< 10^{-7}$ events per flight hour) [7]. On the other hand, multiple fatalities are considered to be of catastrophic severity with a likelihood requirement of 10^{-9} or less [7].

		Catastrophic	Hazardous	Major	Minor	No safety effect
Probable	$> 10^{-5}\,h^{-1}$					
Remote	$< 10^{-5}\,h^{-1}$					
Extremely remote	$< 10^{-7}\,h^{-1}$					
Extremely Improbable	$< 10^{-9}\,h^{-1}$					

Fig. 5.1 Risk reference system for large manned aircraft (the grayed areas signify unacceptable risk). Source: [7]

The risk reference system presented in Fig. 5.1 does not apply to all aircraft and variations exist for smaller or different types of aircraft. This is because it was found that applying certification standards developed for transport category aircraft to smaller ones, led to unrealistically high equipment reliability requirements [8]. In addition to that, the results of accident investigations showed that the main accident cause in manned aviation is pilot error. As such, high equipment reliability would have only a minor effect on overall aviation safety. In 1999, the FAA issued AC 23.1309-1C that contains AMC for aircraft certified based on FAR Part 23. With this AC, four classes of aircraft within that category where defined, each with different acceptable probabilities for failure conditions, as shown in Table 5.1.

5.1.2 Derivation of an ELOS for UAS

Use of the same risk reference system like the one presented in Fig. 5.1 or even Table 5.1 is not straightforward because of the wide range of UAS sizes and characteristics. In addition to that, UAS depend on the on-board flight control system and/or the communication link to operate. This requirement introduces additional failure modes that may increase the total number of accidents for the same reliability requirement. Nevertheless, even if said requirements are adapted as they are to

Table 5.1 FAR Part 23 aircraft classes and corresponding acceptable failure condition probability based on severity, as defined in AC 23.1309-1C. Source: [8]

Aircraft class	Minor	Major	Hazardous	Catastrophic
Class I ($<$ 2,720 kg, SRE)	10^{-3}	10^{-4}	10^{-5}	10^{-6}
Class II ($<$ 2,720 kg, STE, MRE)	10^{-3}	10^{-5}	10^{-6}	10^{-7}
Class III ($>$ 2,720 kg, SRE, MRE, STE, MTE)	10^{-3}	10^{-5}	10^{-7}	10^{-8}
Class IV (commuter)	10^{-3}	10^{-5}	10^{-7}	10^{-9}

SRE: Single Reciprocating Engine, MRE: Multiple Reciprocating Engine, STE: Single Turbine Engine, MTE: Multiple Turbine Engine

a number of UAS classes, they may still lead to unnecessarily high reliability requirements. This is due to the fact that UAS do not carry passengers. As a result, the probability of injuries and fatalities after an accident is greatly reduced, compared with that of general aviation or transport aircraft. The average number and severity of injuries per accident is also expected to be lower.

Since failure frequency requirements prescribed for manned aircraft of the same size can not be used directly, other means to derive such requirements for UAS need be employed. A different approach frequently used in safety engineering, is to define safety constraints for a specific accident based on the desired likelihood of the worst possible outcome [19]. This can in turn be used to determine maximum failure frequency. For UAS operations the worst outcome of most accidents is the occurrence of one or more fatalities and as a result the ELOS can be based on that. Although the ELOS may be based on the frequency of accidents that result in one or more fatality, such an ELOS would also be inadequate since it fails to consider the effect of multiple fatalities.

Although current manned aviation regulations do not impose limits on fatality rates, a statistical analysis of historical data can provide valuable insight on the fatality rates of manned aviation and be the basis for defining the ELOS for UAS. An analysis of NTSB accident data from 1983 to 2006 is presented in Table 5.2. It should be noted that the exact rates may vary depending on the type of aviation (general, regional/commuter, air carrier). Moreover, the period over which the data are averaged can also play a role [1], since there is significant variation from year to year, as shown in Fig. 5.2. This is also evident from a US Navy survey [19] that found an average of 18, 7 and 4.7 ground fatalities per 10 million flight hours for US Navy, Commercial and General Aviation, respectively. The survey included data from 1980 to 1998 for US Navy flights and from 1982 to 1998 for civil aviation. In most cases, an individual, annual risk of fatality of 10^{-6} is considered sufficiently low and below that of other typical everyday activities [21].

Other approaches have employed serious but non-fatal injuries as the basis for their safety assessments, especially when due to the nature of the accident fatalities are unlikely. As an example, [21] proposes the use of casualty limits, the latter defined as the people sustaining injuries of a certain severity and over. A study of NTSB data provided in [21] shows that in aviation, the ratio of casualties to fatalities shows little variation and is close to 2.5. Based on this ratio, casualty limits

Table 5.2 Fatality rates from all accidents based on analysis of NTSB accident data [17] between 1983 and 2006.

Rates per hour	Air Carrier	Commuter	General Aviation	Total
Accident	$2.43 \cdot 10^{-6}$	$2.37 \cdot 10^{-5}$	$8.05 \cdot 10^{-5}$	$5.05 \cdot 10^{-5}$
Fatalities aboard	$8.68 \cdot 10^{-6}$	$1.64 \cdot 10^{-5}$	$2.77 \cdot 10^{-5}$	$2.06 \cdot 10^{-5}$
Ground fatalities	$3.37 \cdot 10^{-7}$	$8.30 \cdot 10^{-6}$	$6.54 \cdot 10^{-7}$	$1.31 \cdot 10^{-6}$

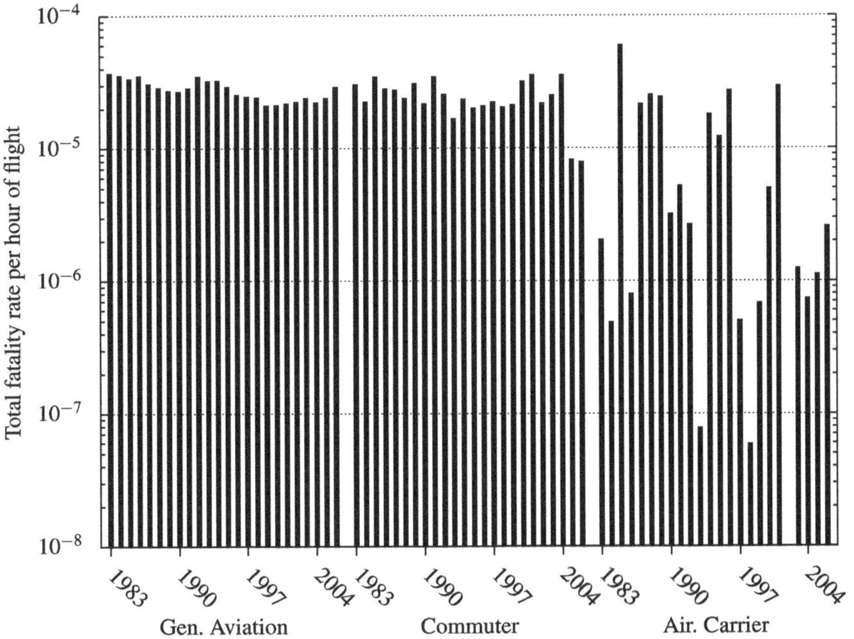

Fig. 5.2 Fatality rates from general aviation, commuter and air carrier accidents as a function of time. Based on analysis of NTSB accident data [17] between 1983 and 2006.

can be derived from those used for fatalities. A common casualty limit is that of 10^{-4} casualties per event, which has been used by a number of agencies like the NASA, the US DoD and the Space Licensing and Safety Office of Australia [21]. The same limit was chosen in [20], complemented by an individual casualty risk probability limit of 10^{-6}. Nevertheless, the use of fatalities has also been advised as a supplemental metric, to better assess the risk involved in an activity [21].

The aforementioned proposed casualty and fatality limits can be contrasted with estimated rates from other sources or activities, provided in Table 5.3, as well as aviation accident statistics given in Table 5.2. When comparing these rates, one must consider that in some activities, a higher risk may be acceptable because of the perceived benefits of participation. It should also be noted, that these rates refer to collective risk [21], i.e. the averaged risk for the entire population. Depending on a person's location, activity and other factors, the actual individual risk may be significantly higher or lower.

Table 5.3 Estimated injury, casualty and fatality rates from different sources or activities.

Activity/Source	Injury rate (h^{-1})	Casualty rate (h^{-1})	Fatality rate (h^{-1})
Motor vehicle accidents (all) [1]	$1.35 \cdot 10^{-5}$	$1.13 \cdot 10^{-6}$	$1.40 \cdot 10^{-7}$
Motor vehicle accidents (occupant) [1]	$8.80 \cdot 10^{-6}$	$6.73 \cdot 10^{-7}$	$5.89 \cdot 10^{-8}$
Pedestrian involved in collision with motor vehicle [1]	$5.10 \cdot 10^{-7}$	$8.92 \cdot 10^{-8}$	$1.04 \cdot 10^{-8}$
Unintentional Falls [1]	$2.45 \cdot 10^{-5}$	$2.20 \cdot 10^{-6}$	$6.06 \cdot 10^{-8}$
Natural Environment [1]	$1.31 \cdot 10^{-7}$	$1.44 \cdot 10^{-8}$	$7.59 \cdot 10^{-9}$
Bicycles & accessories [2]	$1.50 \cdot 10^{-6}$	$8.98 \cdot 10^{-8}$	
Household appliances (ranges, refrigerators, washers) [2]	$4.23 \cdot 10^{-7}$	$1.81 \cdot 10^{-8}$	
Baseball, basketball and football combined [2]	$2.59 \cdot 10^{-5}$	$3.44 \cdot 10^{-7}$	
London Blitz (civilian only) [3]	N/A	$1.04 \cdot 10^{-6}$	$6.22 \cdot 10^{-7}$

[1] The number of injuries and fatalities are from the Web-based Injury Statistics Query and Reporting System available from the Centers for Disease Control and Prevention, National Centers for Injury Prevention and Control and correspond to emergency department admissions in the year 2005. The rates were derived based on an estimated population of 296,410,404 and assuming that every individual (regardless of age, sex or location) is exposed to hazards involving each activity/source an average of 3 h per day. The casualty rate is based on injury cases that required hospitalization or transfer to other facilities such as trauma centers.

[2] The data are from the National Electronic Injury Surveillance System maintained by the US Consumer Product Safety Commission and concern the year 2007. The reported data did not make a distinction between incidents requiring hospitalization and incidents involving deaths. The rates are derived the same way as above, with the exception of the baseball, basketball and football activities, were an average exposure of 3 h per week is assumed instead.

[3] For the London Blitz, the civilian casualties were drawn from historical sources and the rate was obtained by dividing by the population of London in 1939 and by the number of days the Blitz lasted, assuming continuous exposure.

Significant differences can be expected between the effects of various accident types. As a consequence, the ELOS and the risk analysis that will follow need be derived for each accident separately. The following Section provides an overview of the accidents of interest in the case of UAS.

5.2 UAS Accident Types

UAS operations are subject to various hazards that can lead to three primary accidents: unintended or abnormal system mobility operation [24], mid-air collision, and early flight termination [2].

Unintended or abnormal mobility operation refers to accidents that occur when the UAS is still on the ground. In this case the UAS may move unexpectedly, potentially seriously injuring ground crew members. Such accidents usually happen because of operator error. They may occur when the UAS operator does not have a view of the UAS and incorrectly assumes that everyone has cleared the area. This accident type will not be further investigated since the risk can be adequately mitigated with better management of operations and stricter adherence to standard operating procedures.

Mid-air collisions may occur between two UAS systems or between a UAS and a manned aircraft. Depending on the nature of the collision they can result in the loss of one or both of the aircraft. A secondary accident usually following mid-air collisions, is ground impact of debris that may injure people and damage property.

Finally, early flight termination, either controlled or uncontrolled, will result in ground or water impact. Under controlled flight termination it may be feasible to select the point of impact and possibly the speed and orientation of the aircraft. In that case, this capability can be used to reduce the probability of fatalities, as well as damages to property and the aircraft itself.

Potential damages resulting from all these accidents include injury or fatality of people on the ground or on-board another aircraft, damage or loss of the vehicle and damage to property. An indirect damage is environmental pollution either from the payload of the aircraft or as a result of fuel leakage and/or fire following the accident. This is especially important for UAS that will carry chemicals toxic to human beings, for example those used in agricultural applications.

A possible damage that is often ignored is that of societal rejection or outrage that may disrupt future operations. This can occur as a consequence of a high accident rate (even if no injuries occur) or if the accident involves cultural/societal sensitive areas like national parks or monuments, schools and churches. Figure 5.3 summarizes possible accidents and corresponding damages stemming from the operation of UAS in the NAS.

The following sections present models to evaluate UAS risk for the two major accident types; mid-air collisions and ground impact.

5.3 Ground Impact Fatality Risk Modeling

The expected frequency of fatalities (f_F) can be calculated by the following equation:

$$f_F = N_{\text{exp}} P(\text{fatality}|\text{exposure}) f_{\text{GIA}} \tag{5.1}$$

where N_{exp} is the number of people exposed to the accident, $P(\text{fatality}|\text{exposure})$ is the probability a person will suffer fatal injuries given exposure to the accident and f_{GIA} is the rate of ground impact accidents. The $P(\text{fatality}|\text{exposure})$ depends on the resilience of the human body to injury as well as the level of sheltering.

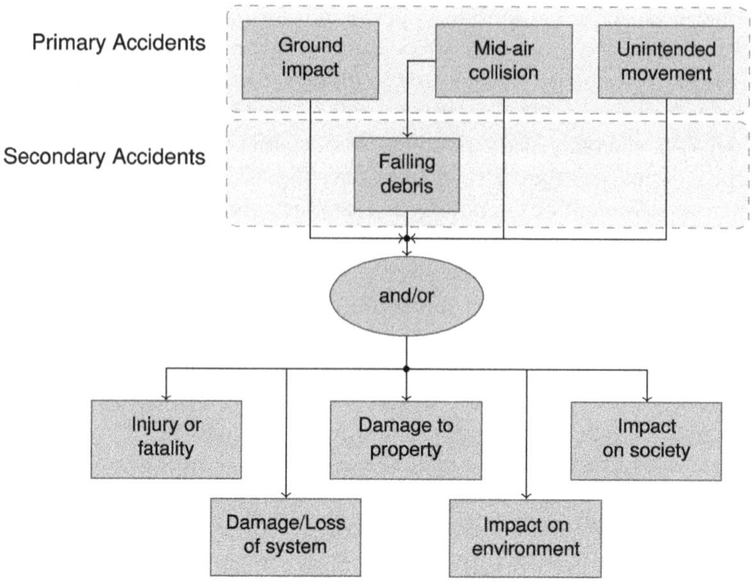

Fig. 5.3 Primary and secondary accidents that can result from the operation of UAS and their possible outcomes.

A similar formulation to the model of eq. (5.1), is provided in [19]:

$$f_F = A_{\text{exp}} \rho P(\text{fatality}|\text{exposure}) f_{\text{GIA}} f_{\text{shelter}} \tag{5.2}$$

In (5.2), N_{exp} is replaced by the product of the lethal area (A_{exp}) by the population density (ρ). An additional term, the sheltering factor f_{shelter} is introduced, that takes values from 0 denoting that everyone is sheltered from the impact, to 1 denoting that nobody is sheltered. This formulation implies a type of "absolute sheltering", where any person considered sheltered, is not affected by the impact.

In certain cases, the exposed population can be divided in groups. Each group is assigned a different probability of fatality given exposure. This can occur when for example, part of the population is inside buildings and part is outside. In this case (5.1) can be expressed as follows:

$$f_F = f_{\text{GIA}} \sum_i N_{i,\text{exp}} P_i(\text{fatality}|\text{exposure}) \tag{5.3}$$

where subscript i refers to the ith group. Although this approach offers better accuracy, it also requires the availability of a library that contains the number of people in each location and the level of their sheltering or at least good estimates thereof [21].

Since the acceptable f_F is provided by the ELOS requirement, it is possible to determine the target level of safety (TLS) for ground impact accidents given by f_{GIA}. Although (5.1) is simple and straightforward, calculation of the terms involved in it is not. In fact, there is a number of options that have been proposed for calculating both N_{exp} and $P(\text{fatality}|\text{exposure})$. The following section will start with the calculation of the f_F term based on the ELOS principle.

5.3.1 Ground Impact ELOS

In determining the fatality rate requirement after ground impacts, special consideration should be given to the fact that UAS are unmanned. This means that only the number of fatalities on the ground are to be taken into account. According to Table 5.2, this number represents only a very small percentage of the total fatalities, about 6%. The ground fatality rate calculated is in the order of $10^{-6}\,\text{h}^{-1}$, although a more conservative ELOS can be derived based on the ground fatality rate of air carriers, which is in the order of $f_F = 10^{-7}\,\text{h}^{-1}$.

It should be noted that Table 5.2 considers all accidents. An alternative analysis can be used by considering only accidents where an in-flight collision with terrain or water occurred (approximately 35% of the total). The updated fatality rates based on NTSB data for the period 1983 to 2006 are presented in Table 5.4. In this case the proposed ELOS would be in the order of $f_F = 10^{-8}\,\text{h}^{-1}$, although it does not include fatalities after emergency landings, ditching and other situations. If the latter are included, the ELOS is closer to $f_F = 10^{-7}\,\text{h}^{-1}$ as shown in Table 5.5.

Table 5.4 Fatality rates for accidents where an in-flight collision with terrain or water occurred. Based on analysis of NTSB accident data [16] between 1983 and 2006.

Rates per hour	Air Carrier	Commuter	General Aviation	Total
Accident	$2.06 \cdot 10^{-7}$	$9.33 \cdot 10^{-6}$	$2.84 \cdot 10^{-5}$	$1.77 \cdot 10^{-5}$
Fatalities aboard	$4.71 \cdot 10^{-6}$	$1.32 \cdot 10^{-5}$	$2.16 \cdot 10^{-5}$	$1.55 \cdot 10^{-5}$
Ground fatalities	$9.84 \cdot 10^{-8}$	$2.86 \cdot 10^{-8}$	$4.46 \cdot 10^{-8}$	$5.99 \cdot 10^{-8}$

Table 5.5 Fatality rates for accidents where one or a combination of in-flight collision with terrain or water, hard/forced landing, runway overrun or ditching occurred. Based on analysis of NTSB accident data [16] between 1983 and 2006.

Rates per hour	Air Carrier	Commuter	General Aviation	Total
Accident	$5.64 \cdot 10^{-7}$	$1.56 \cdot 10^{-5}$	$5.18 \cdot 10^{-5}$	$3.21 \cdot 10^{-5}$
Fatalities aboard	$4.85 \cdot 10^{-6}$	$1.46 \cdot 10^{-5}$	$2.41 \cdot 10^{-5}$	$1.71 \cdot 10^{-5}$
Ground fatalities	$1.01 \cdot 10^{-7}$	$7.63 \cdot 10^{-8}$	$8.43 \cdot 10^{-8}$	$8.89 \cdot 10^{-8}$

For the subsequent analysis and the case studies of Chap. 6, the value for f_F is set to $10^{-7}\,\text{h}^{-1}$, which is the same with that proposed in [21]. However, it should be noted that lower or higher acceptable fatality rates have also been proposed in the past. In [27], although an ELOS of $10^{-7}\,\text{h}^{-1}$ was derived, a target of $10^{-8}\,\text{h}^{-1}$ is proposed instead. This choice was made in an effort to account for the fact that the benefits of UAS operations are not evident to the general public and as a result the tolerance for fatalities will be lower. In [2] analysis is based on multiple acceptable fatality likelihoods ranging from $10^{-6}\,h^{-1}$ to $10^{-9}\,h^{-1}$. The Range Safety Criteria for UAS proposed a fatality rate[1] $10^{-6}\,h^{-1}$ or less based on the US Navy survey discussed previously [19], but their requirements are for military operations that allow higher fatality rates. Finally the NATO USAR adopted a TLS of $10^{-6}\,h^{-1}$ for catastrophic UAS accidents [14], which corresponds to an equal or higher fatality rate.

Although stricter requirements may be attractive, they can seriously impede commercialization of UAS as well as their integration in the NAS. Therefore, a conservative evaluation of the risk from emerging hazards is preferable, since it can be later accommodated as flight hours accumulate and confidence in risk estimates improves.

5.3.2 Exposure to Ground Impact Accidents

Assuming a uniform population density in the area affected by the crash, N_{exp} can be calculated as the product of that area (A_{exp}) by the population density (ρ):

$$N_{exp} = A_{exp} \times \rho \tag{5.4}$$

The population density used in (5.4) is typically estimated using the average population density over the area the UAS will operate. Although use of the actual population density will offer better precision, a standard population density can be used as a reasonable estimate instead. Specifically, EASA has proposed the use of a standard density of 200 people/km^2 [6]. This density was derived taking into account typical civil aviation operations, where a significant percentage of flight time is spent over less densely populated areas. For UAS designed to loiter over populated areas, a higher density will be necessary to avoid underestimating the risk involved. A worst-case scenario of impact at the most densely populated area within the area of operations may also be used to provide a conservative estimate of ρ [19].

There are several ways to determine the A_{exp} based on impact characteristics. For a vertical crash, this area may be approximated by the frontal area of the aircraft augmented by a small buffer to account for the width of an average human [26]. For a gliding descent it can be approximated by (5.5), where the wingspan and length of the aircraft have been increased by the radius of an average person [1]:

$$A_{exp} = W_{aircraft} \left[L_{aircraft} + \frac{H_{person}}{\sin(\text{glide angle})} \right] \tag{5.5}$$

It should be noted, that in some cases instead of the area exposed to the impact, a *casualty* or *lethal area* is mentioned. In this case, attention should be given on

[1] Although [19] employs the word *casualty*, careful review of this document shows that it actually refers to fatalities.

how these areas are defined. This is because in some cases they are the same as A_{exp}, while other times they are defined as the areas within which 100% casualties or fatalities are expected, respectively.

For example, the Range Commanders Council in [21] defines a casualty area as the area where everyone is expected to receive injuries of such severity that they will require hospitalization. On the other hand, the same organization in [19] defines the lethal area merely as an *area of concern*, obtained by:

$$A_{exp} = (L_{aircraft} + 2\,\text{ft})\,(W_{aircraft} + 2\,\text{ft}) \tag{5.6}$$

for a vertically falling vehicle, and:

$$A_{exp} = \left(L_{aircraft} + L_{glide} + L_{stop} + 2\,\text{ft}\right)\,(W_{aircraft} + 2\,\text{ft}) \tag{5.7}$$

where L_{glide} is the gliding distance at an altitude of 6 ft and L_{stop} is the distance required for the aircraft to come to a stop.

5.3.3 Probability of Fatality of Exposed Persons

The human body is capable of sustaining a certain level of force or injury and as a result presence of a person in an area affected by a crash does not guarantee a fatality. Moreover, obstacles such as trees and buildings may provide shelter, thus, increasing the chances of survival. It is evident therefore, that the probability of fatality of a person exposed to a crash need be modeled taking into account the aforementioned factors, namely human vulnerability and sheltering. This Section presents some of the approaches available. Nonetheless, a detailed account of the problem of human vulnerability is beyond the scope of this Section and the interested reader is referred to App. A for more information on the subject.

Despite the observations above, the most commonly used estimate for the probability of fatality of an exposed person is the number one [18]. This is because it is a conservative measure that is not susceptible to criticism. On the other hand it can easily be argued that this measure can be over-conservative especially in the case of small UAS.

According to the Range Commanders Council an exception may be made to the use of unit probability in the case of very light systems [19]. In this case a limit must be defined that divides UAS into two categories, those that do and those that do not cause fatalities. The limit can be based on that used for inert debris, specifically a kinetic energy during impact of 15 J (11 ft lb) for casualty (i.e. reversible injury) and 34 J (25 ft lb) for fatality [21]. The 15 J criterion has been found to be close to the threshold of serious injury for a child [10] and, as a result, it can be stated that it is a conservative measure that protects the entire population.

In an effort to take into account the fact that a person may survive a UAS impact, Weibel et al introduced a penetration factor for calculating the probability of fatality [27]. This factor depends on the characteristics of the UAS and aims at taking into account sheltering. Observing the four example penetration factors provided in [27] it can be argued that Weibel's estimate for smaller vehicles is over conservative. This is because a fatality probability of 5% is assigned to a vehicle that weighs less than 100 g. Due to its low weight, it is impossible to carry a kinetic energy larger than the fatality threshold used in [21]. At the same time, the model seems to underestimate the lethality involved in the impact of a large system. The biggest issue though, is that no method is provided to consistently estimate the penetration factor (penetration parameter in the original) for other UAS, which would make the use of any penetration factor difficult to defend.

In a 1968 study, Feinstein et al investigated the effects of blast, debris and other factors to people [9]. To correlate the probability of fatality with kinetic energy, log-normal curves were derived for impacts to different body parts. Additionally, an averaged curve was proposed, given by:

$$P(\text{fatality}|\text{exposure}) = Z\left(\frac{\ln E_{imp} - \ln a}{b}\right) \tag{5.8}$$

where Z is the cumulative normal distribution whose value is typically obtained from tables, $a = 103\,\text{J}$ is the energy required for a probability of fatality of 50% and $b = 0.538$ is a parameter that affects how fast the probability rises as a function of kinetic energy.

A drawback of Feinstein's model is that it is difficult to adjust it to take into account sheltering. This is because if parameter b was increased so that the curve slope is decreased, the probability of fatality for low kinetic energies would increase. As a result, any change of the b parameter to take into account sheltering, would require an adjustment of the a parameter as well.

The authors have proposed a model to estimate $P(\text{fatality}|\text{exposure})$ as a function of kinetic energy at impact (E_{imp}) that also takes into account sheltering in addition to human vulnerability [4]. The model is presented in (5.9) and it is a variation of the logistic growth model that aligns well with other human vulnerability models. It depends on three parameters, α, β and p_s:

$$P(\text{fatality}|\text{exposure}) = \frac{1}{1 + \sqrt{\frac{\alpha}{\beta}}\left[\frac{\beta}{E_{imp}}\right]^{\frac{1}{4p_s}}} \tag{5.9}$$

The sheltering parameter $p_s \in (0,1]$ determines how exposed is the population to an impact. It is a function of the number of obstacles in the crash trajectory of the aircraft that can absorb impact energy or deflect debris, as well as the ability of people to take shelter behind such obstacles. It takes an average value of 0.5, with higher values meaning better sheltering and a lower probability of fatality for the same kinetic energy. The α parameter is the impact energy required for a fatality probability of 50% when $p_s = 0.5$. Finally, the β parameter is the impact energy threshold required to cause a fatality as p_s goes to zero. Based on the fatality limit of [21], the β parameter can be considered to be a constant with value 34 J.

Further investigation since the publication of the first edition of this book led to the derivation of an improved model to estimate the probability of fatality given exposure to a crash. This improved model is better aligned with other models such as Feinstein's. The model of (5.9) is enhanced with the introduction of a correction factor k, used to improve the estimates given for low kinetic energies, especially those close to, or below the threshold limit of 34 J. In addition to that, the sheltering factor p_s is rescaled, so that its average value matches closer the Feinstein average lethality curve. This model is given by:

$$P(\text{fatality}|\text{exposure}) = \frac{1-k}{1-2k+\sqrt{\frac{\alpha}{\beta}}\left[\frac{\beta}{E_{imp}}\right]^{\frac{3}{p_s}}} \tag{5.10}$$

where $k = \min\left[1, \left(\frac{\beta}{E_{imp}}\right)^{\frac{3}{p_s}}\right]$

The sheltering parameter p_s still determines how exposed is the population to an impact, but now takes values in the range $(0,\infty)$. It is a function of the amount of obstacles in the crash trajectory of the aircraft that can absorb impact energy or deflect debris, as well as the ability of people to take shelter behind such obstacles. It takes an average value of 1, with higher values meaning better sheltering and a lower probability of fatality for the same kinetic energy. The α parameter is the impact energy required for a fatality probability of 50% when $p_s = 6$. Finally the β parameter is the impact energy threshold required to cause a fatality as p_s goes to zero. Based on the fatality limit of [21], the β parameter can be considered to be a constant with value 34 J.

The fatality probabilities as calculated by the model of (5.10) are presented in Fig. 5.4 for different values of the sheltering factor p_s. The Figure also shows the effect of the k factor, which drives the output of the model to zero as the kinetic energy is reduced to the value of β.

The fatality probability models presented in this Section are compared in Fig. 5.5. It is noteworthy that the proposed model can approximate Feinstein's average curve using $\alpha = 32$ kJ and $p_s = 1$ as well as the conservative approach using a low number for p_s. Additional information on this topic is provided in App. A.

5.3.3.1 Kinetic energy estimation

Many human vulnerability models involve the kinetic energy at impact or other functions of impact speed and object mass. This speed may vary depending on the UAS and the descent characteristics and as a result a suitable estimate must be used. A useful conservative substitute for the impact speed is terminal velocity. The latter can be calculated from (5.11), where m is the vehicle mass, g is the acceleration of gravity, ρ_α is the air density, A is the cross-sectional area of the vehicle and C_d is its drag coefficient. The latter two parameters are not always available, since they vary with the orientation of the aircraft during a descent:

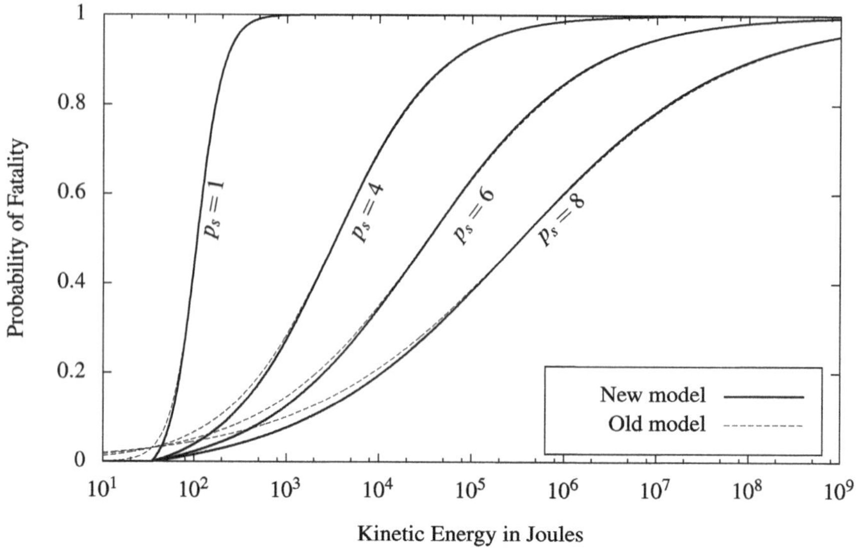

Fig. 5.4 The probability of fatality as a function of kinetic energy impact for the proposed model with $\alpha = 34\,\text{kJ}$, $\beta = 34\,\text{J}$ and for several values of p_s. For comparison purposes the estimates of the older version of the model as presented in [4] are also given.

$$E_{imp} = \frac{m^2 g}{\rho_\alpha A C_d} \tag{5.11}$$

The use of the maximum between the terminal velocity and the *velocity not to exceed* provided by the manufacturer is proposed as an alternative in [19].

In [6, 11, 15], instead of the terminal velocity, the use of the maximum operating velocity (v_{op}) increased by 40% is proposed, instead. This choice overcomes the problem of accurately estimating the parameters required to calculate the terminal velocity, greatly simplifying calculations. The kinetic energy can then be calculated as:

$$E_{imp} = m v_{op}^2 \tag{5.12}$$

It may also be argued, that when the mass of the impacting object is comparable or larger than that of the body part struck, not all of the object's kinetic energy will be absorbed. In fact, after the impact the object will continue to move, in unison with the body, retaining some kinetic energy. The energy absorbed during the collision can be calculated based on momentum conservation by [22]:

$$E = \frac{1}{2} m_1 v_1^2 \left(1 - \frac{m_1}{m_1 + m_2}\right) \tag{5.13}$$

where m_1 and v_1 are the mass and velocity of the object and m_2 refers to the effective mass of the body part struck. As a result when $m_1 \ll m_2$, the effective energy is equal

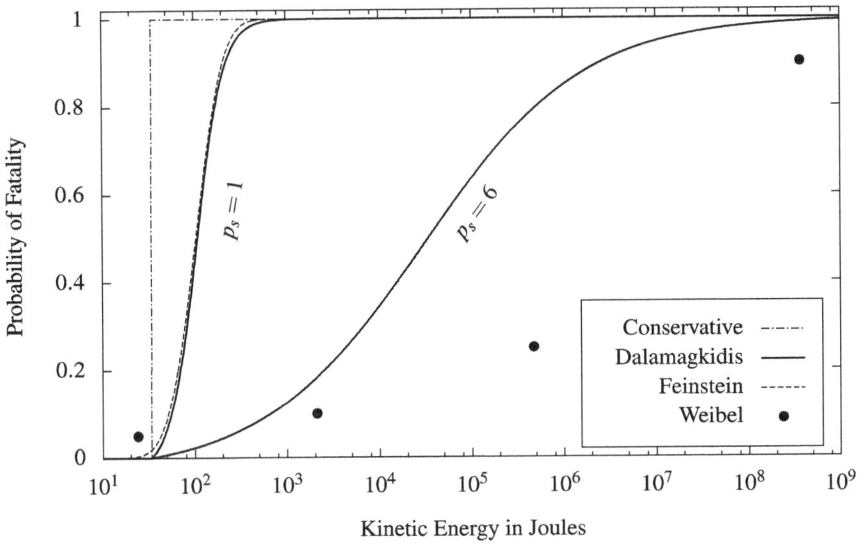

Fig. 5.5 A comparison of the vulnerability models of Feinstein, Dalamagkidis and Weibel, as well as the conservative approach of unit probability above the 34 J kinetic energy threshold.

to the kinetic energy of the projectile. It should be noted that when body movement is constrained (e.g. from a wall), then (5.13) no longer applies and the entire kinetic energy is to be used [22].

Irrespective of which of the aforementioned methods is used to calculate the kinetic energy of the impacting object, if a person is sheltered within a building or a vehicle, some of the kinetic energy will be exhausted to penetrate the shelter. As a result, the energy used in the vulnerability model, may be also adjusted to take into account the effects of sheltering. The following section provides a more detailed account of the factors that need to be considered when incorporating the effects of sheltering; either by kinetic energy adjustments or by estimating the value of a parameter in the vulnerability model.

5.3.3.2 Sheltering estimation

In certain situations, even when some sheltering is available, a conservative estimate of no sheltering may be used [19]. Nevertheless, this could lead to over-conservative results, especially when a significant percent of the population is protected by vehicles and buildings. When estimating the sheltering effect, all the contributing factors need to be evaluated. These factors include the position of the body (standing, sitting or prone), the sheltering provided by buildings or other structures and the level of danger awareness.

Body position is normally taken into account in the vulnerability model. This is typically achieved by averaging the injury expectations calculated for different body positions. It should be noted, that even when a specific area implies that the majority of the exposed population will be in a specific position (e.g. prone for a beach), using just the prone position parameters will be conservative [3]. More information on how the body position is taken into account into human vulnerability modeling is provided in App. A.

Buildings, vehicles and other objects may reduce the risk of injury by absorbing part of or even the entire impact energy. For buildings the kinetic energy required to penetrate the structure needs to be considered [21]. This energy can be as low as 23 J for a roof made by a sheet of 24 gage corrugated aluminum and can exceed 560 J for roofs made of light concrete on 22 gage corrugated steel decking [21]. Although the examples given concern vertical impacts, in most cases, side impacts will also need to be considered [21]. Additionally, if an object manages to penetrate a structure, it is possible for material from the damaged structure to fall, causing secondary hazards [21].

It should be noted that the aforementioned limits concern an irregular, compact, tumbling steel fragment [21]. As a result, a UAS may require additional energy to penetrate the same roof. This is because they are constructed from lighter, less dense materials, because their kinetic energy is going to be distributed over a larger area and because some energy may also be expended for fragmentation instead of penetration. If the types of structures in an area vary significantly it is possible to group buildings into different classes and apply the approach of (5.3), assuming that the distribution of the population within the buildings is known or can be estimated. Accurate prediction of the effect of sheltering with the aforementioned methodology is difficult. This is because it requires precise knowledge regarding the construction of all structures in the affected area, as well as the people and their activities during impact. To overcome this limitation, it is generally acceptable to make estimates based on average or worst case scenarios. Compatible with the worst case scenario is to assume that all buildings offer the sheltering effect of the worst buildings.

Besides the approach of subtracting the energy required to penetrate the structure from the energy available to cause a fatality that was mentioned in the previous Section, a more conservative approach is also possible. The latter assumes that, if an object has sufficient energy to penetrate a structure, everyone inside is going to be a casualty or fatality [21].

Danger awareness may also significantly improve the odds of a person surviving a UAS impact. This is because a person that is aware of the UAS can react sooner and either avoid the UAS entirely or take better shelter. To account for this, as well as the fact that the risk may be taken on a voluntary basis, it has been proposed to use a larger allowable risk of fatality or casualty [21]. On the other hand, it is unlikely that people other than the UAS operating crew will have that benefit.

5.3.4 Frequency of Ground Impact Accidents

The frequency of ground impact accidents (f_{GIA}) is a function of the reliability of the UAS. If a limit for it is known or estimated from (5.1), then the UAS can be designed to meet set TLS. For example, in [21] a limit of 10^{-7} aircraft impacts per mission is proposed that could be used to obtain a limit for f_{GIA}. Alternatively, if the actual f_{GIA} is known, it can be used to determine whether an existing system with known reliability would violate safety limits.

The actual f_{GIA} can be estimated from the history of the UAS itself, provided that a sufficient number of flight hours have accumulated. If the UAS has a long history, the most recent data are to be preferred, since older data may not represent its current performance [19].

An issue arises with newer systems, with few flight hours that may have not exhibited a failure that led to an accident yet. In this case, if the UAS is based on a previous model and shares a significant amount of design characteristics, its reliability can be assumed to be the same with that of the previous model [19]. Otherwise, the f_{GIA} can be estimated using Table 5.6 [19]. The aforementioned table was derived under the assumption of stochastic system behavior, exponential failure distribution and no significant changes in the system or its environment. As a result, care should be taken when using Table 5.6, especially in the case of UAS that are experimental or still in development, since these assumptions may not hold. In that case or when there is a need for a conservative estimate, worst-case assumptions can be made. Such assumptions may take the form of one crash per flight or per flight hour [19].

Table 5.6 Estimate of the expected frequency of ground impact accidents for new UAS. The value shown represents a confidence of 95% that the actual f_{GIA} is less than or equal to $\hat{f}_{GIA}$. Source [19]

Flight hours without crash	$\hat{f}_{GIA}$
10	$3 \cdot 10^{-1}\,\mathrm{h}^{-1}$
30	$1 \cdot 10^{-1}\,\mathrm{h}^{-1}$
100	$3 \cdot 10^{-2}\,\mathrm{h}^{-1}$
300	$1 \cdot 10^{-2}\,\mathrm{h}^{-1}$

If a formal reliability assessment has been carried out for the particular UAS model that included a Failure Mode and Effects Analysis (FMEA) or similar study, then its conclusions can also be used to estimate f_{GIA} [19]. That estimate may be increased by a safety margin to account for failure modes that may have been overlooked or underestimated. Additional information on reliability assessments is provided in Sect. 5.6.

5.4 Mid-air Collision Fatality Risk Modeling

The expected frequency of fatalities (f_F) following mid-air collision accidents can be calculated by the following equation:

$$f_F = E(\text{fatality}|\text{collision}) f_{\text{MaC}} \tag{5.14}$$

$$= N_{\text{exp}} P(\text{fatality}|\text{collision}) P(\text{collision}|\text{CT}) f_{CT} \tag{5.15}$$

where $E(\text{fatality}|\text{collision})$ is the expected number of fatalities after a crash and f_{MaC} is the rate of mid-air collision accidents. The former can be calculated as the product of the number of people exposed to the accident (N_{exp}) by the probability a person will suffer fatal injuries given exposure to the accident ($P(\text{fatality}|\text{collision})$). The f_{MaC} term can be calculated as the product of the probability of collision given that two aircraft are in conflicting trajectories ($P(\text{collision}|\text{CT})$) by the frequency of occurrence of conflicting trajectories (F_{CT}).

As in the case of ground impact accidents, since the acceptable f_F is provided by the ELOS requirement, it is possible to determine the target level of safety (TLS) for mid-air collision accidents given by f_{MaC}. Moreover, if the actual frequency of occurrence of conflicting trajectories is known from historical data, the maximum acceptable value of $P(\text{collision}|\text{CT})$ can be determined, which maps directly to the capabilities of the sense and avoid system.

5.4.1 Mid-air Collision ELOS

In order for an ELOS to be derived, accident statistics involving mid-air collisions are required. The NTSB has defined two categories of relevant accidents; in-flight collisions with obstacles like birds, trees, powerlines and mid-air collisions with other aircraft. The results from the analysis of NTSB data involving these two accidents are tabulated in Tables 5.7 to 5.9.

In contrast to the ground impact accidents, the ELOS for mid-air collision accidents should include the total number of fatalities. This is because on-board fatalities are possible when the accident is between a UAS and a manned aircraft. From the NTSB accident data in Table 5.9 it can be argued that the fatality rate following mid-air collisions with aircraft or other obstacles is in the order of $f_F = 10^{-6}\,\text{h}^{-1}$. A more conservative estimate of $f_F = 10^{-7}\,\text{h}^{-1}$ can be reached from the same Table, if the on-board fatalities after a collision with obstacles other than aircraft are ignored. By deriving the expected number of fatalities after a mid-air collision accident it is then possible to determine the maximum acceptable frequency of such accidents.

Another approach is to assume that in the case of mid-air collisions the fatality expectation is the same, regardless of whether a UAS was involved in the accident. Although this assumption is more conservative, it simplifies subsequent analysis, since one may directly obtain the accident TLS for mid-air collisions. Based on the NTSB data of Table 5.9, the average rate of mid-air collisions involving manned

Table 5.7 Fatality rates for accidents where an in-flight collision with obstacles (e.g. birds, trees, powerlines) occurred. Based on analysis of NTSB accident data [16] between 1983 and 2006.

Rates per hour	Air Carrier	Commuter	General Aviation	Total
Accident	$1.34 \cdot 10^{-7}$	$3.22 \cdot 10^{-6}$	$1.33 \cdot 10^{-5}$	$8.17 \cdot 10^{-6}$
Fatalities aboard	$9.67 \cdot 10^{-7}$	$2.67 \cdot 10^{-6}$	$6.27 \cdot 10^{-6}$	$4.25 \cdot 10^{-6}$
Ground fatalities	$5.97 \cdot 10^{-9}$	$3.81 \cdot 10^{-8}$	$5.73 \cdot 10^{-8}$	$3.93 \cdot 10^{-8}$
Total fatalities	$9.73 \cdot 10^{-7}$	$2.71 \cdot 10^{-6}$	$6.32 \cdot 10^{-6}$	$4.29 \cdot 10^{-6}$

Table 5.8 Fatality rates for accidents where a mid-air collision with another aircraft occurred. Based on analysis of NTSB accident data [16] between 1983 and 2006.

Rates per hour	Air Carrier	Commuter	General Aviation	Total
Accident	None	$2.76 \cdot 10^{-7}$	$5.90 \cdot 10^{-7}$	$3.74 \cdot 10^{-7}$
Fatalities aboard	None	$6.96 \cdot 10^{-7}$	$1.04 \cdot 10^{-6}$	$6.82 \cdot 10^{-7}$
Ground fatalities	None	$1.91 \cdot 10^{-8}$	$2.86 \cdot 10^{-8}$	$1.87 \cdot 10^{-8}$
Total fatalities	None	$7.15 \cdot 10^{-7}$	$1.07 \cdot 10^{-6}$	$7.01 \cdot 10^{-7}$

Table 5.9 Fatality rates for accidents where either a mid-air collision with an object or another aircraft occurred. Based on analysis of NTSB accident data [16] between 1983 and 2006.

Rates per hour	Air Carrier	Commuter	General Aviation	Total
Accident	$1.34 \cdot 10^{-5}$	$3.48 \cdot 10^{-6}$	$1.38 \cdot 10^{-5}$	$8.53 \cdot 10^{-6}$
Total fatalities	$9.73 \cdot 10^{-7}$	$3.42 \cdot 10^{-6}$	$7.40 \cdot 10^{-6}$	$4.99 \cdot 10^{-6}$
Total fatalities[1]	$5.97 \cdot 10^{-9}$	$7.53 \cdot 10^{-7}$	$1.13 \cdot 10^{-6}$	$7.40 \cdot 10^{-7}$

[1] Excluding fatalities aboard after collisions with objects other than aircraft.

aircraft is $7.40 \cdot 10^{-7}$ and under ELOS requirements, a maximum mid-air collision rate of $f_{MaC} = 10^{-7}\,\mathrm{h}^{-1}$ can be proposed for UAS.

Other accident frequency limits proposed for UAS take into account both the manned aircraft threatened and the phase of flight. Such an approach was taken by INOUI, that proposed the limits presented in Table 5.10.

Finally, it should be noted that not all collisions lead to catastrophic accidents. The large variability of aircraft sizes and designs, whether manned or unmanned, and the fact that not all their systems are critical for remaining airborne, means that certain collisions may be survived by one or even both of the aircraft involved. Nevertheless, since it is nearly impossible to account for every possible collision scenario and its effects, every collision is considered a catastrophic accident for both aircraft.

Table 5.10 Maximum acceptable accident frequency depending on ATC type, flight phase and aircraft threatened. The collision accident criteria to be applied corresponds to the one for the highest category of aircraft threatened. Source: [13]

ATC type	Flight phase	>2,730 kg	MEP/SET <2,730 kg	SEP <2,730 kg
Area Control	En route inbound	$3\cdot10^{-9}$	$3\cdot10^{-8}$	$3\cdot10^{-7}$
	En route outbound	$1\cdot10^{-9}$	$1\cdot10^{-8}$	$1\cdot10^{-7}$
	En route transit	$3\cdot10^{-9}$	$3\cdot10^{-8}$	$3\cdot10^{-7}$
Approach	Departure	$1\cdot10^{-9}$	$1\cdot10^{-8}$	$1\cdot10^{-7}$
	Init & interm app	$3\cdot10^{-9}$	$3\cdot10^{-8}$	$3\cdot10^{-7}$
	Final approach	$3\cdot10^{-9}$	$3\cdot10^{-8}$	$3\cdot10^{-7}$
Tower	Landing	$8\cdot10^{-9}$	$8\cdot10^{-8}$	$8\cdot10^{-7}$
	Line-up	$3\cdot10^{-9}$	$3\cdot10^{-8}$	$3\cdot10^{-7}$
	Start-up/push-back	$8\cdot10^{-9}$	$8\cdot10^{-8}$	$8\cdot10^{-7}$
	Take-off	$8\cdot10^{-9}$	$8\cdot10^{-8}$	$8\cdot10^{-7}$
	Taxiing	$6\cdot10^{-8}$	$6\cdot10^{-7}$	$6\cdot10^{-6}$

MEP: Multi Engine Piston, SET: Single Engine Turbine, SEP: Single Engine Piston

5.4.2 Exposure and Risk of Fatality

The number of people exposed to the accident, as well as the probability of them sustaining fatal injuries depends on the aircraft that are involved in the accident and the passengers they carry. As a result it is difficult to get a good estimate without a priori knowledge of all air traffic in the area of operations.

A more general estimate can be derived by noting that the product of N_{exp} and $P(\text{fatality}|\text{collision})$ is in fact the expected number of fatalities per accident. Using the NTSB accident data of Table 5.9, this product is higher for commuter aviation where it takes a value of one, while on average it is closer to 0.58. Moreover, if the on-board fatalities after a collision with obstacles other than aircraft are ignored, the expected number of fatalities per accident drops to below 0.09. It should be noted that this estimate can be considered conservative because in contrast with the accident data it was derived from, the mid-air collisions of interest will always involve at least one aircraft that is unoccupied.

5.4.3 Conflicting Trajectory Expectation

In [27] the mid-air collision risk assessment was based on the use of a gas model of aircraft collisions to estimate the number of expected collisions per hour of flight (f_{MaC}) from:

$$f_{MaC} = \frac{A_{\text{exp}} d}{V \times t} \tag{5.16}$$

where A_{exp} is the exposed area of the threatened aircraft, d is the distance traveled, V is the airspace volume and t is the time required to travel the distance d.

It should be noted that this model estimates the number of mid-air collision hazards due to insufficient spatial and temporal separation given predetermined flight paths or simply the number of *potential* collisions. This is important because in the analysis presented in [27] an unstated assumption is made that all possible collisions are treated as expected collisions for getting the expected level of safety.

An additional term is required to take into account the fact that one or both of the aircraft in a collision course may attempt maneuvers to avoid each other. As a result, the expected number of collisions should be calculated from:

$$f_{MaC} = \underbrace{\frac{A_{\text{exp}} d}{V \times t}}_{E(\text{CT})} \times P(\text{collision}|\text{CT}) \tag{5.17}$$

where CT denotes a conflicting trajectory.

The use of the model in (5.17) to assess $E(\text{CT})$ presents significant difficulties since it requires the exact trajectories (both in space and time) of all air traffic in the area where UAS operations will take place. This requirement is almost impossible to meet because air traffic is dynamic and never identical from day to day. Furthermore, not all traffic is monitored by ATC. Such traffic includes aircraft at low altitudes or aircraft that fly in uncontrolled airspace and are not required to file a flight plan. In addition to that, in the event of a deviation from the predefined trajectory, the number of collision hazards following that event may change. Thus, a worst-case $E(\text{CT})$ may be assumed, instead.

Based on the analysis in [27] high $E(\text{CT})$ is found in proximity of major airways. The highest was found at FL370, where it is approximately 4×10^{-5} CT/hr. Since the results were obtained by averaging data over a 24 hour period, a process that can hide higher peaks, a worst-case $E(\text{CT}) = 10^{-4}$ or even higher can be chosen to also account for future traffic growth.

5.4.4 Collision Probability

Even when two aircraft are on conflicting trajectories, a collision is not guarantied. One or both of the pilots may take action to avoid a hazardous situation by maintaining the required separation between the two aircraft. As a result, the collision probability depends on the collision avoidance capabilities of all the aircraft involved, as well as the measures taken to assure proper separation. If a maximum allowable collision probability is known, it can be used instead of the maximum acceptable accident frequency to determine minimum performance requirements of UAS sense and avoid systems. In fact its use offers advantages because it maps better to what such a system should be capable of accomplishing. To evaluate the actual $P_{\text{max}}(\text{collision}|\text{CT})$ every possible scenario involving a potential conflicting trajectory can be simulated. The results are then aggregated and tested to see whether the target safety levels are accomplished.

It should be noted that when evaluating the sense and avoid system the worst-case $E(\mathrm{CT})$ may be over-conservative. This is because it corresponds to Class A airspace where separation is provided by and is the responsibility of ATC. In the analysis presented in [27], the worst-case conflicting trajectory expectation falls by about an order of magnitude in Class E airspace. If the target safety level is assumed to be a collision rate of $10^{-7}\,\mathrm{h}^{-1}$, then the "see and avoid" capability requirement may be based on achieving $P_{\max}(\mathrm{collision}|\mathrm{CT}) = 1\%$. Nevertheless, the same cannot be assumed for Class G airspace because traffic in that region is not always monitored and accurate estimates on its density are not possible. In addition to that, and especially in very low altitudes the risk of collision with birds, power lines, trees and buildings may be higher than that of a collision with other air traffic.

5.5 Model Choice

Tables 5.11 and 5.12 summarize the parameters involved in modeling risk from UAS ground impact and mid-air collision accidents, respectively, as well as the alternatives presented for estimating their values. It is evident that there are several choices available to an engineer tasked with assessing the risk of UAS operations. The issue at stake and the subject of this Section is what criteria should drive the selection of one alternative over another.

There is a multitude of modeling options available to estimate risk, each with different levels of detail and accuracy. According to [21], any model used for risk/reliability assessment should be based on four basic standards; transparency, clarity, consistency, and reasonableness.

A number of different models may be used in a risk evaluation involving UAS operations depending on the objective and requirements of the risk study in question. A common use of risk models in the UAS domain is for building a safety case that is then used for obtaining a permit to operate in the national airspace system. As a result, the models used must be clearly presented so that they can be reviewed by the regulators and the assumptions and limitations contained therein, must be succinctly expressed. This would lead to compliance with the clarity and transparency standards.

Typically, the regulatory framework does not specify the use of a particular modeling choice over another. Nevertheless, every choice must be defensible. This is achieved by compliance with the consistency and reasonableness standards. The former refers to the use of models that are in use and accepted by the scientific community. The latter means that model selection should be based on rational criteria, risk is not underestimated and a potential review wouldn't raise concerns. As a result, the easier and in fact a common approach is to make conservative estimates.

Conservative estimates are also attractive because of their simplicity and of the associated ease of achieving clarity and transparency. In general, even when high

Table 5.11 A summary of the methods presented to estimate the terms involved in ground impact risk modeling when their values are not known a priori, Eq. (5.1)

Term	Estimate	
f_F	Based on ELOS requirements with typical values in the range of 10^{-6}–$10^{-9}\,h^{-1}$	
N_{exp}	The product of the population density (ρ) and the area affected by the impact (A_{exp})	ρ: (i) Assuming uniform population density, (ii) using a standard population density (e.g. 200 ppl/km^2), (iii) assuming a worst case scenario of impact at the most densely populated area A_{exp}: (i) The area presented by the aircraft perpendicular to its path and augmented by the width of an average person, or (ii) The aforementioned area, including the area the aircraft traverses on the ground until it stops
P(fatality)	(i) Probability of one as a conservative estimate	
	(ii) Zero or one, based on whether the kinetic energy at impact (E_{imp}) exceeds a pre-specified threshold (e.g. 34 J)	E_{imp}: (i) Kinetic energy at terminal velocity, (ii) Kinetic energy at VNE (velocity not to exceed), (iii) Kinetic energy at 140% operational velocity, (iv) The difference between kinetic energy at impact and kinetic energy remaining post-impact (combined human/object), or (v) The kinetic energy calculated with one of the aforementioned means, reduced by the energy required to penetrate sheltering
	(iii) From a vulnerability model based on kinetic energy, e.g. Feinstein et al [9]	E_{imp}: As above
	(iv) From a vulnerability model that also includes the effects of sheltering, e.g. Weibel [25] or Eq. (5.10)	E_{imp}: As above except method (v) Sheltering: (i) Using a conservative value that assumes little or no sheltering, (ii) Based on average sheltering provided by structures, vehicles and other objects, (iii) Based on the building with the worst sheltering, or (iv) Using a database containing structure characteristics and population distribution
f_{GIA}	(i) From previous accident statistics, if sufficient flight hours have accumulated (ii) Assuming an exponential accident distribution for new vehicles without accidents so far (iv) Using a conservative estimate of one crash per flight or per flight hour (v) Based on the results of a formal UAS reliability assessment	

fidelity modeling is possible, a balance must be struck between precision and limitations relevant to cost, resources and time [21].

On the other hand, conservative estimates may lead to irrationally high reliability requirements and/or very strict operational restrictions. This occurs as a consequence of a problem known as compounding conservatism, where use of successive conservative estimates can lead to over-conservative results [21]. To illustrate this

Table 5.12 A summary of the methods presented to estimate the terms involved mid-air collision risk modeling when their values are not known a priori, Eqs. (5.14) and (5.14)

Term	Estimate	
f_F	(i) Based on ELOS requirements with typical values in the range of 10^{-6}–$10^{-8}\,h^{-1}$	
	(ii) As above but considering only fatalities on the ground for calculating ELOS (valid when the accident involves only UAS)	
$E(\text{fatality})$	(i) Estimated from historical data (based on NTSB data from Table 5.9, it takes values in the range 0.02–1)	
	(ii) Estimated from the product of N_{exp} and $P(\text{fatality})$	N_{exp}: The number of people on-board the aircraft involved in the collision, as well as, in the area exposed to debris $P(\text{fatality})$: (i) Use of the number one as a conservative estimate, or (ii) Estimate from historical data, if available
f_{MaC}	(i) From historical data with typical values in the range of 10^{-5}–$10^{-7}\,h^{-1}$	
	(ii) As the product of $P(\text{collision})$ with f_{CT}	$P(\text{collision})$: (i) A conservative estimate of probability one, (ii) Estimated based on the capabilities of the S&A system f_{CT}: (i) Based on the gas model of aircraft collisions using actual traffic data, or (ii) using worst-case air traffic density either at the flight level of operations or the entire airspace

issue, consider the modeling of a ground impact accident scenario using (5.1). In addition, assume that the actual population and area affected by the crash are overestimated by 50% and 20% respectively, and the probability of fatality is considered to be 1 when in fact it is only 20%. In this case, the fatality expectation will be 9 times higher than what it really is. Moreover, if conservative estimates are used for evaluating the possible hazards that may lead to a crash, the reliability requirements for the various aircraft parts can be higher by two or more orders of magnitude. Although such an aircraft would be capable of performing well within the target safety levels, the higher production and maintenance costs could mean that it may never get built.

The problem of compounding conservatism may be addressed by avoiding to take conservative estimates and, instead, opting for the best available estimate [21]. Such an approach has also been advocated by the US Nuclear Regulatory Commision [21]. Of course, when best or mean estimates are used, the uncertainties and possible inaccuracies affecting the final result should be clearly documented [21]. This would also allow adjustments at a later stage to account for newer data as they become available. Moreover, if the uncertainties can be quantified through simulation, sensitivity analysis or other methods, then confidence intervals can be determined. The latter can then be used to adjust results so that proposed reliability targets can be achieved with arbitrary confidence.

As an example, it will be assumed that a probability of fatality of 0.4 was estimated for a particular accident scenario. Further analysis of the model showed that for a confidence level of 95% the actual value is expected to be between 0.3 and 0.5. In this case the 0.5 value may be used instead of 0.4 to determine UAS reliability requirements. If higher confidence is required, the range will be larger and the value used more conservative.

In certain cases —especially when required data are missing— a safety case can be made by comparing the UAS under investigation with a different system already authorized to fly [19]. Qualitative arguments may also be made without a complete analysis of the risk involved [19]. For example the UAS may be too light to cause an injury or the area of operations may be so sparsely populated that the risk to the general public is too low under any conditions.

The modeling techniques described in the previous sections produce an average risk estimate from the operation of a UAS, especially when best estimates are used instead of conservative. What should be mentioned is that in certain cases when using a casualty or fatality metric, particularly catastrophic accidents involving multiple fatalities can occur without violating the target safety levels on average [21]. Such accidents of course are best to be avoided and as a result, it is useful to incorporate catastrophe aversion in the models used [21]. The latter is accomplished by assigning activities that can be particularly dangerous lower acceptable probabilities of occurrence. This in turn may be done either by creating a risk profile if sufficient data are available or easier by using functions of the N_{exp}^k type for expressing the affected population [21]. Of course the latter methodology should be employed with care and for high risk activities only, so as not to unnecessarily inflate the risk in other less dangerous activities. A risk profile gives the function between the number of expected casualties and the expected frequency for each casualty size due to various future incidents [21]. Although this profile is useful for obtaining a better view of the associated risks, it is usually costly and time consuming to obtain.

Regardless of the actual risk model and target safety level chosen, it is necessary to obtain design specifications and requirements on the hardware and software components that comprise the UAS, rather than restrictions on the UAS as a whole. Such requirements can then be readily used for design and procurement decisions. The next Section describes the procedure to derive aforementioned requirements.

5.6 Translating an Accident TLS to System Reliability Requirements

The main issue at stake and the question that needs to be answered, is whether a particular UAS is safe to fly, and if not, what measures need to be taken so that it is. The latter is known as risk management.

The process of risk management of a test range involves four stages [21]:

Mission Definition and Hazard Identification In this stage the characteristics of the mission and of the system are defined and appropriate safety limits are derived. The hazards from operating that system are defined, to be used in the second stage.

Risk Assessment The risk associated with each hazard and possible failure scenario is evaluated. The assessment may be adjusted to incorporate different models and assumptions depending on needs.

Risk Reduction The risk is compared to safety limits and when required it is lowered by adjusting system design or operation, safety limits and/or risk calculation models.

Risk Acceptance If the risk is within acceptable limits, the mission is approved. During the mission the system is still monitor to make sure that risk does not increase.

An alternative method to manage risk is to impose limits on the occurrence rates for each hazard separately, rather than attempt to limit the aggregate risk of fatalities [21]. This method removes some flexibility since it no longer allows the increased risk from some hazards to be balanced out by others. On the other hand, it simplifies the process of assessing risk, since each possible hazard is evaluated separately.

For certain application domains or mission scenarios, risk can be mitigated by controlling the environment within which the UAS operates. For example, by appropriately restricting the areas over which the UAS will fly, the population density can be kept within safe limits. Alternatively the route of the UAS can be adjusted to at least avoid areas of high population density or other high value property [18]. The latter include facilities that could trigger other hazards like a chemical processing plant, cultural or environmental sensitive areas like national monuments or parks, and other areas of high monetary or political cost like a solar farm or a school.

It should be noted however, that even when the area of operations is completely unpopulated, certain failure modes may lead the UAS to exit the restricted area. Sophisticated flight termination systems and pre-programmed commands for emergency situations may help alleviate this issue [21]. The use of procedural safeguards can also reduce risk during UAS operations [19]. Such measures involve the use of safety checklists, the drafting of emergency procedures, as well as operator training. On the other hand, the effectiveness of such measures may be limited in reducing risk to the general public.

Similar measures can be taken to reduce the risk of mid-air collisions. Such a measure involves operating a UAS within restricted airspace where all non-related traffic is banned or notified and separated [18]. The use of secondary aircraft, ground observers, bright markings or radar may also significantly reduce the risk of a mid-air collision, especially in conjunction with pre-determined procedures in the case of potential conflicting trajectories [18].

Of course, besides trying to control the operational environment of the UAS, it is also possible and in certain cases required to design it to meet safety requirements. This is typically accomplished by improving the reliability of the UAS, so that hazard frequency and severity is controlled.

Given an accident TLS, an Operational Safety Assessment (OSA) is required to derive system requirements and verify that the final system complies with set TLS. This assessment is a long procedure that starts at the early stages of the design and involves operation and maintenance of the final system.

The first stage of an OSA is the Operational Environment Definition (OED). This entails a detailed list of the functions that the final system will need to perform, as well as the required hardware and software components. It will also need to include other factors that influence safety like human and environmental factors. For example, higher reliability is required for over-flying high population density areas. The purpose of the OED is to define the range of operating conditions and other factors that can influence safety, to be used in the second stage of the OSA, the Function Hazard Assessment (FHA).

The FHA examines every aspect of the system under investigation, in an effort to determine all possible hazards that can lead to failure conditions. Typically this is accomplished by dividing the system functions into broad categories and subdividing each category into lower level functions, until a list of basic functions is derived. In the case of UAS, care should be given in considering all possible functions, including those that reside with the ground control station or a launch and retrieval system.

As an example in the FHA presented in [12], the high level UAS functions were defined as aviate (fly the plane), navigate, communicate and mitigate hazards. Each of these functions were then divided into lower level functions. For example, aviate was divided into:

1. Control flight path;
2. Control ground path;
3. Control air to ground transition;
4. Command and control between ground control station and UAS, and,
5. Control UAS subsystems.

Each of these functions were further subdivided into more basic functions.

After the basic system functions have been identified, failure conditions are derived for each of these functions. These conditions may range from total loss of the function, to capability deterioration and misleading information (e.g. false positive failure identification). Each failure condition is then assigned one of the following severities: no safety effect, minor, major, hazardous and catastrophic. This assignment is based on the expected outcome following the failure and the possible outcomes defined in the risk reference system used. In cases where the assignment is not obvious, a conservative choice is usually taken, under the condition that it will not create unreasonable requirements. Since each of these severity levels has been assigned a TLS in the risk reference system, a TLS is now available for each failure condition.

Table 5.13 provides an excerpt from the FHA of [12] with respect to the "convey flight path state" and "determine guidance command" functions, belonging to the "control flight path" function which in turn belongs to the "aviate" category. It should be noted that the aforementioned FHA was done on a generic UAS and changes may be required for an actual UAS depending on its type and configuration.

Table 5.13 Excerpt from a generic UAS Preliminary Functional Hazard Assessment, showing failure conditions and corresponding severity for two low-level functions. Source: [12]

Number	Function	Failure Condition	Criticality
1.1.1-a	Convey flight path state	Detected loss of function	hazardous
1.1.1-b	Convey flight path state	Display freezes with old information	hazardous
1.1.1-c	Convey flight path state	Incorrect information	catastrophic
1.1.2-a	Determine guidance command	Detected loss of function (alternate means to provide function exist)	minor
1.1.2-b	Determine guidance command	Undetected loss of function (alternate means to provide function exist)	major
1.1.2-c	Determine guidance command	Incorrect command determined (alternate means to provide function exist)	major
1.1.2-d	Determine guidance command	Loss of function (no alternate means exist), "soft landing" function available	hazardous
1.1.2-e	Determine guidance command	Loss of function (no alternate means exist), "soft landing" not function available	catastrophic

The next step in the FHA is to determine the hazards associated with each system function that can lead to each potential failure condition. During this analysis several factors need to be taken into account:

- The effect of a single malfunction to other system functions performed by the same component.
- The effect of a failure in one system function to other related functions.
- The effect of possible undetected, low-severity failures to the onset of new failures.
- The effect of alleviating factors, like the presence of backup systems or alternative means to provide the function.
- The effect of intensifying factors, like increased workload, adverse environmental conditions or the absence of failure notification.

This analysis will result in determining a TLS for each hazard associated with the operation of the final system. The OSA then proceeds with evaluating the effect of component failures using the Failure Modes and Effects Analysis (FMEA) and Fault Tree Analysis (FTA). The final goal is to determine appropriate component reliability levels as well as other means to control hazards, such that the determined TLS for each hazard is not violated. Special notice should be taken to the fact that in certain cases, manned aviation safety assessments have considered the pilot as an additional source of redundancy, an assumption that may not always hold for UAS.

Although it always preferable to design for minimum risk, in some cases safety or warning devices may need to be retrofitted or when even that is not possible the risk will need to be managed through training and procedures [23]. In any case, even in these latter cases the associated hazards will need to be identified. In the approach suggested by the Range Commanders Council in [19], the goal is to identify the hazards inherent in operating/testing a particular UAS, rather than designing a new system. In this case the different hazardous conditions are identified along with the possible causes, and measures are taken when required to reduce the risk to acceptable for the range levels. Example hazardous conditions and their causes that are provided in [19] are presented in Tables 5.14 to 5.18.

Table 5.14 Hazards that may result in uncontrolled flight. Source: [19]

Hazardous condition	Possible Causes
Loss of propulsion	(i) engine failure, (ii) fuel starvation, (iii) stuck throttle, and (iv) icing / weather
Loss of lift	(i) structural failure, and (ii) icing / weather
Loss of heading / attitude / position information	(i) heading / attitude system failure, and (ii) navigation system failure
Unplanned loss of link	(i) radio frequency interference, (ii) flight beyond horizon, (iii) antenna masking, (iv) loss of ground control station, (v) software interrupt between ground control station and air vehicle, (vi) atmospheric attenuation, (vii) inadvertent deactivation of autopilot, and (viii) loss of satellite link
Loss of control surface performance	(i) stuck servo, (ii) autopilot failure, and (iii) icing / damage to control surface
Loss of UAV electrical power	(i) generator failure, (ii) backup battery failure, and (iii) excessive load from payload
Loss of ground control station (GCS)	(i) Loss of GCS power, (ii) GCS transmitter/ receiver / antenna failure, and (iii) GCS computer failure

Table 5.15 Hazards that may result in a ground impact accident. Source: [19]

Hazardous Condition	Cause
Mission planning error or operator error	(i) flight below minimum enroute altitude, and (ii) undetected man-made obstacles (towers, cables)
Attitude error	(i) incorrect barometer setting, and (ii) inadequate alert for altitude deviation
Navigation error	(i) nav system failure, (ii) nav system discrepancy (INS vs. GPS), and (iii) map display inaccuracy
Failure to see and avoid terrain	(i) no capability, and (ii) autonomous operation
Loss of link "fly home" mode	(i) mission planning error for loss of link mode

Table 5.16 Hazards that may result in a mid-air collision. Source: [19]

Hazardous condition	Cause
Navigation error	(i) nav system failure, (ii) nav system discrepancy (INS vs. GPS), and (iii) map display inaccuracy
Altitude error	(i) incorrect barometer setting, and (ii) inadequate alert for altitude deviation
Unable to "see-and-avoid"	(i) limited capability, and (ii) autonomous operation
Mission planning error	(i) inadvertent flight into established routes of other aircraft
Not seen by other aircraft	(i) strobe / position lights inadequate or fail, (ii) IFF failure, (iii) TCAS failure, and (iv) ATC/UAV operator comm link failure

Table 5.17 Hazards that may result in an accident during take-off or landing. Source: [19]

Hazardous condition	Cause
Pilot induced oscillation	(i) system latency
Automatic landing system failure	(i) RFI, (ii) handoff errors, and (iii) missed approach procedures
Operator error	(i) outside weather / wind limits, and (ii) internal pilot / external pilot handoff errors

It should be noted that there are in fact a multitude of safety analysis tools available that are applicable to different requirement, design/development stages and for different types of systems. For civil aircraft the reader is referred to the most recent iteration of the following documents:

- Air Force Pamphlet 91-215 — "Operational Risk Management Guidelines and Tools"
- EUROCONTROL DAP/SSH/091 — "Safety Case Development Manual".
- FAA System Safety Handbook
- FAA AC 23.1309
- RTCA DO-160 — "Environmental Conditions and Test Procedures for Airborne Equipment".
- DO-264/ED-78A — "Guidelines for Approval of the Provision and Use of Air Traffic Services Supported by Data Communications"
- MIL-STD-882D — "Standard Practice for System Safety"
- SAE ARP 4761 — "Guidelines and Methods for Conducting the Safety Assessment Process on Civil Airborne Systems and Equipment".
- SAE ARP 926 — "Fault/Failure Analysis Procedure".
- SAE ARP 4753 — "Certification Considerations for Highly Integrated or Complex Aircraft Systems".

Table 5.18 Factors that may contribute to aircraft loss. Source: [19]

Contributing factor	Cause
Inadequate operator response	(i) failure to recognize flight critical situation, and (ii) flight-critical information missing, erroneous, or ambiguous delays in information flow
Incorrect inputs of flight critical parameters	(i) operator entry errors
Operator information overload	(i) tasking Vs time available, (ii) sensory overload over time
Critical information unavailable, inadequate, blocked, etc.	(i) design dependent
Latency of flight control commands	(i) operator far removed from control loop, (ii) non-deterministic software, and (iii) control link through satellite
Operator fatigue	(i) inadequate crew rest, (ii) task saturation, and (iii) long / boring mission
Control of multiple UAVs	(i) workload issues
Software paths to unsafe state	(i) unexpected reboot, and (ii) inadequate software safety process

5.7 Risk Mitigation

Ideally a UA will be designed in such a way that safety requirements are met for every conceivable application. In reality though there are currently many systems in operations that may violate these requirements under certain operating scenarios or during certain phases of flight. When a risk assessment indicates such an issue, there are several alternatives besides not proceeding with a flight.

When the flight authorization is requested for a specific scenario, the affected population may be controlled to reduce the risk of fatality. The latter can be accomplished by selecting a different route that avoids the more densely populated areas or when possible by notifying people of the flight and/or restricting access to certain areas [19]. Another possibility is to select a day and time of flight that will ensure lower population density or on average better sheltering of the people in the area of operations.

Although fail-safe systems may or may not be considered in the risk assessment depending on their characteristics, they may be considered as risk mitigation measures. Parachutes to reduce kinetic energy at impact, pyrotechnics to decrease the mass of the impacting objects and crash-ports to contain the energy of the impact can be used to reduce the effects of a hazard rendering continued flight impossible. Recovery mechanisms and pre-programmed "return home" functions can also be used [19] to reduce the risk of an accident in lost-link situations or for containing the aircraft in the area of operations. Special care should be given in identifying and avoiding "high-value/high-consequence" areas such as those mentioned in the previous Section.

With respect to the risk of mid-air collisions, several measures may be taken to detect other traffic and recognize threats to the UAS safety. Such measures may involve additional on-board equipment, the use of observers, cooperation with ATC and the use of other ground-based assets (e.g. radar) [19]. Moreover, preprogrammed maneuvers may be stored to provide the necessary commands when a threat has been detected [19].

Although the aforementioned measures may not be suitable for obtaining airworthiness certificate, they can be used for obtaining permission for flying under certain conditions and/or scenarios. It is also a fact, that many of these measures may significantly increase the cost of an operation. As a result, it may still be necessary to abort the flight altogether.

References

1. Clothier R, Walker R (2006) Determination and evaluation of UAV safety objectives. In: Proc. 21st International Unmanned Air Vehicle Systems Conference, Irvine, CA, USA, pp 18.1–18.16
2. Clothier R, Walker R, Fulton N, Campbell D (2007) A casualty risk analysis for unmanned aerial system (UAS) operations over inhabited areas. In: Proc. 12th Australian International Aerospace Congress and 2nd Australasian Unmanned Air Vehicles Conference, Melbourne, Australia
3. Cole JK, Young LW, Jordan-Culler T (1997) Hazards of falling debris to people, aircraft, and watercraft. Sandia Report SAND97-0805, Sandia National Laboratories
4. Dalamagkidis K, Valavanis K, Piegl L (2008) Current status and future perspectives for unmanned aircraft system operations in the US. Journal of Intelligent and Robotic Systems 52(2):313–329
5. European Aviation Safety Agency (2009) Airworthiness certification of Unmanned Aircraft Systems (UAS). Policy Statement, E.Y013-01
6. European Aviation Safety Agency (EASA) (2005) A-NPA, No. 16/2005, policy for unmanned aerial vehicle (UAV) certification
7. European Aviation Safety Agency (EASA) (2007) Certification specification 25 (CS25). Amendment 3
8. Federal Aviation Administration (1999) Equipment, systems and installations in part 23 airplanes. AC 23.1309-1C
9. Feinstein DI, Haugel WF, Kardatzke ML, Weinstock A (1968) Personnel casualty study. Tech. Rep. Project No. J 6067, Illinois Institute of Technology Research Institute
10. Haber JM, Linn AM (2005) Practical models of human vulnerability to impacting debris. In: Proc. First IAASS Conference: "Space Safety, a New Beginning", Nice, France, ESA SP-599, pp 543–548
11. Haddon DR, Whittaker CJ (2002) Aircraft airworthiness certification standards for civil UAVs. UK Civil Aviation Authority

12. Hayhurst KJ, Maddalon JM, Miner PS, Szatkowski GN, Ulrey ML, Dewalt MP, Spitzer CR (2007) Preliminary considerations for classifying hazards of unmanned aircraft systems. Tech. Rep. NASA TM-2007-214539, National Aeronautics and Space Administration, Langley Research Center, Hampton, Virginia
13. INnovative Operational UAS Integration (INOUI) (2009) Proposal for the integration of UAS into non-segregated airspace. Booklet
14. Joint Capability Group on Unmanned Aerial Vehicles (2007) STANAG 4671 - Unmanned Aerial Vehicle Systems Airworthiness Requirements (USAR). draft, NATO Naval Armaments Group
15. Joint JAA/Eurocontrol Initiative on UAVs (2004) A concept for european regulations for civil unmanned aerial vehicles (UAV). Final Report
16. National Transportation Safety Board (NTSB) (2008) Accident database and synopses. Online, [online] http://www.ntsb.gov/ntsb/query.asp
17. National Transportation Safety Board (NTSB) (2008) Aviation accident statistics. Online, [online] http://www.ntsb.gov/aviation/Stats.htm
18. Range Safety Group, Range Commanders Council (1999) Range safety criteria for unmanned air vehicles. Document 323-99
19. Range Safety Group, Range Commanders Council (1999) Range safety criteria for unmanned air vehicles - rationale and methodology supplement. Supplement to document 323-99
20. Range Safety Group, Range Commanders Council (2007) Common risk criteria standards for national test ranges. Document 321-07
21. Range Safety Group, Range Commanders Council (2007) Common risk criteria standards for national test ranges: Supplement. Supplement to document 321-07
22. Sturdivan LM, Viano DC, Champion HR (2004) Analysis of injury criteria to assess chest and abdominal injury risks in blunt and ballistic impacts. Journal of Trauma 56(3):651–663
23. US Department of Defense (2000) Standard practice for system safety. MIL-STD-882D
24. US Department of Defense (2007) Unmanned systems safety guide for DoD acquisition. First Edition (Version.96)
25. Weibel RE (2005) Safety considerations for operation of different classes of unmanned aerial vehicles in the national airspace system. Master's thesis, Department of Aeronautics & Astronautics, Massachusetts Institute of Technology
26. Weibel RE, Hansman RJ (2003) Safety considerations for operation of small unmanned aerial vehicles in civil airspace. Presented in MIT Joint University Program Quarterly Meeting
27. Weibel RE, Hansman RJ (2004) Safety considerations for operation of different classes of UAVs in the NAS. In: Proc. AIAA 4th Aviation Tehcnology, Integration and Operations Forum and AIAA 3rd Unmanned Unlimited Technical Conference, Workshop and Exhibit, Chicago, IL, USA

Chapter 6
Case Studies

Traditional scientific method has always been at the very best 20-20 hindsight. It's good for seeing where you've been. It's good for testing the truth of what you think you know, but it can't tell you where you ought to go.

Robert M. Pirsig
(Zen and the Art of Motorcycle Maintenance, 1974)

Using the methodologies described in Sect. 5.3 it is possible to derive the reliability requirements with respect to ground impact for various types of UAS and under different scenarios.

To illustrate the differences between vulnerability models, three cases were investigated using twelve UAS, seven fixed-wing and five rotary-wing. The systems were chosen to span all sizes and their basic characteristics are shown in Table 6.1. The cases involve an optimistic, a pessimistic and an average scenario, each featuring different population densities affected and different levels of sheltering expected. A description of each case and the parameters used are provided in Table 6.2.

Although the results are subject to the uncertainties inherent in the parameters and the models themselves, they should be accurate in terms of order of magnitude and for comparisons between different vulnerability models, as well as between UAS types and sizes.

Table 6.1 Characteristics of seven fixed wing and five rotary-wing UAS of various sizes, used for the case analysis. Source: [3, 4]

	Weight (kg)	Dimensions (m)	Oper. Speed (m/s)	Oper. Altitude (ft)
RQ-4 Global Hawk	11,612	35.4 (wingspan)	177	65,000
MQ1 Predator	1,021	14.8 (wingspan)	70	20,000
RQ-2 Pioneer	205	5.2 (wingspan)	41	15,000
Neptune	36	2.1 (wingspan)	43	8,000
Aerosonde	15	2.9 (wingspan)	42	12,000
RQ-11 Raven	1.9	1.3 (wingspan)	15	1,000
Mosquito	0.5	0.34 (wingspan)	1.25	400
RQ-6 Fire Scout	1,157	8.4 (rotor diameter)	65	20,000
CL-327 Guardian	350	4.0 (rotor diameter)	44	18,000
Rmax IIG	94	3.12 (rotor diameter)	5.6	500
Vario XLV	22	2.5 (rotor diameter)	16	500
Maxi Joker 2	8	1.8 (rotor diameter)	20[a]	400

[a] guesstimated

K. Dalamagkidis et al., *On Integrating Unmanned Aircraft Systems into the National Airspace System*, Intelligent Systems, Control and Automation: Science and Engineering 54, DOI 10.1007/978-94-007-2479-2_6,

For each case, the probability of fatality was first estimated. To illustrate the different results that may be reached due to the choice of a model, four different modeling options were employed:

i. The probability of fatality was based on the vulnerability model of (5.10) using a kinetic energy estimate based on (5.12).
ii. Using the same model as above, but with a more conservative kinetic energy estimate based on terminal velocity.
iii. Using the older model of (5.9) with the conservative kinetic energy estimate. This was chosen to evaluate the effects of the correction factor k.
iv. The probability of fatality is 0 or 1, based on the kinetic energy threshold of 34 J, initially presented in Sect. 5.3.3. The kinetic energy is calculated using the terminal velocity.

For the first three options and under all scenarios, parameter α was chosen to be 100 kJ and β equal to 34 J.

In addition to the probability of fatality, the required system reliability was also calculated for each of the aforementioned fatality probability models, as well as for the conservative estimate of probability of 1. The UAS reliability requirement has been given in minimum hours between ground impact accidents and its calculation is based on a target level of safety of 10^{-7} fatalities per hour of flight. The system reliability requirement was derived since it allows a comparison with the current performance of manned and unmanned aviation. The results for each UAS and each case are summarized in Tables 6.3 to 6.8.

In the pessimistic scenario, with the exception of the Mosquito micro UAS, the probability of fatality associated with each UAS is 100%. As a result, there are no differences between the reliability requirements calculated from the different models. Considering that current manned aviation accident rates are in the order of $10^{-7}\,h^{-1}$ for air carriers and $10^{-5}\,h^{-1}$ for general aviation (Table 5.4), it is obvious that for operations in high population density areas, certain UAS will need to exceed this performance. It is also noteworthy that using a fatality probability of 1, would impose an unrealistic requirement on the Mosquito micro UAS, despite the fact that normally no fatality risk is expected. The usefulness of the correction factor k is also evident in this case, since model (5.9) would still impose a requirement of 181 h.

In the optimistic scenario, the most striking differences between vulnerability models can be seen! This is due to the effect of the sheltering factor, which is not taken into account when using a threshold kinetic energy. Smaller systems feature fatality probabilities of 10–25% that are further reduced to 4–10% when using a less conservative estimate for the kinetic energy at impact. This is also evidenced in the system reliability requirement, which is at least an order of magnitude smaller compared to that obtained using the threshold function. The benefits are evidenced in larger systems as well, where the reliability requirement is lower by a factor up to 3.

In the average scenario, the fatality probabilities of the smaller systems are in most cases over 50%, and for the larger systems they approach 100%. Nevertheless,

Table 6.2 The parameters used for each test case and a description of a possible corresponding scenario.

Scenario	Pop. Density (ppl/km^2)	p_s	Description
1 - Optimistic	50	7	Low population density area. It is also assumed that people are afforded significant sheltering either by natural obstacles (e.g. trees) or they can be trained to avoid or take cover when required. This scenario may correspond to surveillance of a remote military installation or to a forest monitoring application.
2 - Average	200	3	The population density is equal to the standard population density of 200 ppl/km^2 [2]. This scenario aims at simulating operations in suburban regions. The sheltering factor assumes that there is some sheltering, but it does not offer significant protection.
3 - Pessimistic	5,000	1	This scenario features very high population density. Additionally the sheltering factor used corresponds to no protection from sheltering at all. This case corresponds to the scenario of a search and rescue operation in a metropolitan area, where several people are in open areas preoccupied with other tasks.

the lower reliability requirements involved compared with an over-conservative approach allow for systems that will be significantly easier and cheaper to develop.

A similar analysis can be done to determine regions that can be safely overflown given the UAS reliability. Results of such an analysis are given in Tables 6.9 to 6.11 for the United States, Europe and Australia, respectively, for half of the UAS of Table 6.1. The differences are mainly due to the differences in population distribution.

It is noteworthy that a large UAS like the RQ-4 Global Hawk can safely loiter over only 38.8% of the US area and there will still be 20% of that area that would be unreachable even if it reaches the accident rate of general aviation. Even for smaller systems like the RQ-2 Pioneer and the RQ-6 Fire scout and even if they reach the 100,000 T_{GI} limit, many areas, mostly in and around major cities, remain out-of-bounds.

The figures are even worse for Europe, where because of the relatively high population density, only about 20% is currently safe to loiter over with a RQ-4 Global Hawk. Loitering over metropolitan centers will require significant reliability —beyond current capabilities— to maintain the target safety levels. On the other hand, Australia, with the large swaths of sparsely populated land presents an easier target. Nevertheless previous observations regarding metropolitan areas hold true in this case as well.

Small UAS will be safe to operate over most areas without high reliability requirements. It is noteworthy that with even relatively low reliability a large per-

Table 6.3 Fatality probability with respect to ground impact accidents for twelve UAS under the pessimistic scenario. Four different means of obtaining the fatality probability are used.

UAS Model	$P(\text{fatality}\|\text{exposure})$ Eq. (5.10)[1]	Eq. (5.10)[2]	Eq. (5.9)[2]	34 kJ limit[2]
RQ-4 Global Hawk	100.0%	100.0%	100.0%	100.0%
MQ1 Predator	100.0%	100.0%	100.0%	100.0%
RQ-2 Pioneer	100.0%	100.0%	100.0%	100.0%
Neptune	100.0%	100.0%	100.0%	100.0%
Aerosonde	100.0%	100.0%	100.0%	100.0%
RQ-11 Raven	97.4%	100.0%	100.0%	100.0%
Mosquito	0.0%	0.0%	0.2%	0.0%
RQ-6 Fire Scout	100.0%	100.0%	100.0%	100.0%
CL-327 Guardian	100.0%	100.0%	100.0%	100.0%
Rmax type IIG	100.0%	100.0%	100.0%	100.0%
Vario XLV	100.0%	100.0%	100.0%	100.0%
Maxi Joker	99.6%	100.0%	100.0%	100.0%

[1] Using vehicle kinetic energy estimated from (5.12).
[2] Worst-case vehicle kinetic energy estimate.

Table 6.4 Reliability requirement for twelve UAS with respect to ground impact accident under the pessimistic scenario. Five different fatality probability estimates are used.

UAS Model	Required time between ground impact accidents in hours Eq. (5.10)[1]	Eq. (5.10)[2]	Eq. (5.9)[2]	34 kJ limit[2]	$P = 1$
RQ-4 Global Hawk	28,002,000	28,002,000	28,002,000	28,002,000	28,002,000
MQ1 Predator	7,879,500	7,879,500	7,879,500	7,879,500	7,879,500
RQ-2 Pioneer	1,738,500	1,738,500	1,738,500	1,738,500	1,738,500
Neptune	507,000	507,000	507,000	507,000	507,000
Aerosonde	646,000	646,000	646,000	646,000	646,000
RQ-11 Raven	280,485	287,987	287,987	288,000	288,000
Mosquito	0	0	181	0	100,800
RQ-6 Fire Scout	3,421,626	3,421,626	3,421,626	3,421,626	3,421,626
CL-327 Guardian	874,737	874,737	874,737	874,737	874,737
Rmax type IIG	566,023	566,069	566,069	566,069	566,069
Vario XLV	388,748	388,768	388,767	388,772	388,772
Maxi Joker	227,655	228,497	228,497	228,512	228,512

[1] Using vehicle kinetic energy estimated from (5.12).
[2] Worst-case vehicle kinetic energy estimate.

centage of the area of the United States, Europe and Australia will be available to these systems. Nevertheless, although low reliability may be permissible in terms of security, it may be unacceptable due to other factors, like cost.

To further illustrate these results, Figs. 6.1 to 6.9 present graphically the areas of United States, Europe and Australia where these six types of UAS can safely loiter over, depending on their respective T_{GI}. These are followed by similar maps for the state of Texas and the Dallas-Fort worth Metroplex (Figs. 6.10 to 6.15), Colorado

Table 6.5 Fatality probability with respect to ground impact accidents for twelve UAS under the optimistic scenario. Four different means of obtaining the fatality probability are used.

UAS Model	$P(\text{fatality}\mid\text{exposure})$ Eq. (5.10)[1]	Eq. (5.10)[2]	Eq. (5.9)[2]	34 kJ limit[2]
RQ-4 Global Hawk	94.5%	95.1%	95.0%	100.0%
MQ1 Predator	75.4%	81.1%	80.9%	100.0%
RQ-2 Pioneer	49.0%	72.5%	72.2%	100.0%
Neptune	31.7%	50.7%	50.7%	100.0%
Aerosonde	23.5%	29.2%	29.7%	100.0%
RQ-11 Raven	3.6%	10.7%	12.0%	100.0%
Mosquito	0.0%	0.0%	1.3%	0.0%
RQ-6 Fire Scout	57.9%	75.2%	75.0%	100.0%
CL-327 Guardian	48.1%	56.3%	56.2%	100.0%
Rmax type IIG	9.8%	26.5%	27.1%	100.0%
Vario XLV	10.2%	13.0%	14.1%	100.0%
Maxi Joker	5.2%	10.1%	11.4%	100.0%

[1] Using vehicle kinetic energy estimated from (5.12).
[2] Worst-case vehicle kinetic energy estimate.

Table 6.6 Reliability requirement for twelve UAS with respect to ground impact accident under the optimistic scenario. Five different fatality probability estimates are used.

UAS Model	Required time between ground impact accidents in hours Eq. (5.10)[1]	Eq. (5.10)[2]	Eq. (5.9)[2]	34 kJ limit[2]	$P = 1$
RQ-4 Global Hawk	264,481	266,239	266,006	280,020	280,020
MQ1 Predator	59,394	63,909	63,736	78,795	78,795
RQ-2 Pioneer	8,514	12,598	12,558	17,385	17,385
Neptune	1,609	2,571	2,570	5,070	5,070
Aerosonde	1,517	1,884	1,919	6,460	6,460
RQ-11 Raven	102	309	346	2,880	2,880
Mosquito	0	0	13	0	1,008
RQ-6 Fire Scout	19,825	25,728	25,648	34,216	34,216
CL-327 Guardian	4,210	4,926	4,917	8,747	8,747
Rmax type IIG	554	1,501	1,536	5,661	5,661
Vario XLV	396	504	550	3,888	3,888
Maxi Joker	119	232	262	2,285	2,285

[1] Using vehicle kinetic energy estimated from (5.12).
[2] Worst-case vehicle kinetic energy estimate.

and the Denver metropolitan area (Figs. 6.16 to 6.21) and finally for the state of Florida and the county of Hillsborough (Figs. 6.22 to 6.27).

Table 6.7 Fatality probability with respect to ground impact accidents for twelve UAS under the average scenario. Four different means of obtaining the fatality probability are used.

UAS Model	$P(\text{fatality}\vert\text{exposure})$ Eq. (5.10)[1]	Eq. (5.10)[2]	Eq. (5.9)[2]	34 kJ limit[2]
RQ-4 Global Hawk	100.0%	100.0%	100.0%	100.0%
MQ1 Predator	100.0%	100.0%	100.0%	100.0%
RQ-2 Pioneer	99.5%	99.9%	99.9%	100.0%
Neptune	97.4%	99.6%	99.5%	100.0%
Aerosonde	93.6%	96.6%	96.5%	100.0%
RQ-11 Raven	17.9%	66.5%	66.3%	100.0%
Mosquito	0.0%	0.0%	0.8%	0.0%
RQ-6 Fire Scout	99.8%	100.0%	100.0%	100.0%
CL-327 Guardian	99.4%	99.7%	99.7%	100.0%
Rmax type IIG	61.7%	95.4%	95.3%	100.0%
Vario XLV	63.8%	75.6%	75.3%	100.0%
Maxi Joker	30.3%	63.6%	63.4%	100.0%

[1] Using vehicle kinetic energy estimated from (5.12).
[2] Worst-case vehicle kinetic energy estimate.

Table 6.8 Reliability requirement for twelve UAS with respect to ground impact accident under the average scenario. Five different fatality probability estimates are used.

UAS Model	Required time between ground impact accidents in hours Eq. (5.10)[1]	Eq. (5.10)[2]	Eq. (5.9)[2]	34 kJ limit[2]	$P = 1$
RQ-4 Global Hawk	1,120,073	1,120,074	1,120,074	1,120,080	1,120,080
MQ1 Predator	315,066	315,128	315,127	315,180	315,180
RQ-2 Pioneer	69,177	69,504	69,504	69,540	69,540
Neptune	19,743	20,189	20,188	20,280	20,280
Aerosonde	24,184	24,949	24,934	25,840	25,840
RQ-11 Raven	2,057	7,665	7,641	11,520	11,520
Mosquito	0	0	34	0	4,032
RQ-6 Fire Scout	136,552	136,814	136,813	136,865	136,865
CL-327 Guardian	34,792	34,896	34,895	34,989	34,989
Rmax type IIG	13,967	21,607	21,589	22,643	22,643
Vario XLV	9,922	11,751	11,715	15,551	15,551
Maxi Joker	2,774	5,816	5,799	9,140	9,140

[1] Using vehicle kinetic energy estimated from (5.12).
[2] Worst-case vehicle kinetic energy estimate.

Table 6.9 The percentage of the US area over which each UAS can loiter without violating set TLS requirement, based on exhibited reliability. The bold column represents the reliability of manned general aviation. Population density data: [1].

	T_{GI} in hr				
	10^2	10^3	10^4	10^5	10^6
RQ-4 Global Hawk	0.4%	7.1%	38.8%	**79.5%**	96.6%
MQ1 Predator	2.5%	25.6%	64.2%	**93.8%**	99.0%
Neptune	43.8%	83.9%	97.2%	**99.9%**	100.0%
RQ-6 Fire Scout	7.7%	40.8%	81.4%	**96.8%**	99.8%
Rmax type IIG	55.9%	91.5%	98.5%	**100.0%**	100.0%
Maxi Joker	89.4%	98.1%	100.0%	**100.0%**	100.0%

Table 6.10 The percentage of the area of Europe over which each UAS can loiter without violating set TLS requirement, based on exhibited reliability. The bold column represents the reliability of manned general aviation. Population density data: [1].

	T_{GI} in hr				
	10^2	10^3	10^4	10^5	10^6
RQ-4 Global Hawk	4.2%	6.2%	19.3%	**50.5%**	93.6%
MQ1 Predator	5.8%	12.0%	36.4%	**82.6%**	98.7%
Neptune	22.7%	56.9%	95.5%	**99.8%**	100.0%
RQ-6 Fire Scout	7.0%	20.5%	52.9%	**94.3%**	99.6%
Rmax type IIG	31.4%	74.0%	98.0%	**99.9%**	100.0%
Maxi Joker	67.8%	97.2%	99.9%	**100.0%**	100.0%

Table 6.11 The percentage of the area of Australia over which each UAS can loiter without violating set TLS requirement, based on exhibited reliability. The bold column represents the reliability of manned general aviation. Population density data: [1].

	T_{GI} in hr				
	10^2	10^3	10^4	10^5	10^6
RQ-4 Global Hawk	30.0%	72.7%	92.6%	**98.9%**	99.7%
MQ1 Predator	61.9%	87.0%	97.8%	**99.5%**	99.9%
Neptune	94.0%	99.2%	99.8%	**100.0%**	100.0%
RQ-6 Fire Scout	73.1%	93.0%	99.0%	**99.7%**	100.0%
Rmax type IIG	97.1%	99.4%	99.9%	**100.0%**	100.0%
Maxi Joker	99.3%	99.9%	100.0%	**100.0%**	100.0%

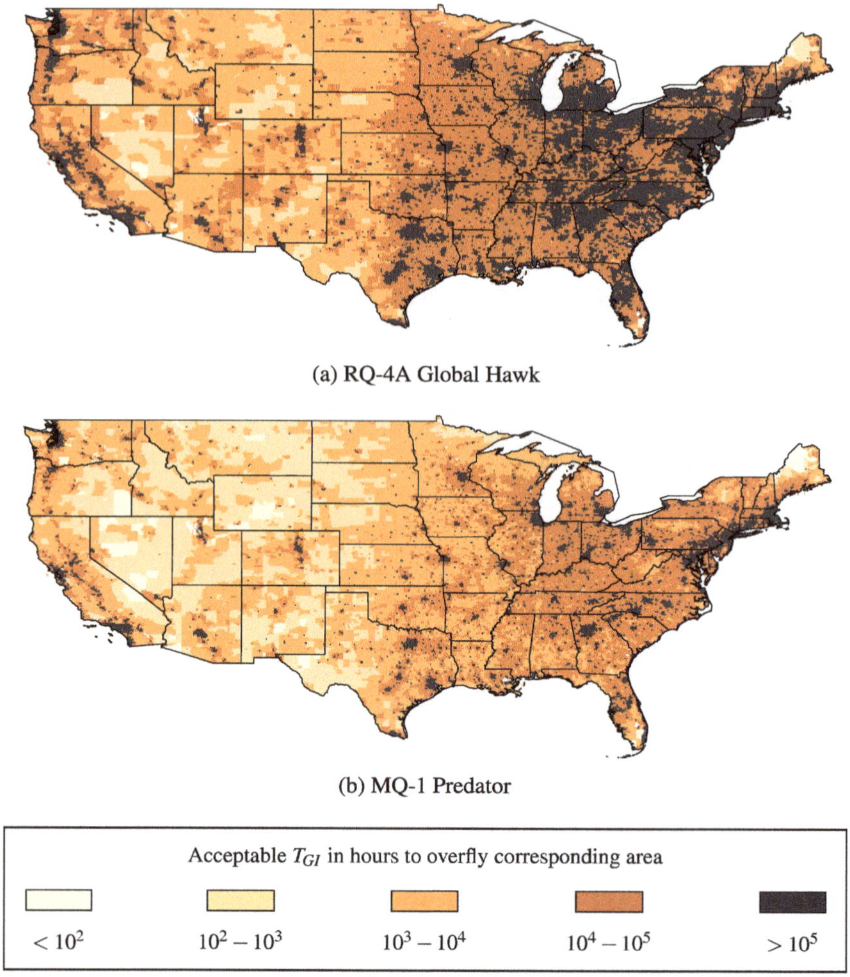

Fig. 6.1 The areas of the US, the RQ-4A Global Hawk and the MQ-1 Predator UAS are allowed to loiter over based on their reliability with respect to ground impact occurrence frequency.

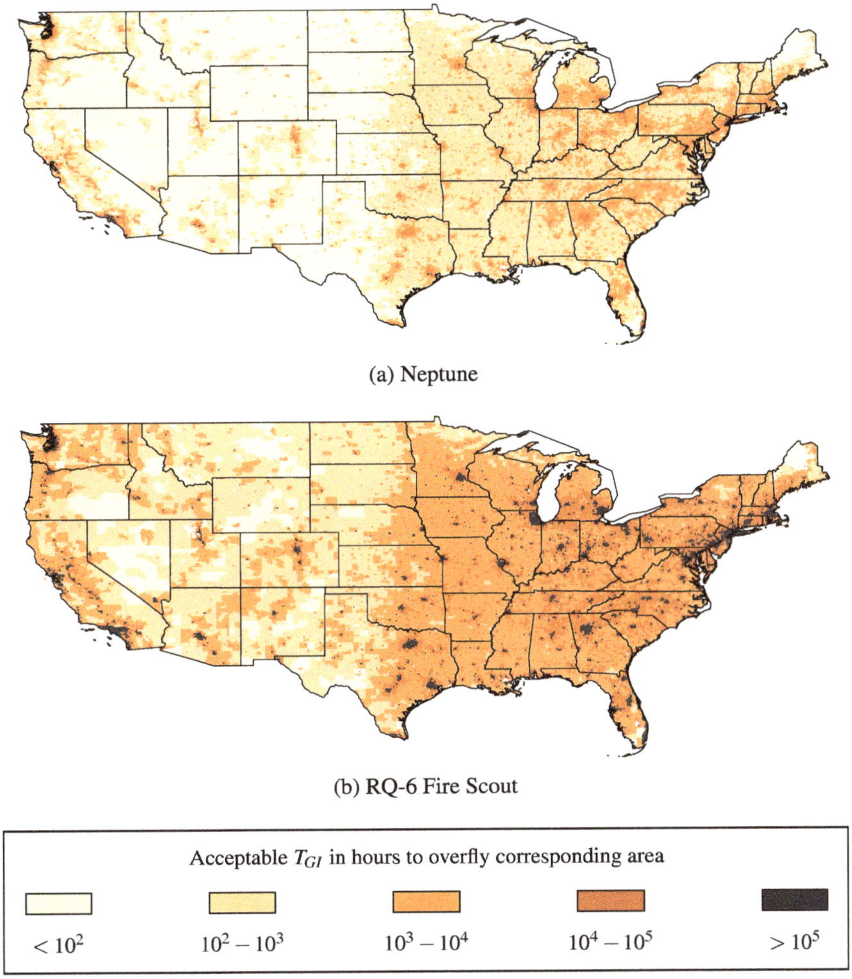

(a) Neptune

(b) RQ-6 Fire Scout

Fig. 6.2 The areas of the US, the Neptune and the RQ-6 Fire Scout UAS are allowed to loiter over based on their reliability with respect to ground impact occurrence frequency.

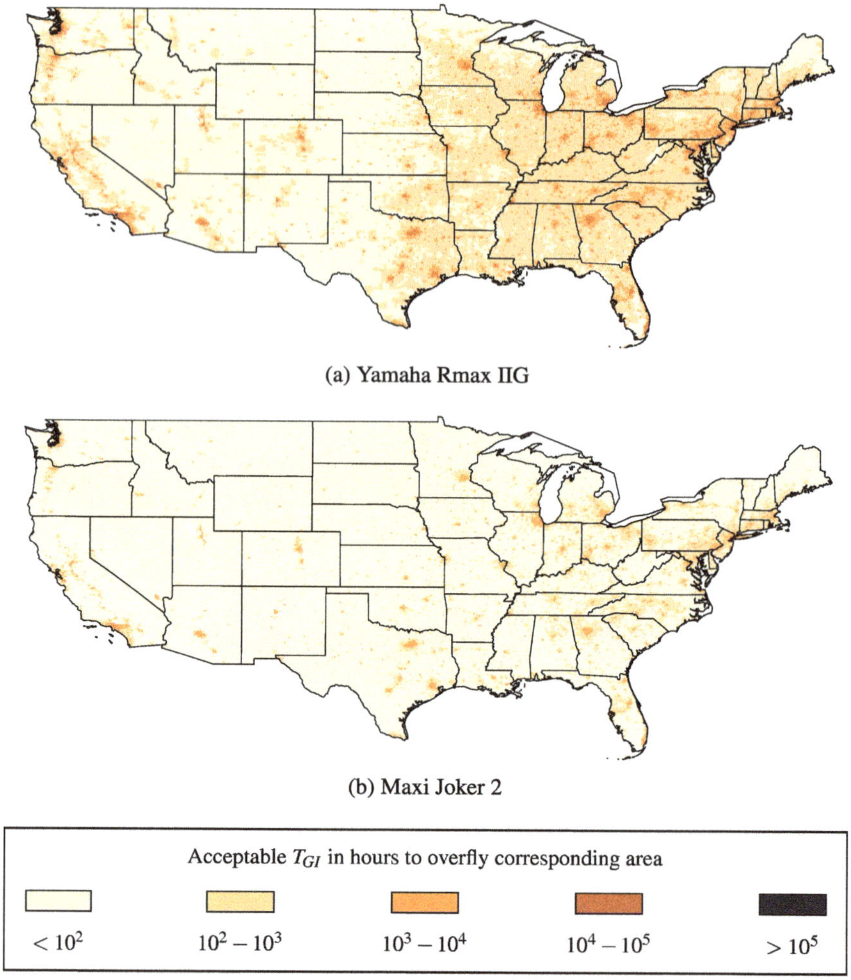

Fig. 6.3 The areas of the US, the Yamaha Rmax IIG and Maxi Joker 2 helicopters are allowed to loiter over based on their reliability with respect to ground impact occurrence frequency.

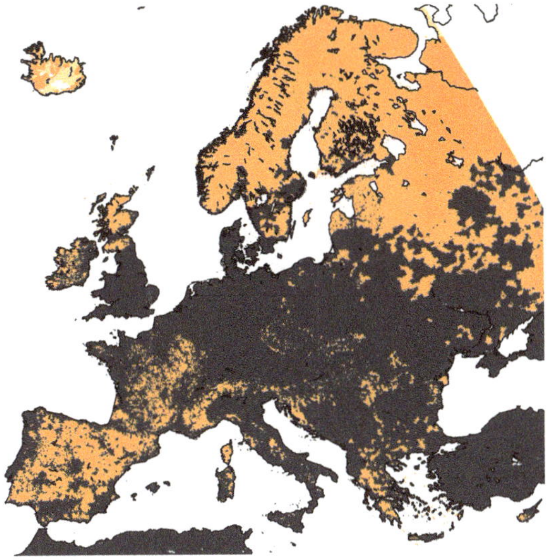

(a) RQ-4A Global Hawk

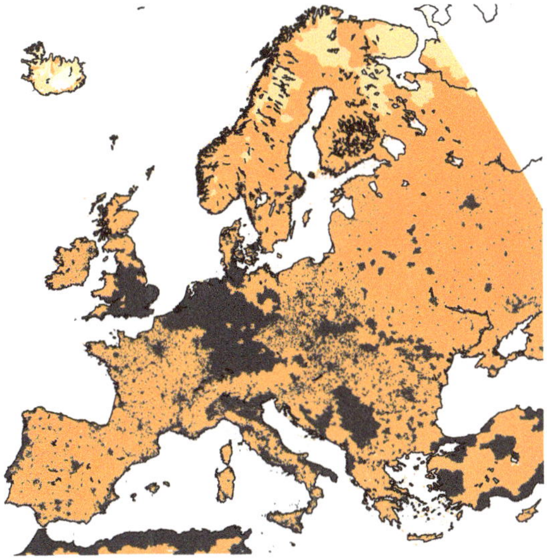

(b) MQ-1 Predator

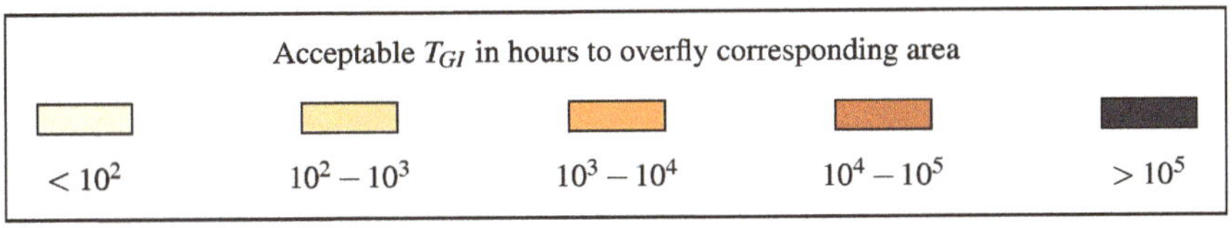

Fig. 6.4 The areas of Europe, the RQ-4A Global Hawk and the MQ-1 Predator UAS are allowed to loiter over based on their reliability with respect to ground impact occurrence frequency.

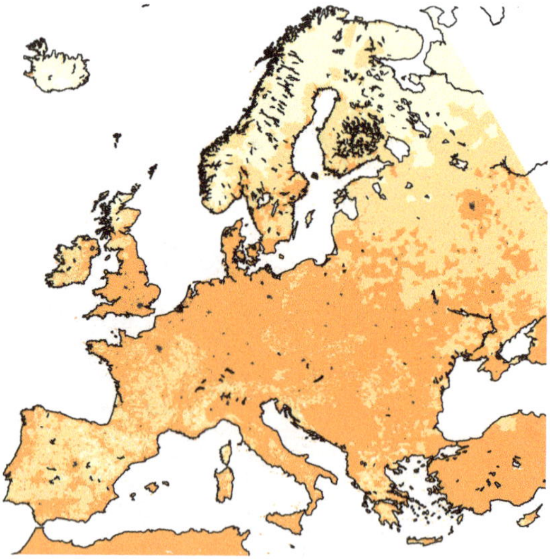

(a) Neptune

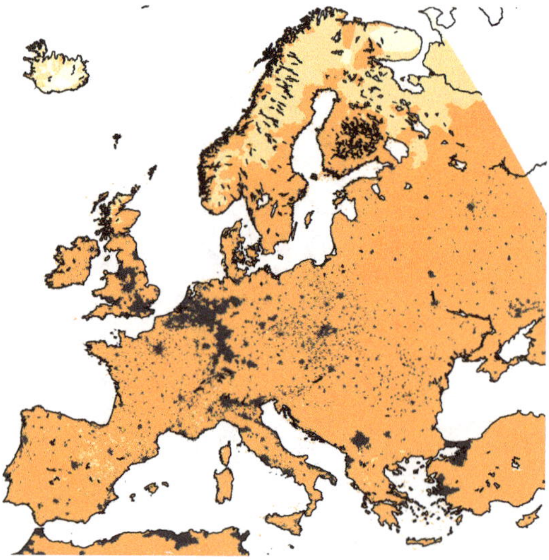

(b) RQ-6 Fire Scout

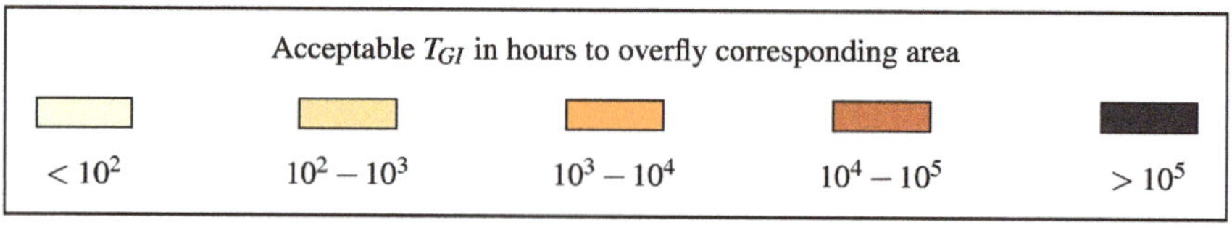

Fig. 6.5 The areas of Europe, the Neptune and the RQ-6 Fire Scout UAS are allowed to loiter over based on their reliability with respect to ground impact occurrence frequency.

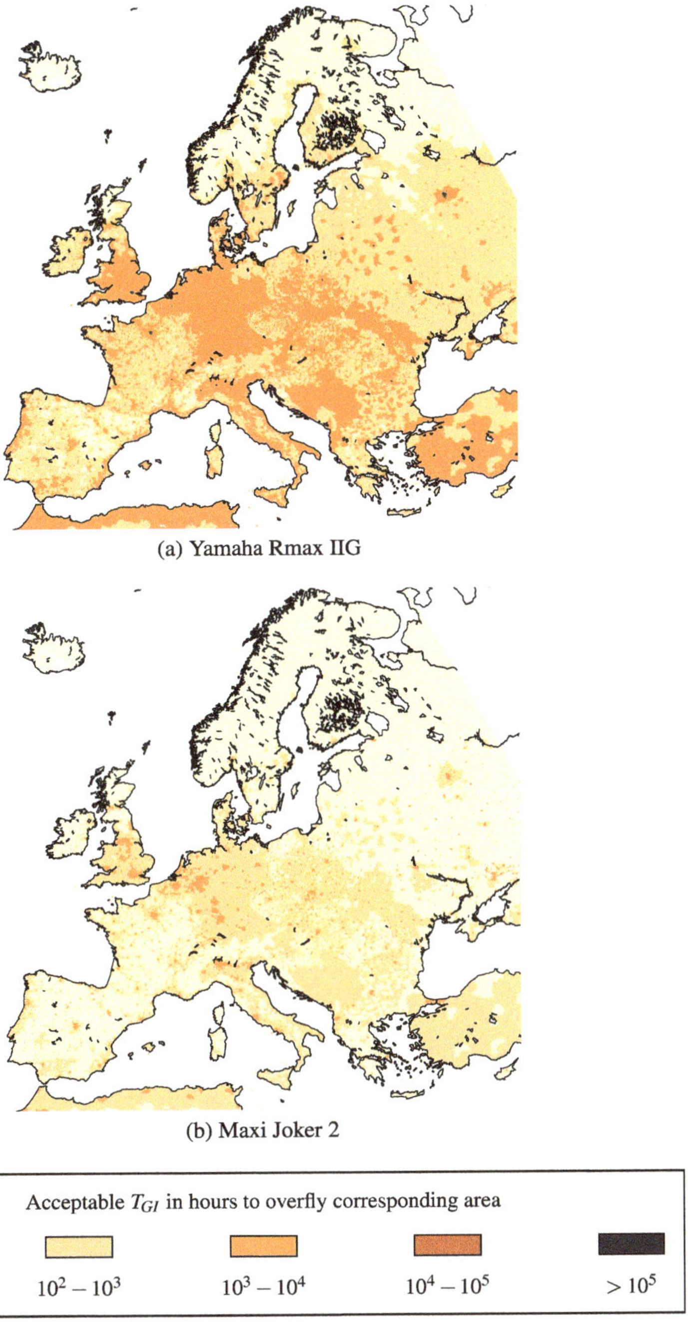

Fig. 6.6 The areas of Europe, the Yamaha Rmax IIG and Maxi Joker 2 helicopters are allowed to loiter over based on their reliability with respect to ground impact occurrence frequency.

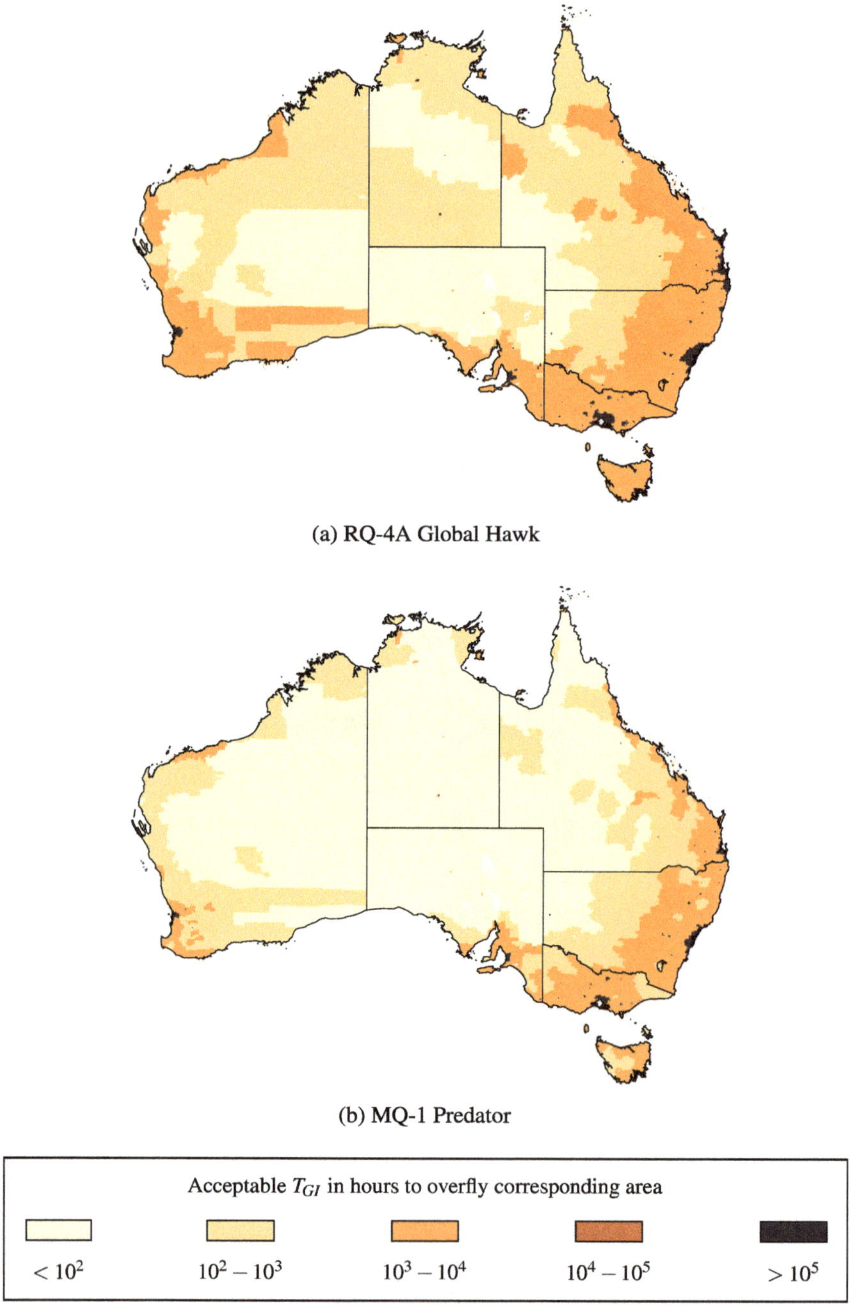

Fig. 6.7 The areas of Australia, the RQ-4A Global Hawk and the MQ-1 Predator UAS are allowed to loiter over based on their reliability with respect to ground impact occurrence frequency.

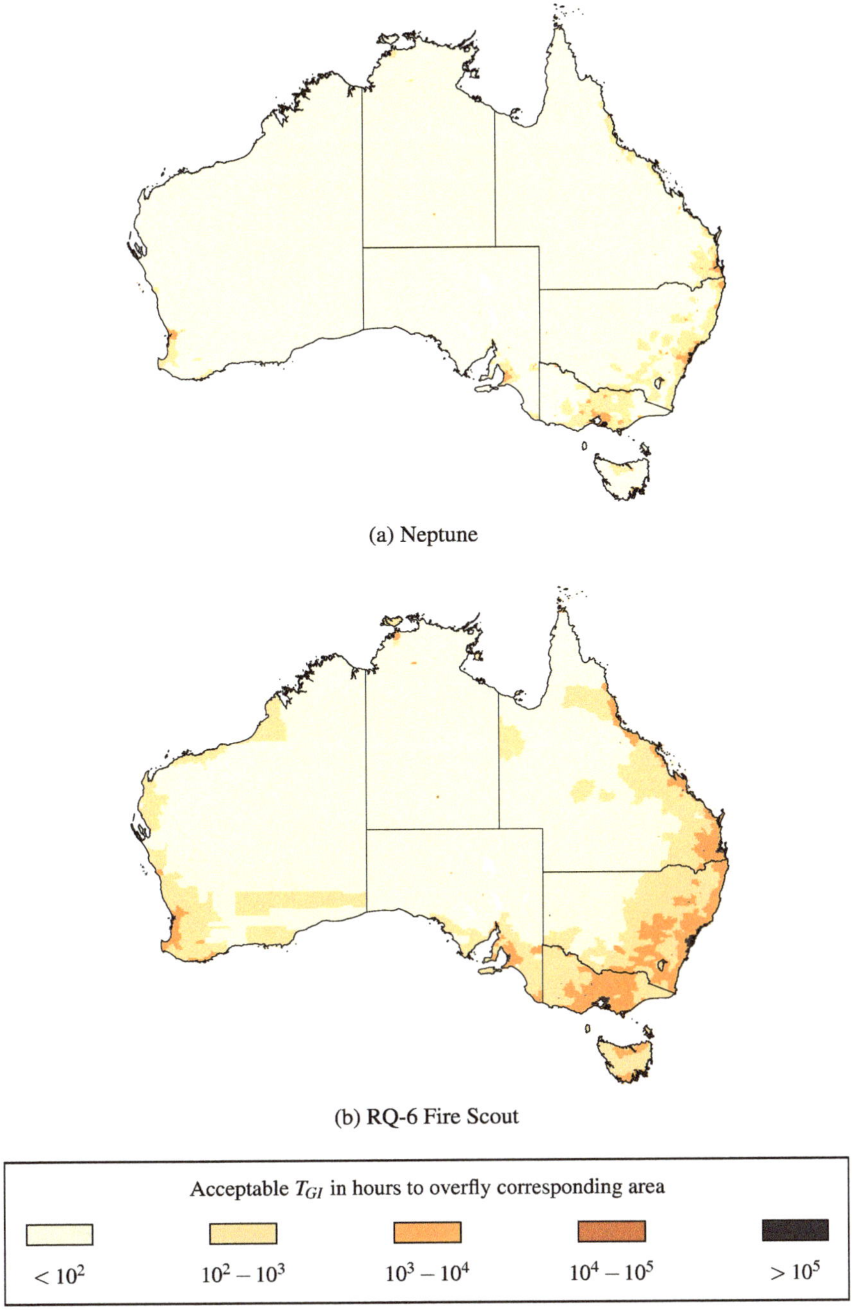

Fig. 6.8 The areas of Australia, the Neptune and the RQ-6 Fire Scout UAS are allowed to loiter over based on their reliability with respect to ground impact occurrence frequency.

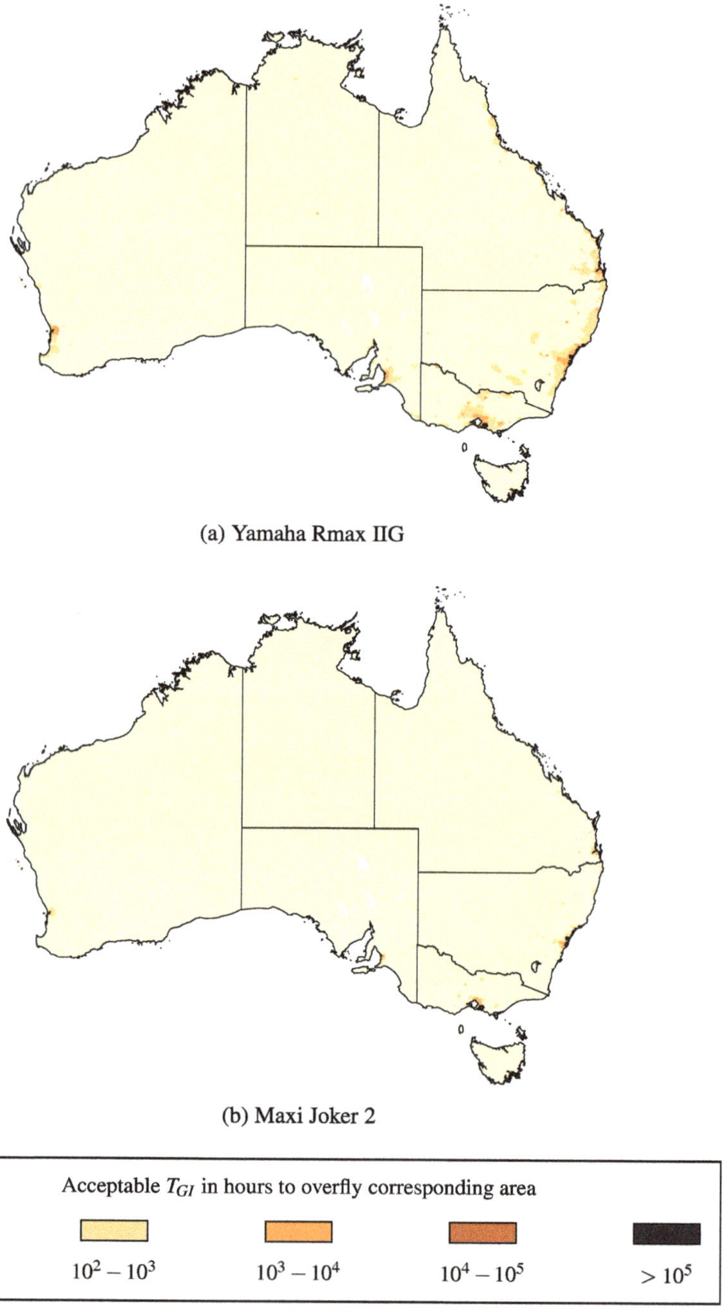

Fig. 6.9 The areas of Australia, the Yamaha Rmax IIG and Maxi Joker 2 helicopters are allowed to loiter over based on their reliability with respect to ground impact occurrence frequency.

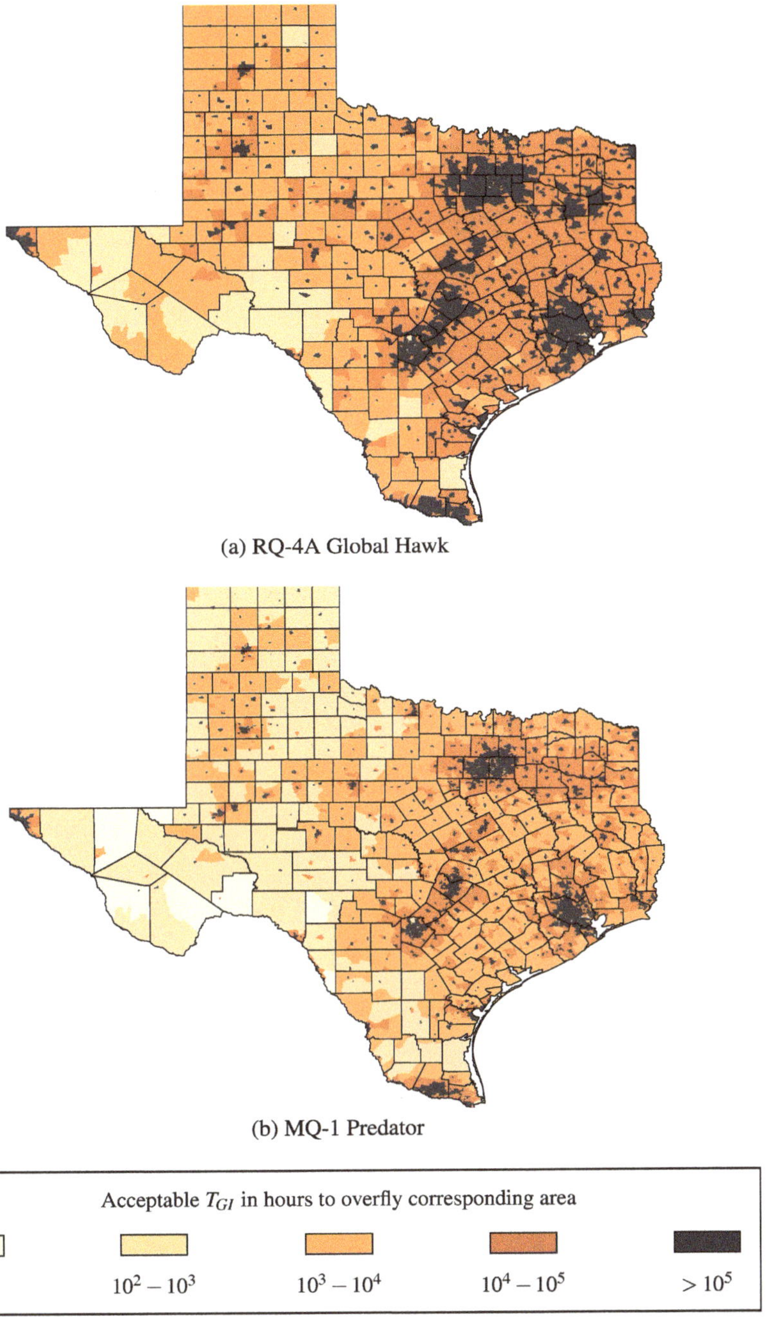

Fig. 6.10 The areas of the State of Texas, the RQ-4A Global Hawk and the MQ-1 Predator UAS are allowed to loiter over based on their reliability with respect to ground impact occurrence frequency.

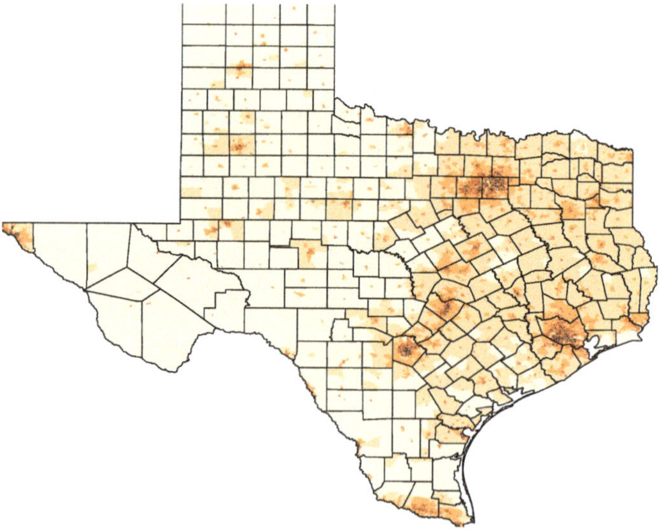

(a) Neptune

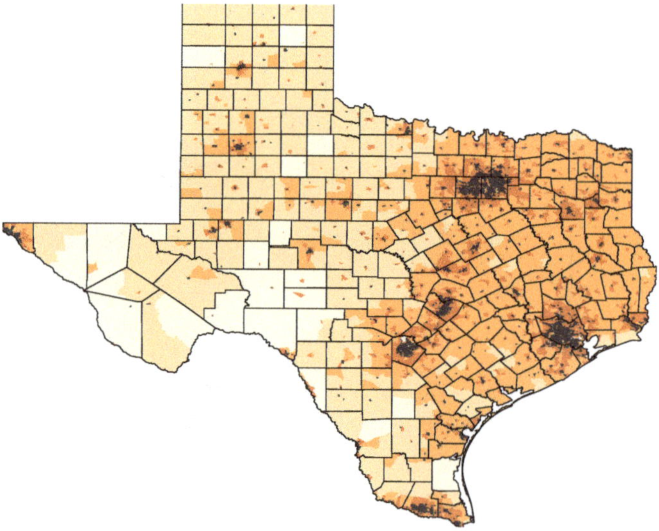

(b) RQ-6 Fire Scout

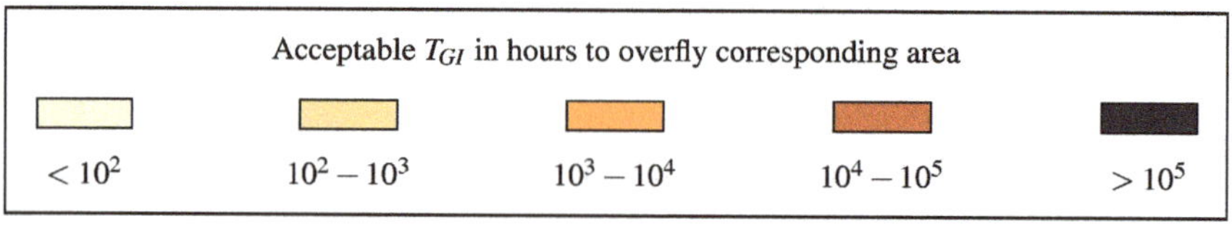

Fig. 6.11 The areas of the State of Texas, the Neptune and the RQ-6 Fire Scout UAS are allowed to loiter over based on their reliability with respect to ground impact occurrence frequency.

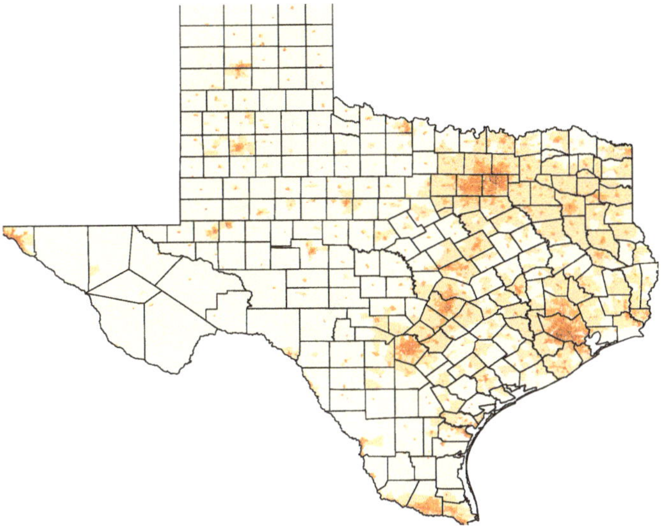

(a) Yamaha Rmax IIG

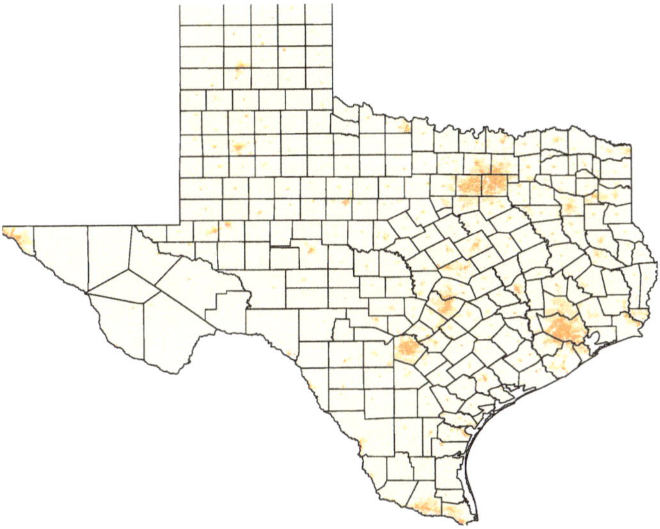

(b) Maxi Joker 2

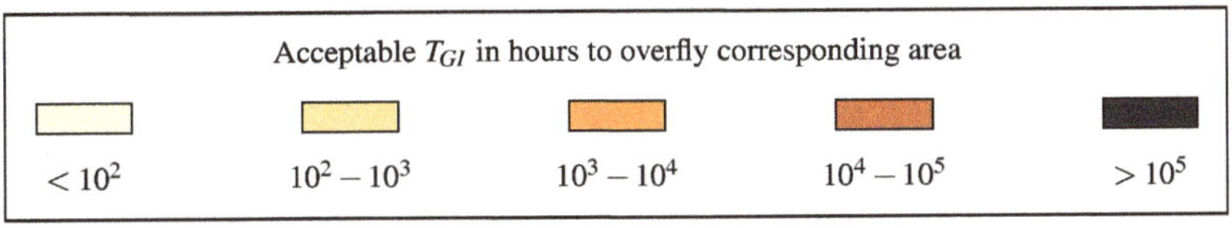

Fig. 6.12 The areas of the State of Texas, the Yamaha Rmax IIG and Maxi Joker 2 helicopters are allowed to loiter over based on their reliability with respect to ground impact occurrence frequency.

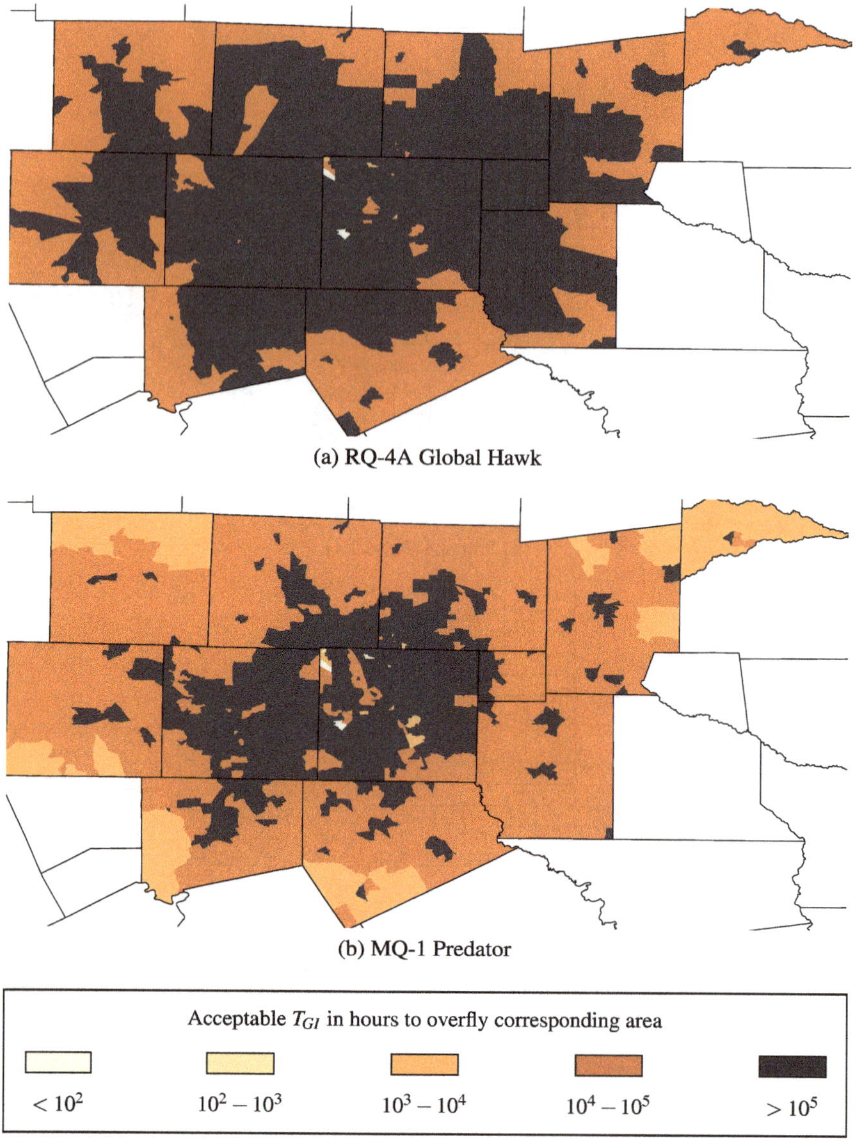

Fig. 6.13 The areas of the Dallas - Fort Worth, TX metropolitan area, the RQ-4A Global Hawk and the MQ-1 Predator UAS are allowed to loiter over based on their reliability with respect to ground impact occurrence frequency.

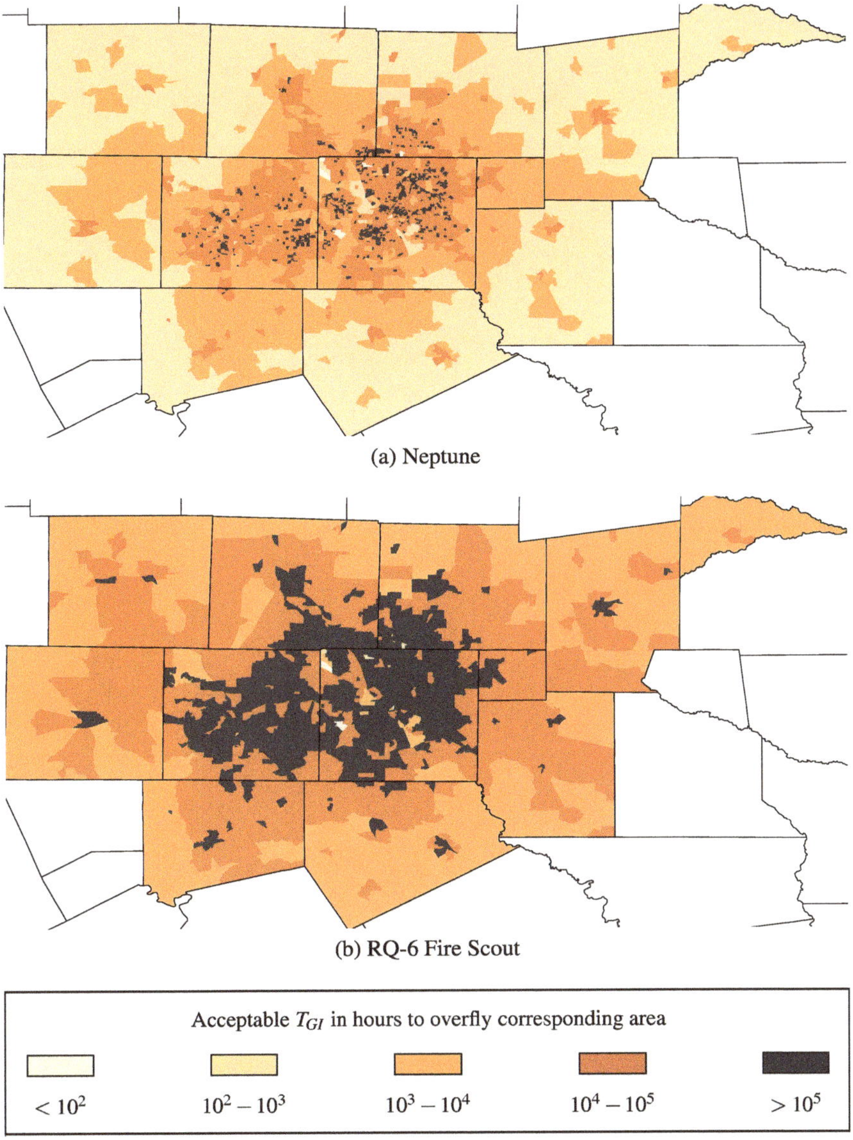

Fig. 6.14 The areas of the Dallas - Fort Worth, TX metropolitan area, the Neptune and the RQ-6 Fire Scout UAS are allowed to loiter over based on their reliability with respect to ground impact occurrence frequency.

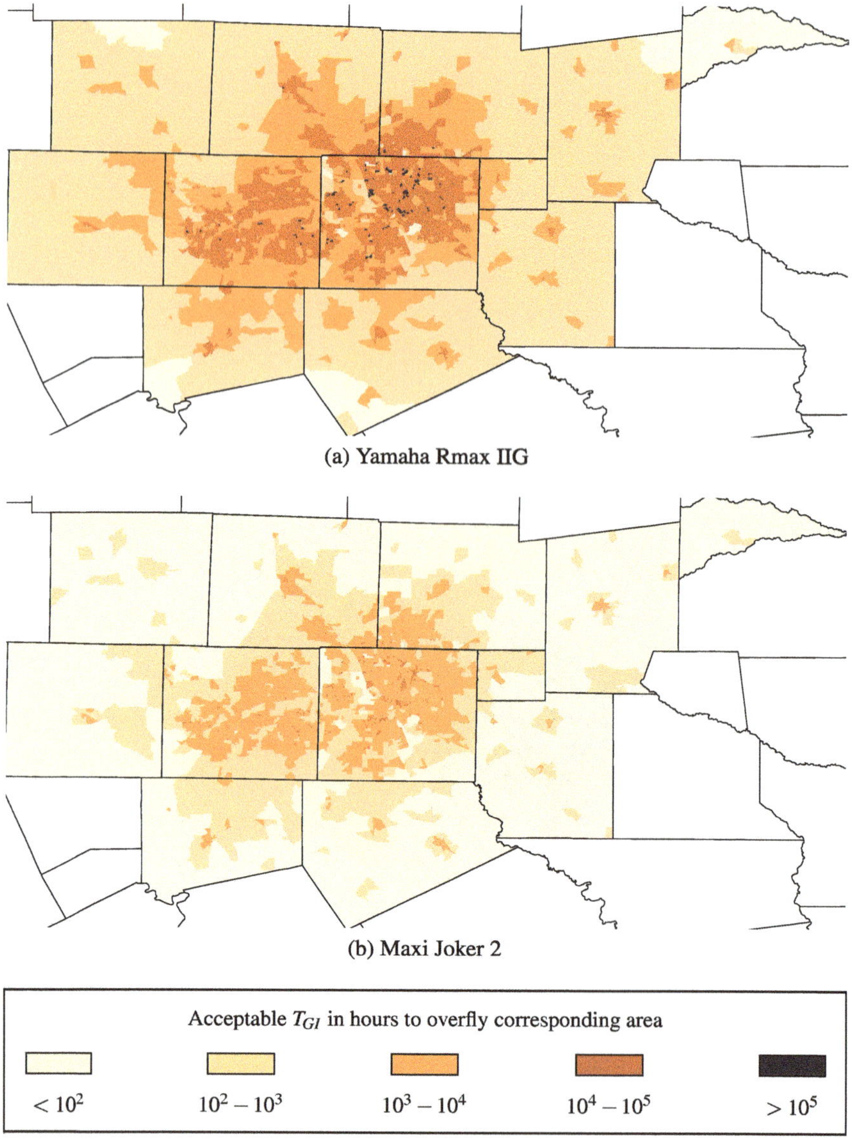

Fig. 6.15 The areas of the Dallas - Fort Worth, TX metropolitan area, the Yamaha Rmax IIG and Maxi Joker 2 helicopters are allowed to loiter over based on their reliability with respect to ground impact occurrence frequency.

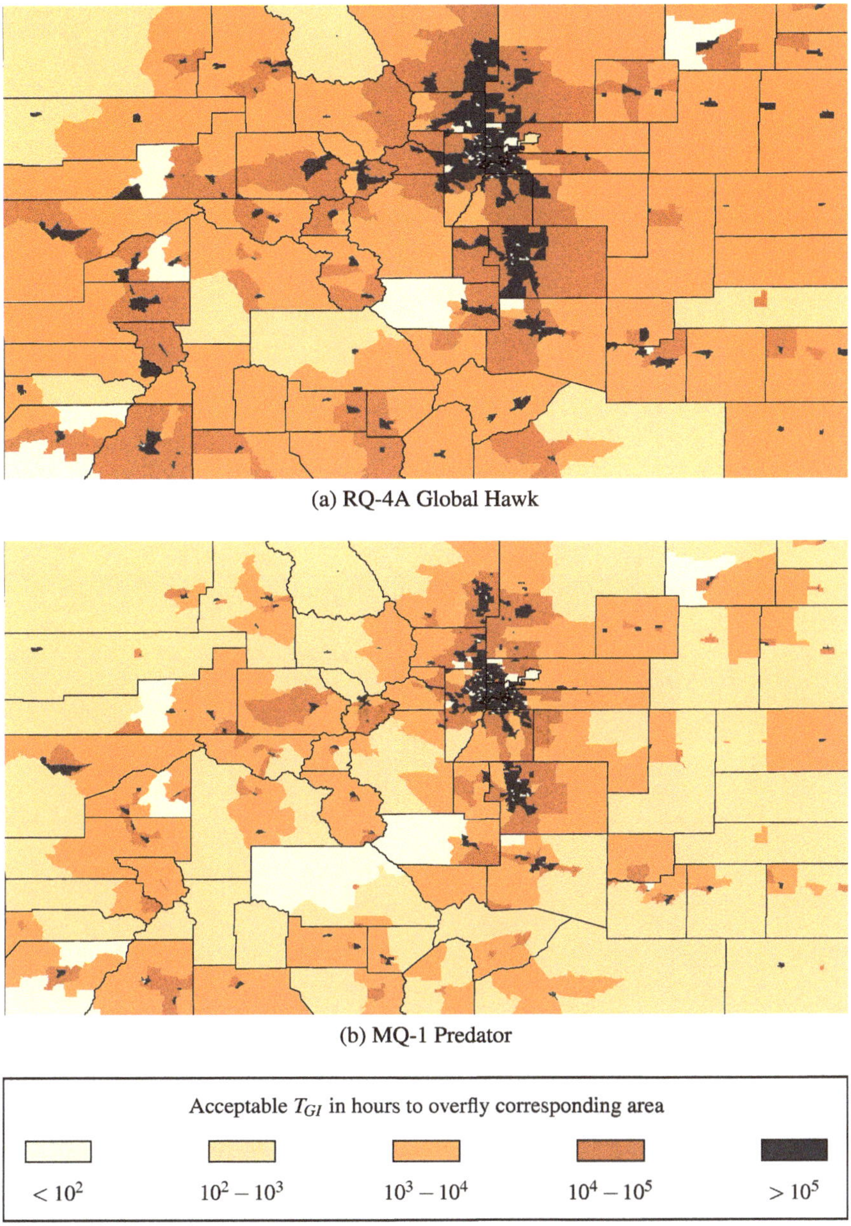

Fig. 6.16 The areas of the State of Colorado, the RQ-4A Global Hawk and the MQ-1 Predator UAS are allowed to loiter over based on their reliability with respect to ground impact occurrence frequency.

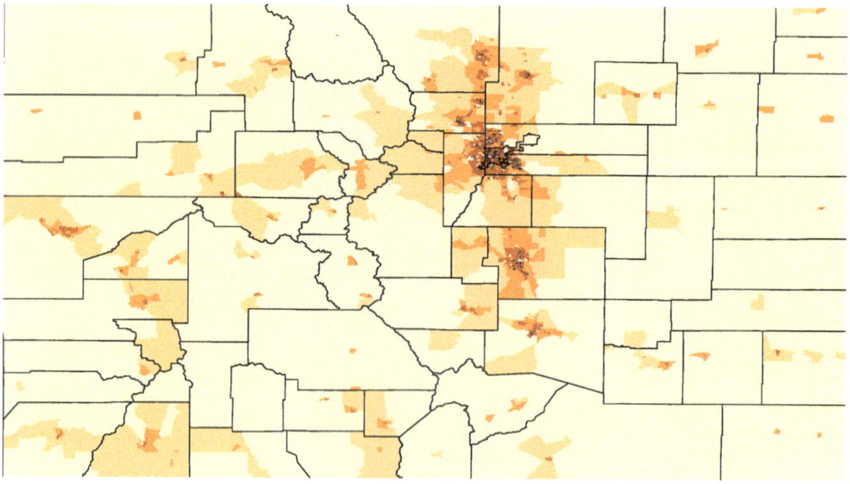

(a) Neptune

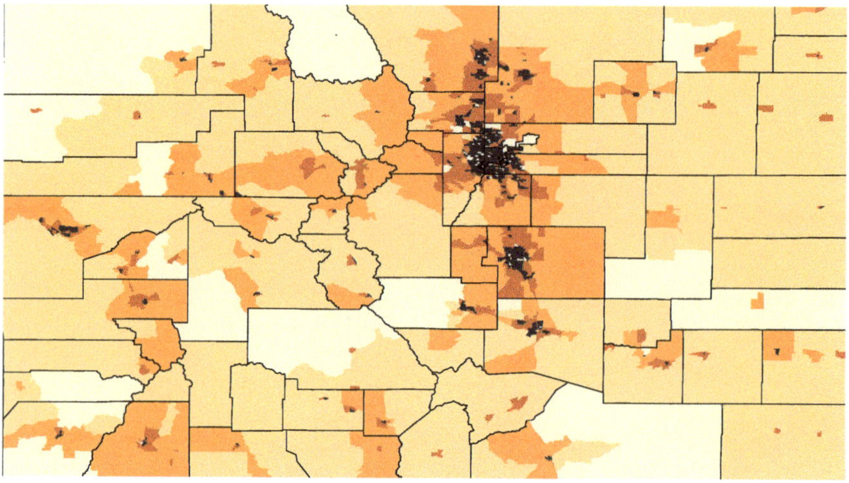

(b) RQ-6 Fire Scout

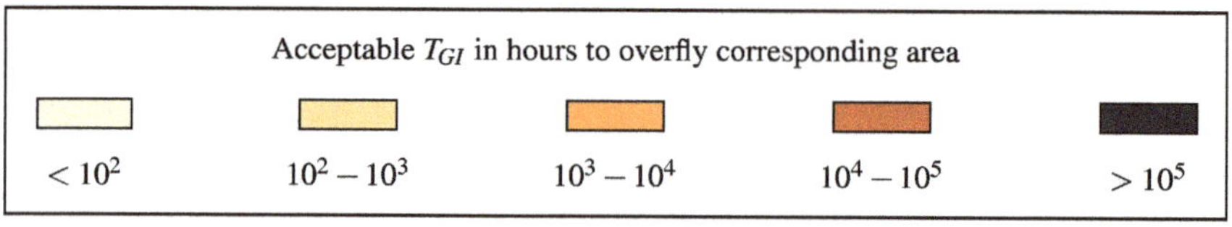

Fig. 6.17 The areas of the State of Colorado, the Neptune and the RQ-6 Fire Scout UAS are allowed to loiter over based on their reliability with respect to ground impact occurrence frequency.

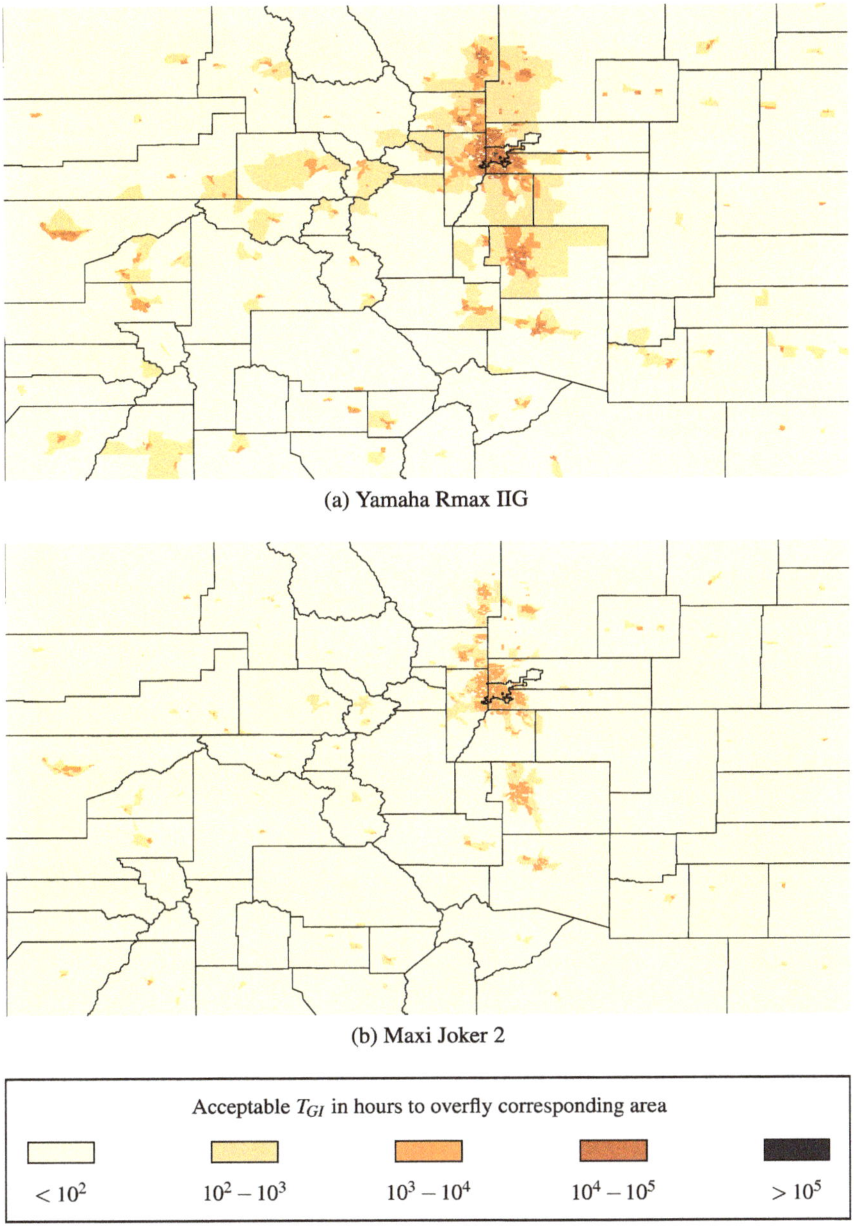

Fig. 6.18 The areas of the State of Colorado, the Yamaha Rmax IIG and Maxi Joker 2 helicopters are allowed to loiter over based on their reliability with respect to ground impact occurrence frequency.

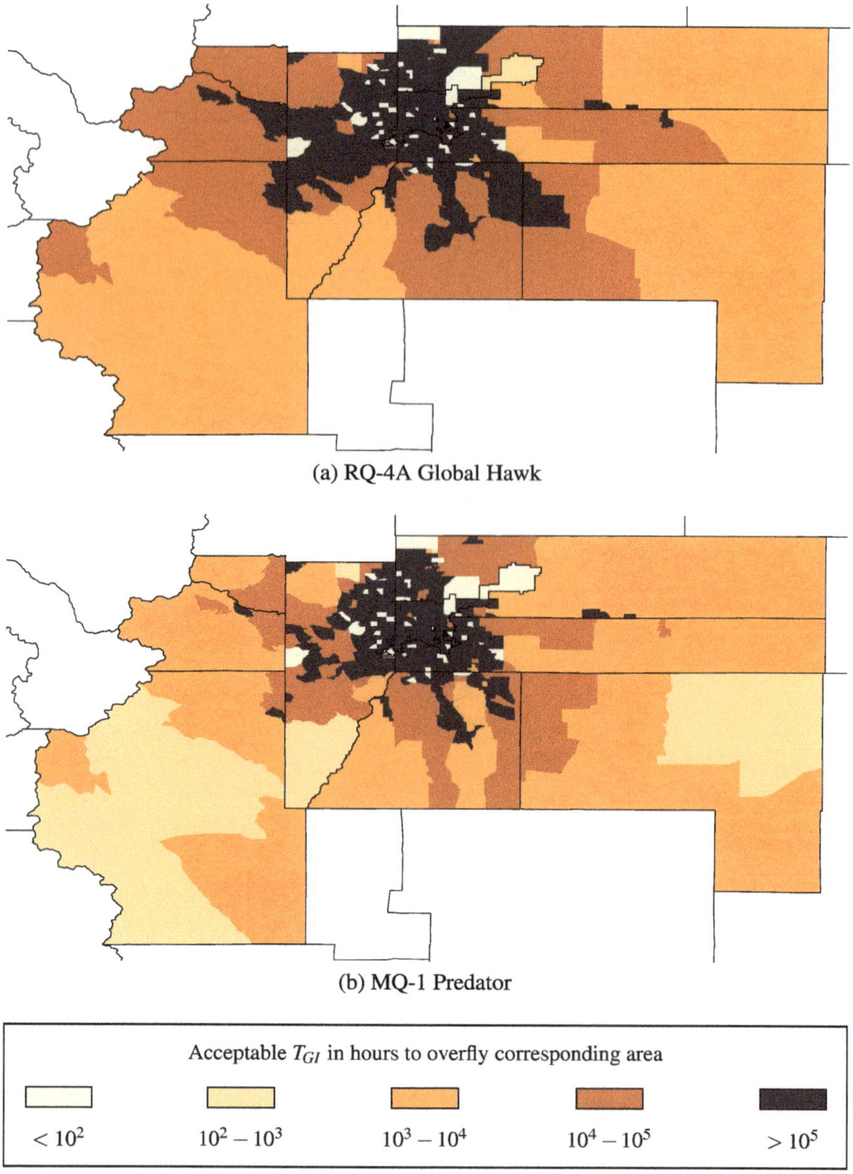

(a) RQ-4A Global Hawk

(b) MQ-1 Predator

Fig. 6.19 The areas of the Denver, CO metropolitan area, the RQ-4A Global Hawk and the MQ-1 Predator UAS are allowed to loiter over based on their reliability with respect to ground impact occurrence frequency.

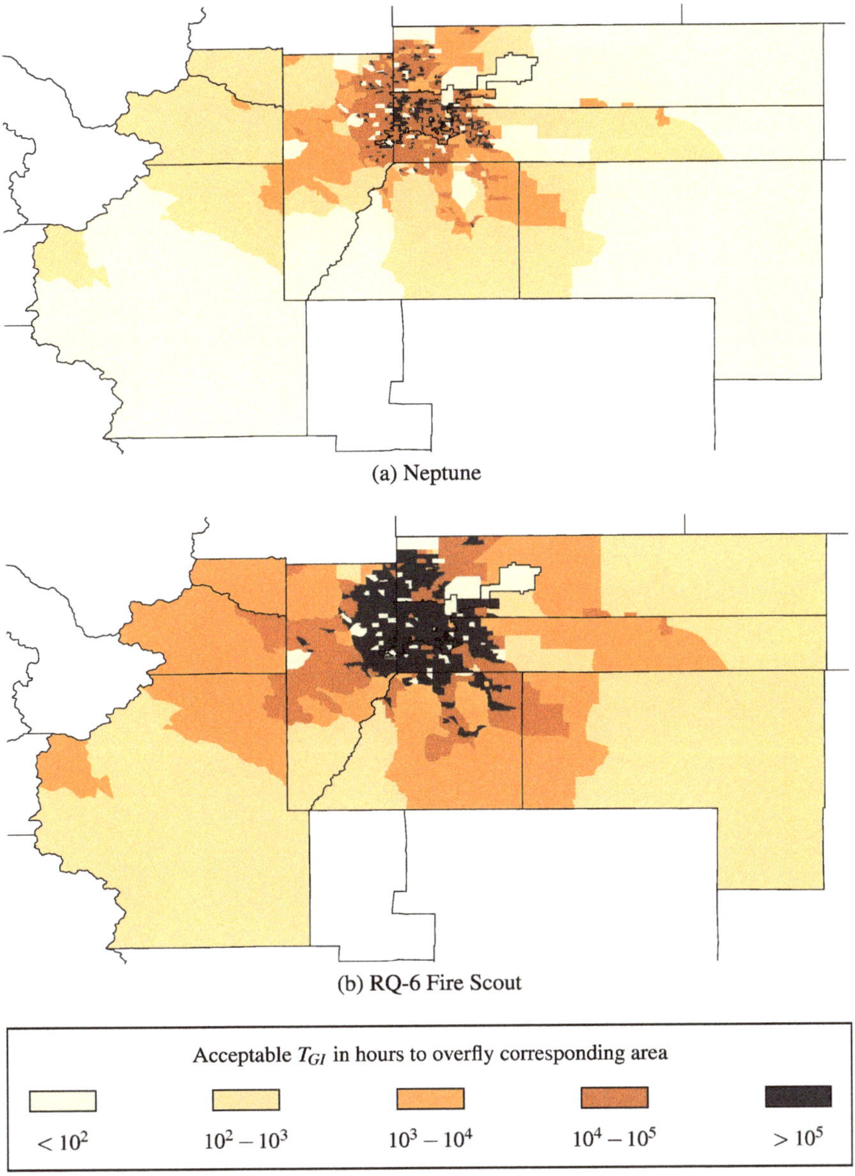

Fig. 6.20 The areas of the Denver, CO metropolitan area, the Neptune and the RQ-6 Fire Scout UAS are allowed to loiter over based on their reliability with respect to ground impact occurrence frequency.

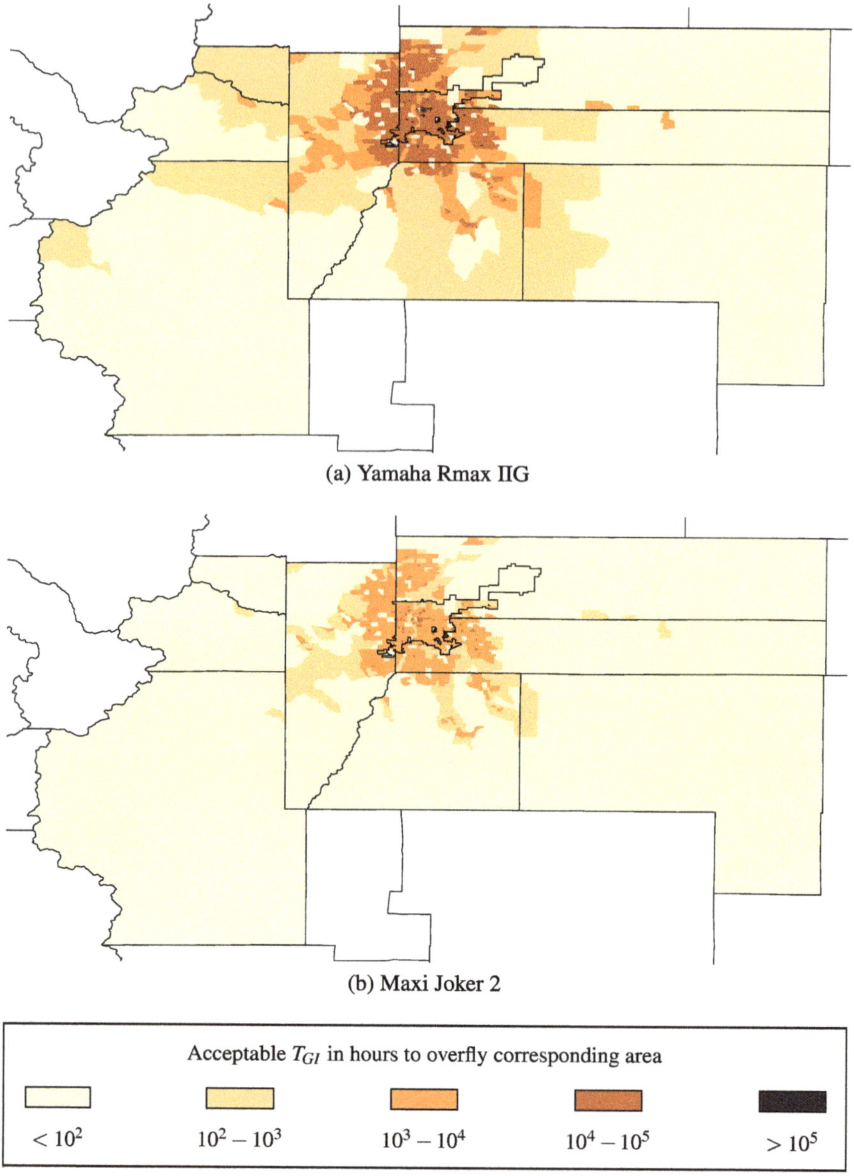

Fig. 6.21 The areas of the Denver, CO metropolitan area, the Yamaha Rmax IIG and Maxi Joker 2 helicopters are allowed to loiter over based on their reliability with respect to ground impact occurrence frequency.

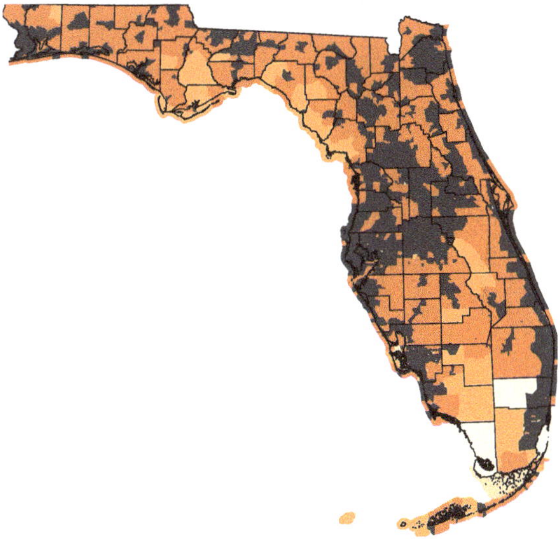

(a) RQ-4A Global Hawk

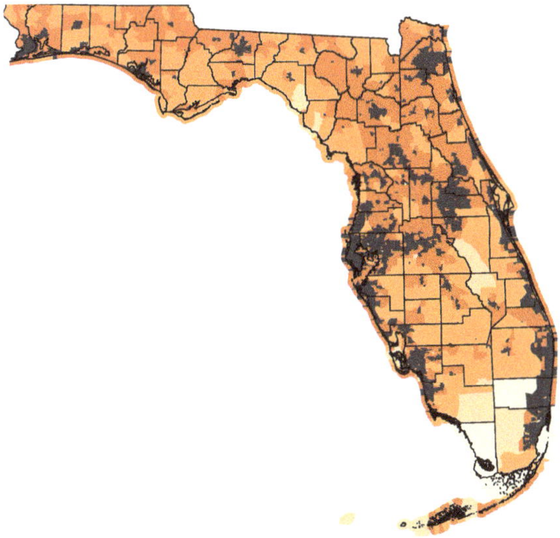

(b) MQ-1 Predator

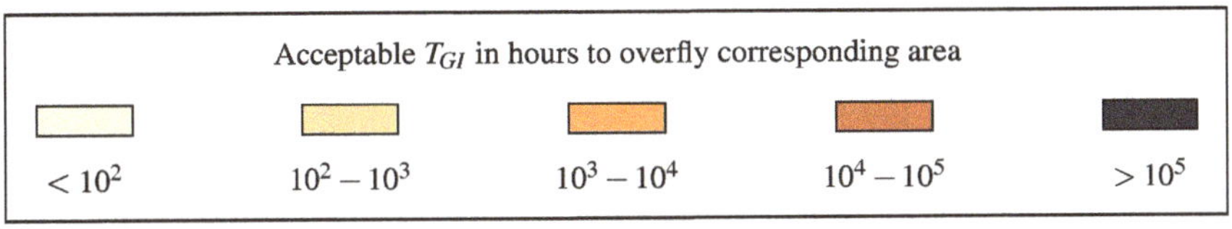

Fig. 6.22 The areas of the State of Florida, the RQ-4A Global Hawk and the MQ-1 Predator UAS are allowed to loiter over based on their reliability with respect to ground impact occurrence frequency.

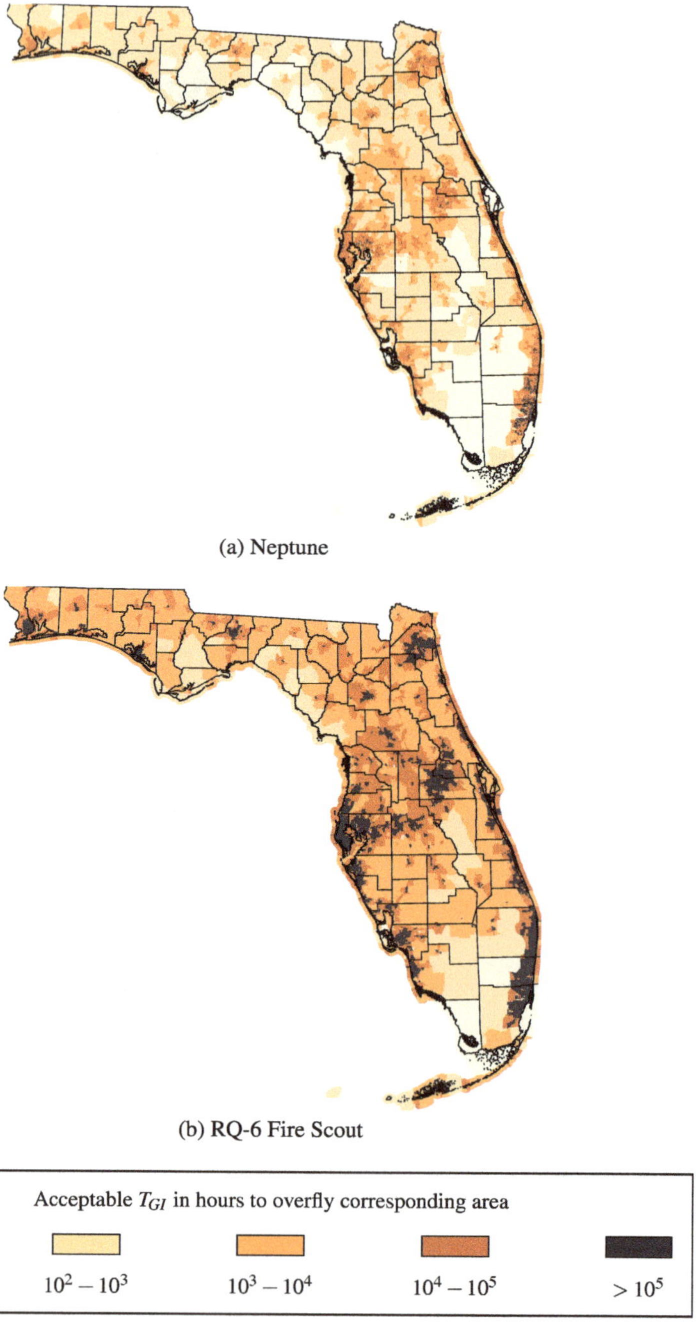

Fig. 6.23 The areas of the State of Florida, the Neptune and the RQ-6 Fire Scout UAS are allowed to loiter over based on their reliability with respect to ground impact occurrence frequency.

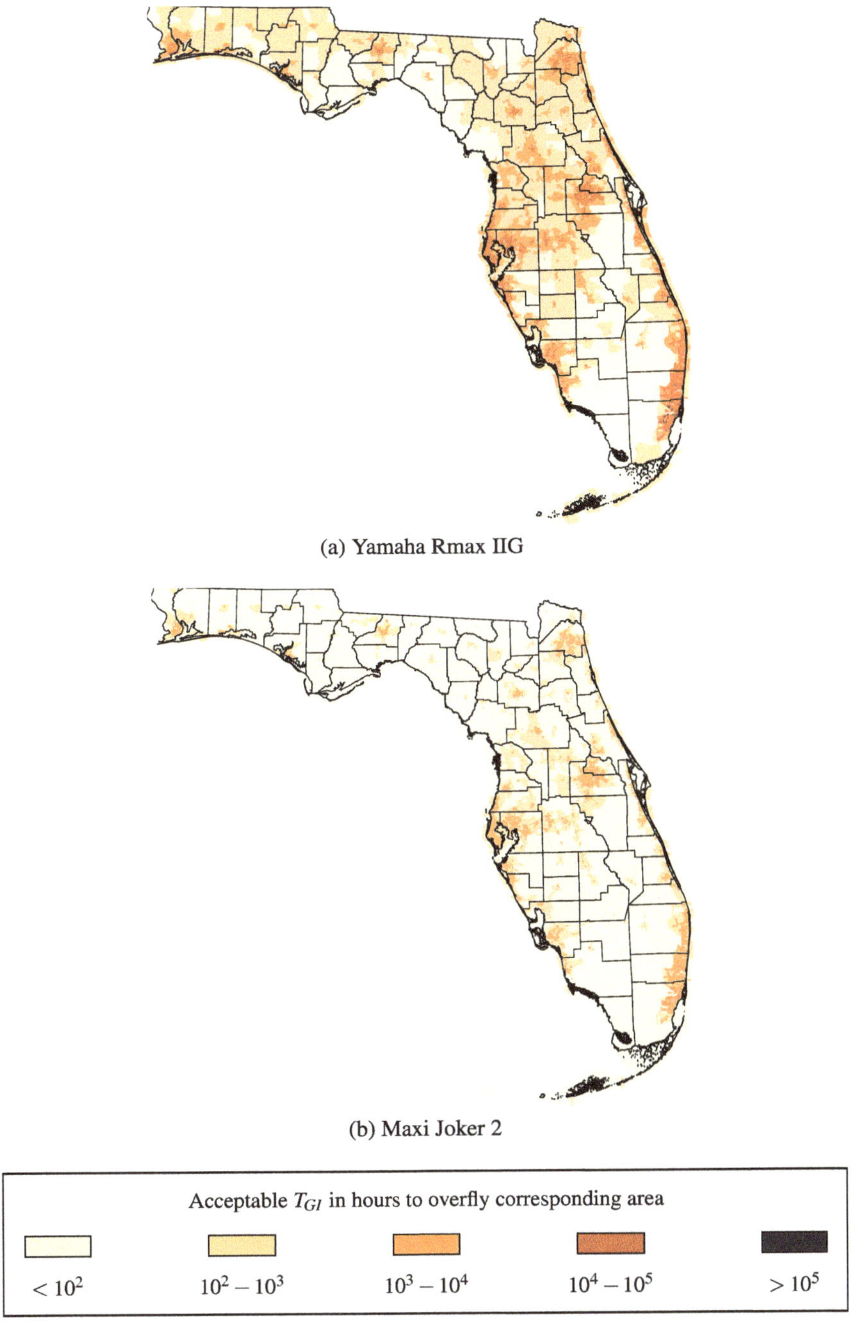

Fig. 6.24 The areas of the State of Florida, the Yamaha Rmax IIG and Maxi Joker 2 helicopters are allowed to loiter over based on their reliability with respect to ground impact occurrence frequency.

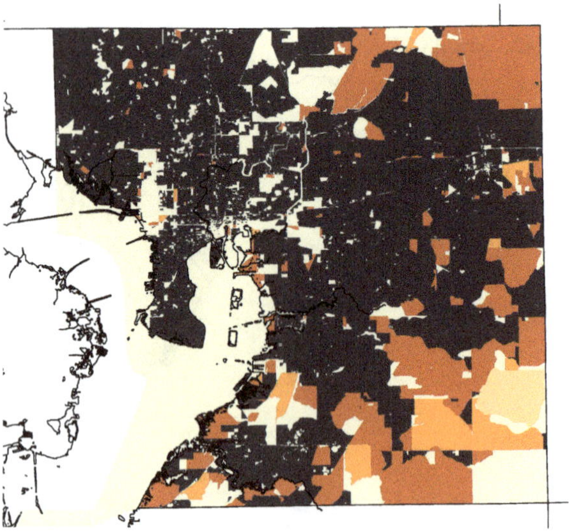

(a) RQ-4A Global Hawk

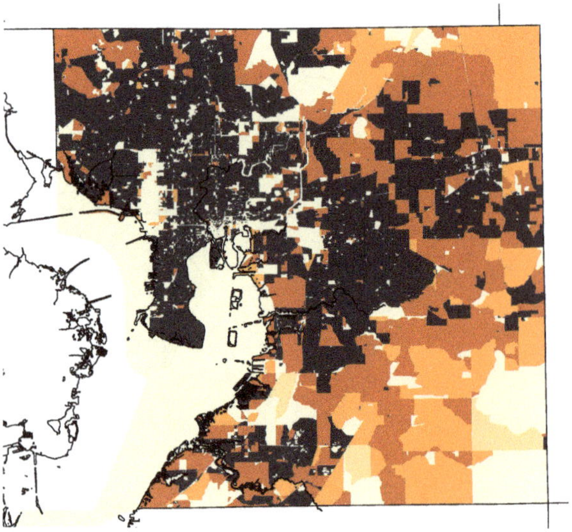

(b) MQ-1 Predator

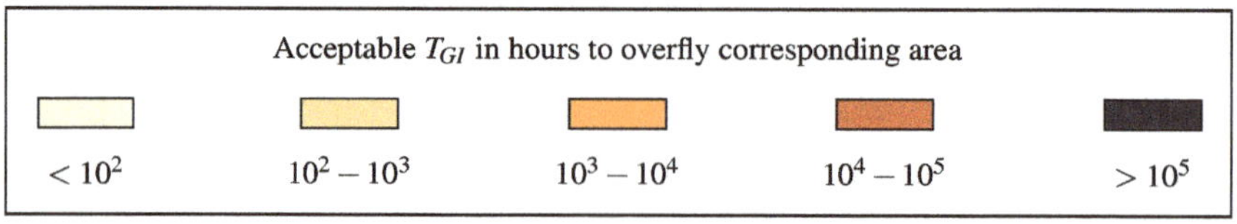

Fig. 6.25 The areas of the Hillsborough County, FL, the RQ-4A Global Hawk and the MQ-1 Predator UAS are allowed to loiter over based on their reliability with respect to ground impact occurrence frequency.

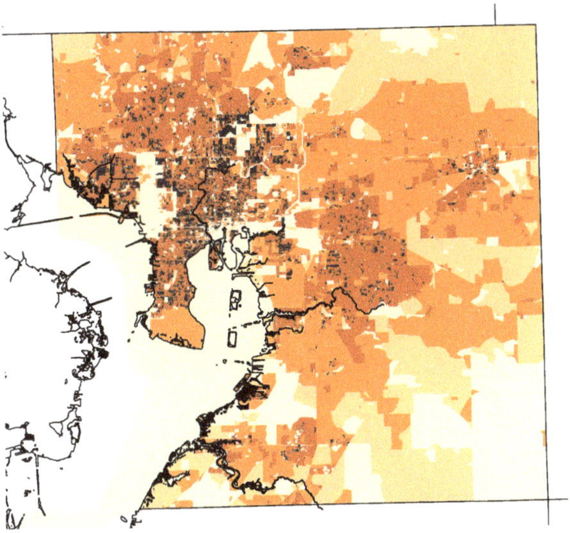

(a) Neptune

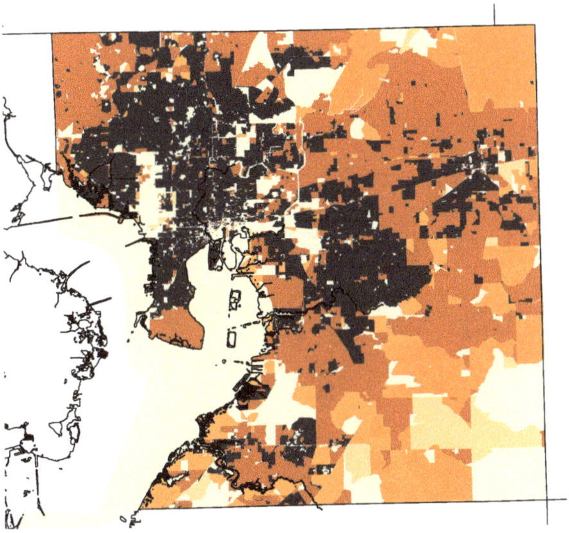

(b) RQ-6 Fire Scout

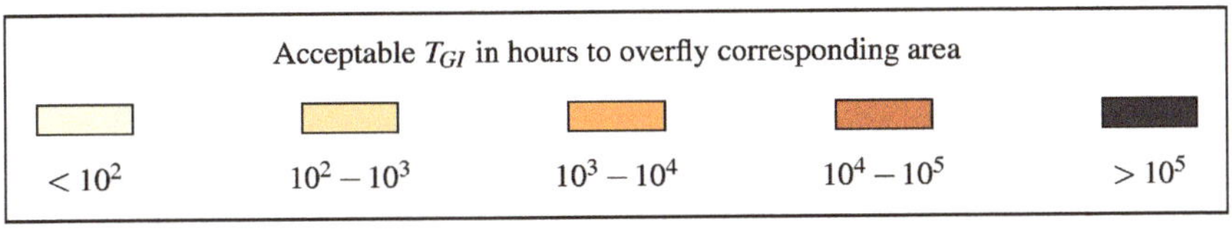

Fig. 6.26 The areas of the Hillsborough County, FL, the Neptune and the RQ-6 Fire Scout UAS are allowed to loiter over based on their reliability with respect to ground impact occurrence frequency.

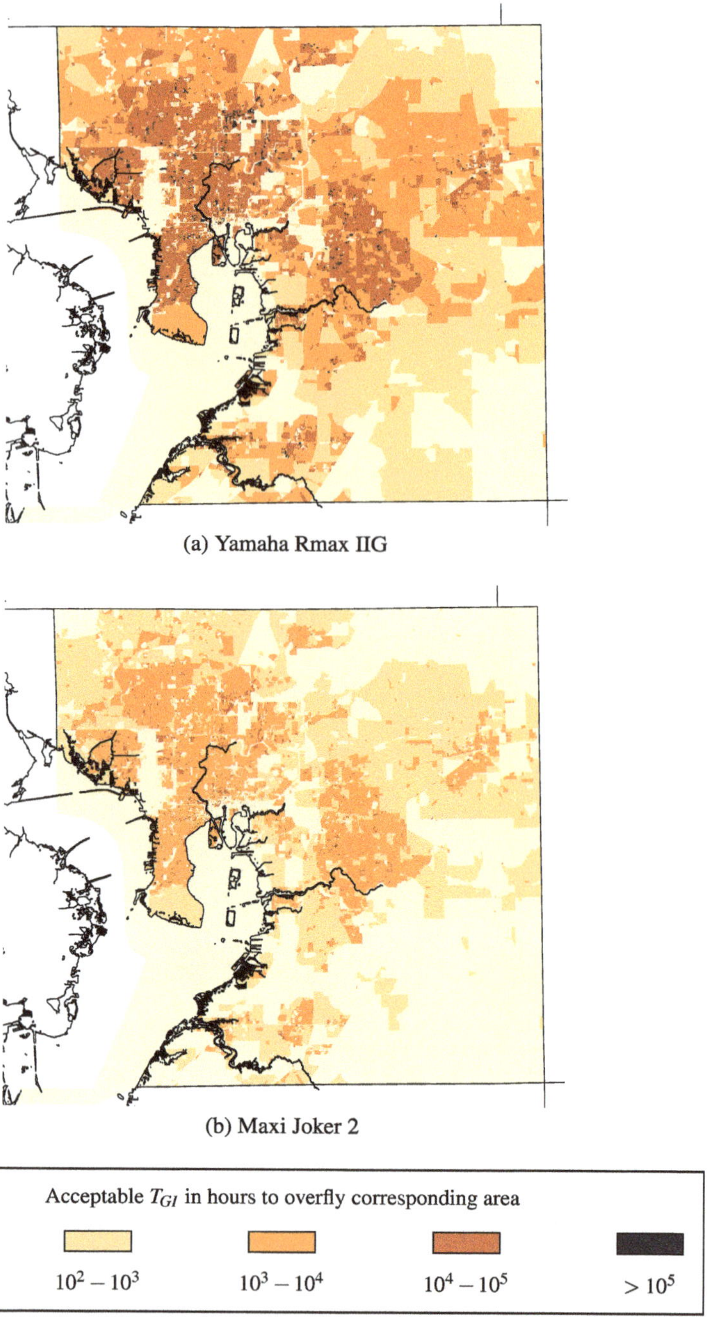

Fig. 6.27 The areas of the Hillsborough County, FL, the Yamaha Rmax IIG and Maxi Joker 2 helicopters are allowed to loiter over based on their reliability with respect to ground impact occurrence frequency.

References

1. Center for International Earth Science Information Network (CIESIN), Columbia University, Centro Internacional de Agricultura Tropical (CIAT) (2005) Gridded Population of the World Version 3 (GPWv3): Population Density Grids. Palisades, NY: Socioeconomic Data and Applications Center (SEDAC), Columbia University, [online] http://sedac.ciesin.columbia.edu/gpw
2. European Aviation Safety Agency (EASA) (2005) A-NPA, No. 16/2005, policy for unmanned aerial vehicle (UAV) certification
3. FSF editorial staff (2005) See what's sharing your airspace. Flight Safety Digest 24(5):1–26
4. US Department of Defense Office of the Secretary of Defense (2005) Unmanned aircraft systems roadmap 2005-2030. Report

Chapter 7
Thoughts and Recommendations on a UAS Integration Roadmap

A law, to be respectable, should match and protect human life, freedom and progress. If this is not the case, then it is simply an order backed by violence.

Paul Rosenberg
(Mindless Slogans, 2007)

This Chapter aims at discussing elements of a viable roadmap leading to integration of UAS into the NAS. This roadmap must address several issues regarding ATM integration and operations. A list of the basic requirements that UAS will eventually need to adhere to, includes [6]:

- Airworthiness
- Operator certification (pilot in command must be licensed)
- Aircraft documentation (manuals, charts)
- 2-way communication with ATC
- Navigation according to airspace
- Surveillance according to airspace (ADS-B, TCAS)
- Sense and avoid collisions with aircraft, obstacles or terrain
- Predictable behavior

Although this list is simplistic, it underscores two key points; UAS must comply with similar requirements with respect to airworthiness, operator training and documentation and they must also behave indistinguishably from manned aircraft with respect to ATM.

In this Chapter, first the key differences between manned and unmanned aviation are presented, since they can be expected to drive the regulation development. Following that, information and recommendations are provided on key issues like the development of a risk reference system for UAS, classification for regulatory purposes and certification of systems and operators. The Chapter concludes with a discussion of technology-related issues that will need to be adequately resolved before UAS can enjoy unrestricted access to the NAS. When applicable, existing regulations for manned aviation are adapted; however, the need for new rules, procedures and regulations is also essential.

The account given is by no means an exhaustive or authoritative answer to the issue of UAS integration in the NAS; its purpose is to be as informative as possible and, hopefully, lay the foundation for the development of a detailed roadmap. Regardless, the fact is that sooner or later UAS will be allowed to fly in parallel to manned aviation; therefore a well thought roadmap is imperative.

K. Dalamagkidis et al., *On Integrating Unmanned Aircraft Systems into the National Airspace System*, Intelligent Systems, Control and Automation: Science and Engineering 54, DOI 10.1007/978-94-007-2479-2_7,

7.1 Regulation Development

The JAA UAV Task Force in its document presented four basic principles to be employed in the development of UAS regulations; fairness, equivalence, responsibility and transparency [31]. To ensure fairness, applicable manned aviation regulations should apply as they are instead of developing new, tailor-made regulations for UAS. The principle of equivalence was discussed in Sect. 5.1 and besides equivalence in risk levels also mandates equivalence in operation requirements. Responsibility refers to maintaining a chain of accountability regardless of how or where the UAS is operated. Finally transparency is achieved by making the UAS transparent to other airspace users and the ATC.

Recent results from the development of regulation for the LSA category aircraft were very encouraging. Within a few years and with substantially reduced cost, the appropriate regulation was developed and practice showed no reduction in safety. Nevertheless, UAS technology is considered far from mature and the FAA, as well as other national aviation agencies, is taking a careful, conservative approach.

A major question in the development of UAS regulation is whether it can be based on the current FAR or whether UAS have substantially different characteristics that warrant new regulation. Although the FAR does not specifically mention UAS, regulations may be assumed to apply equally to all aircraft categories [1]. However, several paragraphs pertaining to passenger comfort and safety will need to be removed and others that assume a pilot onboard will need to be adapted.

The FAA CoE for General Aviation Research (CGAR) has conducted related research into UAS regulation to determine how applicable is current regulation to UAS. In 2007, a regulatory review was presented at the CGAR Annual Meeting where it was found that only 30% of current manned aviation regulation applies as is to UAS; 54% may apply or may require revisions and 16% does not apply. The study concluded that there are significant gaps that remain to be addressed.

Adaptation of current regulation has the advantage of utilizing the experience and expert knowledge gained from years of manned aviation operations. This is reflected also in the literature where there seems to be a consensus on basing UAS regulation on that of manned aircraft of the same category, the latter defined primarily by MTOW [11, 16, 20, 31]. This is achieved by removing the non-applicable paragraphs and adding any additional requirements where needed, just like other special aircraft categories. Of course airworthiness regulations are typically general and refer to existing standards rather than present technical specifications themselves [29]. This allows easier revising by updating the standard rather than the long legal path of updating the regulations [29]. As a result, the current lack of technical standards for areas that are specific to UAS is a significant obstacle in regulation development.

Regardless, applying current aviation regulation to UAS is not without problems. This is due to a number of differences, some of which may require a different approach towards aviation safety and a new way of thinking. The following sections summarize some of the most important differences.

7.1.1 Applications

Traditionally, safety levels have been considered under the assumption that the vast majority of manned aircraft fly in point-to-point operations, transporting people or goods. This implies that a significant portion of their flight time is spent over less densely populated areas an assumption that has been taken into account in the drafting of aviation safety regulations [16].

This same assumption does not hold for UAS. This is especially true for surveillance/patrolling applications, where UAS are required to loiter over specific areas. It is obvious that if such areas under consideration have very low population (borders, forests, etc.), then, the safety level requirement obtained for manned aviation would be over conservative; on the contrary, if the UAS is required to loiter over a metropolitan area, this safety level would be inadequate.

7.1.2 Flight Characteristics

Although UAS come in a wide range of designs and types, many systems are typically lighter and less maneuverable than the average manned aircraft [46]. This difference introduces certain issues from the ATM standpoint. The reduced maneuverability may result in UAS that are not capable of following the climb/descend and turning profiles required for well established departure and approach procedures [14]. This in turn means that the transparency principle will need to be violated, since ATC will need to take into consideration the reduced performance and/or new procedures will need to be developed. This may also increase the overall workload of ATC controllers [46]. In addition to that, their lower weight make them particularly susceptible to the wake turbulence of larger aircraft, thus requiring larger separation [14].

7.1.3 Sacrificability

Early, abnormal flight termination, regardless of whether it is controlled or not, is a major concern for manned aviation due to the high probability of fatalities associated with it. Any failure condition that can lead to such an accident is typically considered catastrophic and the lowest probability of occurrence needs to be demonstrated. This also entails imposing the strictest reliability requirements on related flight safety equipment.

In the case of UAS, although an uncontrolled ground impact may still be considered a catastrophic accident, it is acceptable —and possibly desirable— to allow a UAS to crash in a controlled manner, in order to minimize the risk to people and property. In fact, in its recommendations, the sUAS ARC proposed a general rule that "*during an emergency, the safety of people on the surface must be given priority over the sUAS*."

7.1.4 Pilot Physically Removed from Cockpit

An aircraft pilot is intimately aware of the surroundings, as well as the performance of the aircraft. Vibrations, smells, noise, controller feel and other indicators of aircraft performance and possible failures are available. The pilot in command is also the ultimate authority of the aircraft, being able to assume full control of every aspect of its operation.

On the other hand a UAS operator, being physically removed from the cockpit, has limited perception of the aircraft state. This is because one relies only on data sent back from the UAS, which may not provide important or needed information and lack the aforementioned sensory indicators [1, 21]. For remotely operated UAS, this separation has the added side-effect that there may be a communications lag between the UA and its pilot. It is noteworthy, that in the case of satellite communications the latency for the signal to travel between satellite and UA pilot can be 250 ms [38]. As a result, half a second is required between sensing something and executing a correction, even if instantaneous pilot reaction is assumed. Although during cruising this may not be a problem, in critical phases of the flight like landing it may have catastrophic consequences. As a result, it is currently standard practice for UAS operating in Iraq to fly based on a satellite link and switching to a local channel for take-off and landing [34].

Furthermore, the authority of controlling the UA may reside fully or partially with the onboard control system, while the operator is limited to observing and providing only high-level commands. Specifically, depending on the level of autonomy, the operator may need little to no training to operate the UA, or may in fact operate more than one at the same time [21]. As a result, it is possible that full control of the aircraft may not be available.

In addition to that, it has been suggested that since the pilot is in a safe environment and life-threatening consequences from mistakes are not expected, pilot errors may be more frequent [24]. Similarly, maintenance personnel may become complacent and negligent in their duties [24].

7.1.5 UAS as Systems

Due to the nature of a UAS, flight critical systems are now separated from the aircraft itself. This subsystems include the ground control station, the communication link and potentially other facilities provided from the ground, such as see and avoid. As a result, the following requirement taken from FAR Part 25 airworthiness standards, would apply to those subsystems, as well:

> Failure of any single element, component or connection should not prevent *continued safe flight and landing* or significantly reduce the capability of the airplane to cope with failure conditions.

Based on the above, the safety of the ground control station is of paramount importance, since it is a vital part of a UAS, replacing the role of the cockpit in manned aircraft. Natural disasters affecting installations housing the ground control station may result in catastrophic accidents, by either causing damage to vital equipment or disabling the UA operator. Interference from other activities may also degrade the quality of the communication link, possibly leading to complete loss of communication with the aircraft. The issue of security against malicious intent of the control station as well as the communications link between it and the UA are also of particular importance since it may be possible for an external entity to disrupt or even assume effective control of the UA.

7.1.6 Take-off Weight

Manned aircraft have a wide range of take-off weights, starting at about 100 kg for ultralight, unpowered vehicles and about twice that for powered. General aviation aircraft are typically several times heavier and the largest transport aircraft have reached 600,000 kg (Airbus A380).

On the other hand, UAS span the entire spectrum from a few grams and up to —currently— 12,000 kg. This means that UAS cover a number of current manned aviation aircraft classes and several FAR parts. In addition to that, there is a class of vehicles, lighter than 100 kg, for which there is no equivalent manned aviation regulation. It is reiterated here that AC 21-97 is applicable to recreational R/C modeling only and not to UAS, according to the FAA, and similar regulations are in force in many other countries.

This gap is of particular importance when considering the fact that there is already a significant interest for small UAS. Such systems are considered the entry point for civil commercial applications in the future because of their low cost, portability and smaller associated risks [51]. They are also considered by the military as a key instrument in support of urban warfare operations. These systems also have the added advantage of not requiring extensive safety modeling and testing, due to their low range and low kinetic energy [37].

7.1.7 Passengers and Cargo

Although currently UAS are designed to fly without people or cargo on-board, it is conceivable in the future that passengers and/or cargo will be transported aboard unmanned aircraft. Regardless, the procedures relating to the crash of an aircraft and especially those concerning Search and Rescue operations will need to be adapted to accommodate UAS [32]. This entails measures and procedures so that at the time of a crash it is known whether a specific UAS carried passengers and what wass their number or whether it was transporting dangerous cargo and what was the nature of that cargo [32].

7.1.8 Payload

Instead of cargo, several UAS applications like weather monitoring, communications relaying and law enforcement will require the use of sophisticated sensors, communication devices or other equipment. This payload may be intricately connected to the flight control system and capable of changing high level mission commands or generating new waypoints.

As a result, new hazards emerge because of the possible control failure induced by the payload [21]. On the other hand the payload may also be used to provide redundancy for main aircraft sensors thus mitigating other hazards [21].

7.1.9 Operational Security

Several UAS designs are light, cheap, can be made with off-the-shelf components and don't require any infrastructure to be present in order to operate. As a result, it is possible for people to operate such systems without any certification or training, virtually undetected [14]. This raises the issue of not only developing the regulations, but consistently enforcing them, which may become an issue, especially for the lighter UAS classes.

7.2 Operational Risk Reference System

The risk reference system used in manned aviation has defined severity categories to classify failure conditions based on the expected outcome. Each of these categories caries a different TLS, thus, ensuring that requirements are proportional to the perceived risks.

However, the aforementioned categories —as currently defined— do not apply to UAS. This is because they are concerned with the health and comfort of the passengers and crew onboard the aircraft. New severity descriptions have been proposed in [30] and [22] that aim to address the differences between UAS and manned aviation. A comparison of the severity definitions from [19, 22, 30] is provided in Tables 7.1 and 7.2.

Table 7.1 A comparison of the definitions of failure condition severity categories between those currently used for manned aviation and those proposed for unmanned systems. Where appropriate differences have been italicized.

Severity	AC 23.1309-1C [19]	Hayhurst et al. [22]	STANAG 4671 [30]
No Safety Effect	Failure Conditions that would have no affect on safety (that is, Failure Conditions that would not affect the operational capability of the airplane or increase crew workload).	Failure Conditions that would have no affect on safety (that is, Failure Conditions that would not affect the operational capability of the airplane or increase *flight* crew workload).	Failure Conditions that have no affect on safety.
Minor	Failure Conditions that would not significantly reduce airplane safety and involve crew actions that are well within their capabilities. Minor Failure Conditions may include a slight reduction in safety margins or functional capabilities, a slight increase in crew workload (such as routine flight plan changes), or some physical discomfort to passengers or cabin crew.	Failure Conditions that would not significantly reduce *UAS* safety and involve *flight* crew actions that are well within their capabilities. Minor Failure Conditions may include a slight reduction in safety margins or functional capabilities, or a slight increase in *flight* crew workload (such as routine flight plan changes).	Failure Conditions that do not significantly reduce *UAV System* safety and involve *UAV* crew actions that are well within their capabilities. These conditions may include a slight reduction in safety margins or functional capabilities, and a slight increase in *UAV* crew workload.
Major	Failure Conditions that would reduce the capability of the airplane or the ability of the crew to cope with adverse operating conditions to the extent that there would be a significant reduction in safety margins or functional capabilities; a significant increase in crew workload or in conditions impairing crew efficiency; or a discomfort to the flight crew or physical distress to passengers or cabin crew, possibly including injuries.	Failure Conditions that would reduce the capability of the *UAS* or the ability of the *flight* crew to cope with adverse operating conditions to the extent that there would be a significant reduction in safety margins or functional capabilities; a significant increase in *flight* crew workload or in conditions impairing *flight* crew efficiency; or a discomfort to the flight crew, possibly including injuries; *or a potential for physical discomfort to persons.*	Failure Conditions that either by themselves or in conjunction with increased crew workload, result in a worst credible outcome of an emergency landing of the UAV on a predefined site where it can be reasonably expected that a serious injury will not occur; or failure conditions which could potentially result in injury to UAV crew or ground staff.

Table 7.2 (cont'd from Table 7.1). A comparison of the definitions of failure condition severity categories between those currently used for manned aviation and those proposed for unmanned systems. Where appropriate differences have been italicized.

Severity	AC 23.1309-1C [19]	Hayhurst et al. [22]	STANAG 4671 [30]
Hazardous	Failure Conditions that would reduce the capability of the airplane or the ability of the crew to cope with adverse operating conditions to the extent that there would be the following: 1. A large reduction in safety margins or functional capabilities; 2. Physical distress or higher workload such that the flight crew cannot be relied upon to perform their tasks accurately or completely; or 3. Serious or fatal injury to an occupant other than the flight crew.	Failure Conditions that would reduce the capability of the *UAS* or the ability of the *flight* crew to cope with adverse operating conditions to the extent that there would be the following: 1. A large reduction in safety margins or functional capabilities; 2. Physical distress or higher workload such that the *UAS* flight crew cannot be relied upon to perform their tasks accurately or completely; or 3. *Physical distress to persons, possibly including injuries.*	Failure conditions that either by themselves or in conjunction with increased crew workload, result in a worst credible outcome of a controlled-trajectory termination or forced landing potentially leading to the loss of the UAV where it can be reasonably expected that a fatality will not occur; or failure conditions which could potentially result in serious injury to UAV crew or ground staff
Catastrophic	Failure Conditions that are expected to result in multiple fatalities of the occupants, or incapacitation or fatal injury to a flight crewmember normally with the loss of the airplane.[1,2]	Failure Conditions that are expected to result in *one or more fatalities or serious injury to persons, or the persistent loss of the ability to control the flight path of the aircraft normally with the loss of the aircraft.*[1]	Failure Conditions that result in a worst credible outcome of at least uncontrolled flight (including flight outside of pre-planned or contingency flight profiles/areas) and/or uncontrolled crash, which can potentially result in a fatality; or failure conditions which could potentially result in a fatality to UAV crew or ground staff.

[1] The phrase "are expected to result" is not intended to require 100% certainty that the effects will always be catastrophic. Conversely, just because the effects of a given failure, or combination of failures, could conceivably be catastrophic in extreme circumstances, it is not intended to imply that the failure condition will necessarily be considered catastrophic.

[2] The term "Catastrophic" was defined in previous versions of the rule and the advisory material as a Failure Condition that would prevent continued safe flight and landing.

Hayhurst et al. [22], tried to adopt the severity categories of AC 23.1309 with a minimal amount of changes to the wording. The term "airplane" was substituted with "UAS", "crew" with "flight crew" and references to passengers were removed. The catastrophic category was re-written to include loss of trajectory control and loss of the aircraft. It can be argued that although their work tried to remain close to the word of AC 23.1309, it isn't true to its spirit. This is because the same failure condition can have different outcomes depending on whether a UAS or a manned aircraft is involved.

As an example, although in manned aviation loss of an aircraft would invariably mean a high probability of multiple fatalities, in the case of UAS this is not necessarily true. As a result the catastrophic category is probably overreaching by including conditions that are more likely of hazardous or lower severity (i.e. there is no distinction between the controlled vs the uncontrolled loss of the aircraft).

The severity definitions given in the AMC section of the STANAG 4671 [30] are completely different than those of AC 23.1309 and are more focused on the expected outcome of each failure condition. As a result the use of these definitions would be preferable.

Nevertheless, the decision whether to base the TLS for each failure severity on AC 23.1309 or AC 25.1309 rules is a more difficult one. This is because as mentioned in Sect. 5.1.2, the failure modes of UAS are different in number, nature and severity. It is indicative that the EASA has considered applying one or the other or a combination of both depending on the UAS [16]. Similarly the FAA has recently asserted that UAS are very complex systems for which the assumptions used to derive AC 23.1309 levels does not hold and as such require the use of 25.1309 [13].

It should be noted that although the severity categories are typically linked to the effect on people (i.e. the number of casualties and fatalities), other effects may also be taken into account, like those considered in [40] and presented in Table 7.3.

Table 7.3 RCC323 Hazard Severity Categories. Source: [40]

Description	Effect on people	Effect on property	Environmental effects
Catastrophic	death, permanent disability	greater than \$1 million	severe
Critical	severe injury, permanent partial disability, hospitalization for 5 or more people	\$200,000 to \$1 million	major
Marginal	minor injury, 1 or more lost workdays	\$10,000 to \$ 200,000	minor
Negligible	less than minor injury	less than \$10,000	less than minor

Similarly to the severity categories, the frequency definitions must also be clearly given. The NATO FINAS group used the number of events per hour, similarly to what is used by the US Navy [40]. Nevertheless, in the former remote probability is defined as a frequency of occurrence of less than $10^{-4}\,h^{-1}$, while in the latter the same label is given to frequencies in the range 10^{-5}–$10^{-6}\,h^{-1}$. A different approach is taken in MIL-STD-882D, where the definitions are given with respect to the life of the aircraft. For example, the remote probability is defined as *"unlikely but possible to occur in the life of an item, with a probability of occurrence less than* 10^{-3} *but greater than* 10^{-6} *in that life"* [47]. In MIL-STD-882D, the requirements are also extended to the lifetime of the entire fleet, where the remote probability is defined as *"unlikely, but can reasonably be expected to occur"* [47].

Based on the analysis in Chap. 5, it is possible to determine an appropriate risk reference system by going back from an accident (or fatality) requirement. This is accomplished by performing an OSA for a typical system for each UAS class similar to the one presented in [22]. As a result before determining TLS for each failure condition severity category, an appropriate UAS classification is required. Aircraft classification is important to ensure appropriate design, construction, maintenance and operation requirements. Applying different requirements for different classes can ensure compliance with target levels of safety without imposing unreasonably high requirements that can increase costs and impede commercialization.

7.3 UAS Classification

A major step in facilitating UAS regulation development is to determine an appropriate UAS classification, with each class carrying different requirements based on perceived associated risks. There are a number of metrics that have been used for UAS classification, including MTOW, size, operating conditions, capabilities or any combination of these and other characteristics. It should be noted that some of these metrics have minimal effect on the safety performance requirements of the system. A comprehensive classification of UAS demonstrating the wide variety of UAS systems and capabilities is presented in Table 7.4. A simpler classification scheme used by the USAF and relies on take-off weight, operating altitude and airspeed is shown in Table 7.5.

7.3.1 Classification Based on Ground Impact Risk

As mentioned in Chap. 5, the primary measure proposed for aircraft classification is the MTOW. In fact, AC 23.1309 already divides FAR Part 23 aircraft into 4 classes based on MTOW and engine type, with different target levels of safety for each one. Let it be stated that the UAS take-off weight range extents down to a few grams and as a result the whole spectrum of UAS cannot fit in the manned aviation classes. Therefore, a need arises to determine appropriate UAS classes for regulatory purposes.

MTOW is a good metric to classify aircraft since it correlates well with the expected kinetic energy imparted at impact, which in turn is considered to be the primary factor affecting the probability of fatalities [16, 20, 31, 40, 41, 53].

Table 7.4 UAS categorization for differentiation of existing systems. Source: [2]

	Mass (kg)	Range (km)	Flight Alt. (m)	Endurance (h)
Micro	<5	<10	250	1
Mini	<20/25/30/150[a]	<10	150/250/300	<2
		Tactical		
Close Range (CR)	25-150	10-30	3.000	2-4
Short Range (SR)	50-250	30-70	3.000	3-6
Medium Range (MR)	150-500	70-200	5.000	6-10
MR Endurance (MRE)	500-1500	>500	8.000	10-18
Low Altitude Deep Penetration (LADP)	250-2500	>250	50-9.000	0.5-1
Low Altitude Long Endurance (LALE)	15-25	>500	3.000	>24
Medium Altitude Long Endurance (MALE)	1000-1500	>500	3.000	24-48
		Strategic		
High Altitude Long Endurance (HALE)	2500-5000	>2.000	20.000	24-48
Stratospheric (Strato)	>2.500	>2.000	>20.000	>48
Exo-Stratospheric (EXO)	TBD	TBD	>30.500	TBD
		Special Task		
Unmanned combat AV (UCAV)	>1.000	1.500	12.000	2
Lethal (LET)	TBD	300	4.000	3-4
Decoys (DEC)	150-250	0-500	50-5.000	<4

[a] Varies with national legal restrictions

UAS Category	MTOW	Operating Altitude	Speed (KIAS)
Group 1	up to 9 kg	<1,200 ft AGL	<100 kts
Group 2	9–25 kg	<3,500 ft AGL	<250 kts
Group 3	600 kg	<18,000 ft MSL	<250 kts
Group 4	>600 kg	<18,000 ft MSL	any
Group 4	>600 kg	>18,000 ft MSL	any

Table 7.5 The Joint UAS Group Classification. Source: [45]

Of course there are other factors that can influence the risk of fatalities and the required target levels of safety. These factors include:

- *Sheltering:* Buildings, trees, vehicles and other obstacles can shelter a person from the impact, thus, reducing the probability for a serious injury or fatality.
- *Population density:* Areas with lower population density result in a smaller number of people exposed to the crash. This number can be three or more orders of magnitude higher in metropolitan areas compared to rural regions, thus, significantly influencing the TLS.

- *Aircraft dimensions:* Larger aircraft can affect larger areas, thus, increasing the number of people exposed to a crash.
- *Percentage of voluntary versus involuntary exposure:* Voluntary exposure relates to people involved in the operation of the UAS who are more aware of the risks and can take steps to avoid them. Although this factor has only a minor effect on the risk, it can affect the level of risk that is acceptable. This is because higher fatality rates may be acceptable for voluntary exposure, since the operators can justify the increased risk by way of the benefits of using the technology [7]. For example a farmer may accept an increased level of risk, in order to achieve higher efficiency and reduce costs of agricultural operations. In fact, in [41] an increase by a factor of 10 is proposed in the level of risk acceptable for voluntarily exposed persons.

Based on the analysis carried out in Sect. 5.3.1, the reliability requirement for 43 UAS of various types and sizes was calculated using (5.10). This requirement is given in the minimum hours between ground impact accidents needed to maintain an expected number of fatalities of less than $10^{-7}\,\text{h}^{-1}$. The kinetic energy at impact was calculated using the maximum estimate from (5.11) and (5.12). The other parameters of the fatality probability model were assigned average values ($\alpha = 100\,\text{kJ}$), $\beta = 34\,\text{J}$ and $p_s = 6$). A standard population density of $200\,\text{ppl/km}^2$ has been assumed, which corresponds to the average scenario used to evaluate manned aircraft [16]. The reliability requirement for each UAS is plotted against its MTOW and presented in Fig. 7.1. The existence of an approximately linear relationship between the MTOW and the aforementioned requirement is evident.

Using Fig. 7.1 a natural classification of UAS may be based on the order of magnitude of their MTOW, where each subsequent class will require an accident rate an order of magnitude smaller than the previous ones. Such a classification is presented in Table 7.6 where the MTOW-based classes were derived using the linear interpolation mentioned earlier, including a 250% safety margin to ensure a conservative estimate. The reliability target represents catastrophic accidents, so less strict targets may be acceptable for accidents of lower severity.

It should be noted that the reliability targets in Table 7.6 were derived for a standard population density of $200\,\text{ppl/km}^2$. For operation over metropolitan areas, this number can be significantly higher and even prohibitive for large systems as demonstrated in Chap. 6. A classification of larger systems could be then based on their area of operations, allowing for lower reliability for systems that will be operated away from major population centers. Nevertheless such a classification would in effect define a new airspace class over major cities, imposing further load on ATM and as a result it is unlikely.

Although the actual classes may vary depending on the model parameters used, such a classification is important because of the significant differences in risk presented by aircraft of different classes. For example, the Micro and Mini class UAS, as defined in Table 7.6, are so light that it is almost impossible for a fatality or serious injury to occur after a ground impact. They are also unlikely to cause problems to other aviation, provided that they operate with sufficient clearance from airports, due to their typically low operating altitudes.

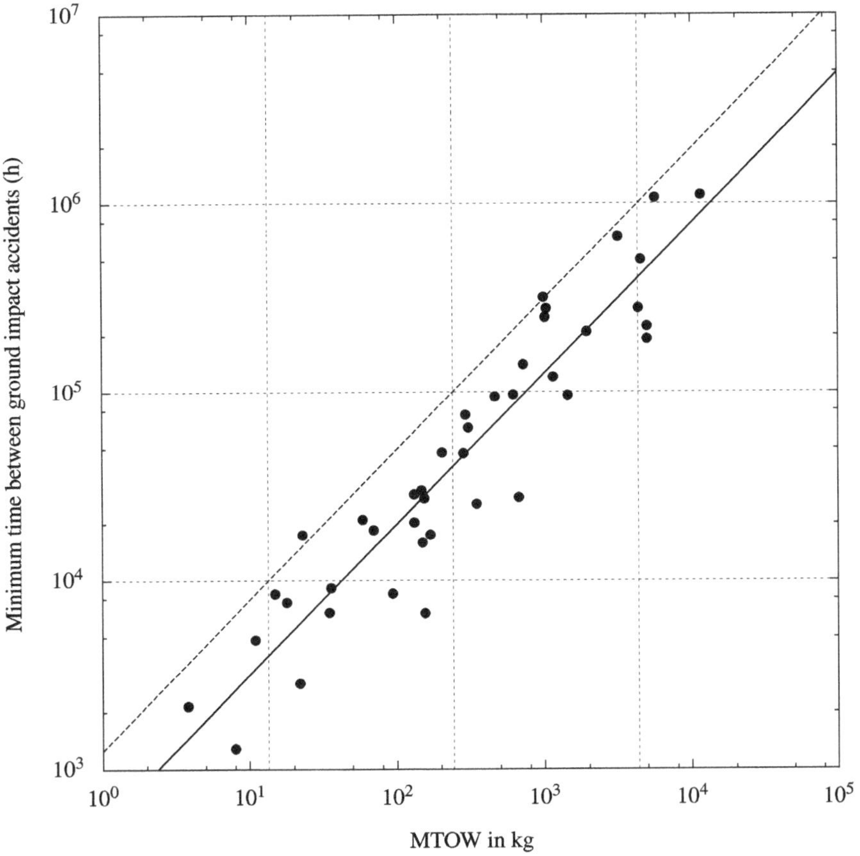

Fig. 7.1 The calculated time between ground impact accidents requirement versus the corresponding MTOW for 43 UAS of different types and sizes. The calculations are done for a population density of 200 people per km^2, $\alpha = 100$kJ, $\beta = 34$J and $p_s = 6$ using (5.10). The relationship is approximately linear with respect to their logarithms. The dotted line corresponds to the requirement derived with a 250% safety margin added.

7.3.2 Classification Based on Mid-air Collision Risk

Although MTOW provides a good basis to classify aircraft based on the risk they present to people and property after a ground impact, UAS classes based on altitude may also be of interest since they will dictate to a degree collision avoidance requirements [56]. A simple classification is proposed below:

1. *Very low altitude (VLA)* operating in Class G airspace and typically in altitudes less than 400–500 ft.
2. *Medium altitude (MA)* operating in Class A through E airspace.
3. *Very high altitude (VHA)* operating in Class E airspace above FL600.

Table 7.6 Proposed UAS classification for certification purposes.

Number	T_{GI}	Category MTOW	Name	Notes
0	10^2	less than 1 kg	Micro	Most countries don't regulate this category since these vehicles pose minimal threat to human life or property.
1	10^3	up to 1 kg	Mini	These two categories correspond to converted R/C model aircraft. The operation of the latter is based on AC91-57 which the FAA has decided is not applicable for UAS.
2	10^4	up to 13.5 kg	Small	
3	10^5	up to 242 kg	Light/Ultralight	Airworthiness certification for this category may be based either on ultralights (FAR Part 103), LSA (Order 8130) or normal aircraft (FAR Part 23).
4	10^6	up to 4,332 kg	Normal	Based on MTOW these vehicles correspond to normal aircraft (FAR Part 23).
5	10^7	over 4,332 kg	Large	These vehicles best correspond to the transport category (FAR Part 25).

The VLA class of UAS will operate in uncontrolled airspace and at such altitudes that it will mostly encounter uncooperative aviation (ultralights, parachutists, etc.) as well as stationary obstacles like buildings, trees and power lines. This class can be divided into two subclasses, based on whether they are operated within line-of-sight (LOS) or beyond line-of-sight (BLOS), each with different collision avoidance requirements:

- *VLA/LOS* where the operator has always visual contact with the aircraft and is responsible for collision avoidance, and,
- *VLA/BLOS* where a sophisticated collision avoidance system that incorporates sense and avoid will be required. This requirement may be waived under special circumstances and provided that other means for collision avoidance are provided (e.g. observers).

Collision avoidance in aeromodeling has been based on the operator and LOS operation. Despite the fact that R/C model airplanes have been suggested to present a mid-air collision risk to other aircraft [52], there is only a small number of incidents reported in the Aviation Safety Reporting System (ASRS) database, all of which occurred between 1993 and 1998. Furthermore, incident occurrence was either due to model operators violating restrictions or because the pilot of the manned aircraft was unaware of authorized R/C model activity. As such, LOS operation can be considered adequate to ensure appropriate levels of safety for people and property, without additional onboard equipment.

On the other hand VLA/BLOS presents significant challenges. This is due to a large number of obstacles stationary or moving, some of which will be difficult to detect in time. The latter include power lines or other small obstacles, other systems that may enter the airspace from a stationary position on the ground and obstacles that may be occluded by buildings or trees until the last moment. As a result routine VLA/BLOS operations are not to be expected before the S&A technology matures.

UAS in the MA class will operate in controlled airspace. As a result to mitigate the risk of mid-air collisions they will require a cooperative collision avoidance system in addition to an efficient S&A system for uncooperative aviation. Operation in controlled airspace will also impose a requirement for two-way communication with ATC and the capability to timely comply with ATC instruction.

A subclass of the MA class may be considered for UAS that will operate exclusively in Class A airspace. Since separation in Class A airspace is provided by the ATC, the S&A requirement can be relaxed. Nevertheless, this requirement cannot be waived completely since all aircraft operating in Class A will need to transition there through other airspace classes.

Finally, the VHA class is reserved for UAS operating above class A airspace. Such a UAS is the RQ-4 Global Hawk with normal cruising altitude of 65,000 ft. Currently, traffic at these altitudes is scarce and mostly public or military in nature. Although more relaxed requirements may seem attractive for this class of UAS, it is very unlikely that they will be very different than those of the MA class. This is because traffic in these altitudes is expected to rise due to the operation of UAS and because this traffic will need to traverse other controlled airspace to reach that altitude.

It should be noted that some UAS operators may be able to use restricted airspace to reach class A or above class A airspace with out traversing other airspace classes. In this case the certifying authority may issue a waiver of certain requirements for that particular UAS or operator.

The proposed classification is summarized in Table 7.7.

Table 7.7 Proposed classification based on class of airspace used.

Class	Airspace Class	S&A	Transponder	2-way ATC communication
VLA/LOS	Class G	Not Required	Not Required	Not Required [a]
VLA/BLOS	Class G	Required [b]	Required [b]	Not Required [a]
MA	Class A-E	Required	Required	Required
MA/A	Class A	Required [b]	Required	Required
VHA	Above Fl600	Required [b]	Required	Required [b]

[a] Communication with ATC before operation may still be required.
[b] May be waived for certain types of operations or under certain conditions.

7.3.3 Classification Based on Autonomy

Another way to categorize UAS that is also of interest for certification purposes is based on their level of autonomy. Already UAS are exhibiting autonomy in certain functions and this trend is expected to increase, especially if one operator is to control more than one UA at the same time [39]. In 2005 the Autonomous Control Levels (ACL) where proposed to measure autonomy. More specifically, ten (10) such levels were proposed based on requirements like situational awareness, analysis, coordination, decision making and operational capability [8]. A list of the ACL is presented in Table 7.8.

Table 7.8 Autonomous control levels. Source: [8]

ACL	Level descriptor
0	Remotely piloted vehicle
1	Execute preplanned mission
2	Changeable mission
3	Robust response to real-time faults/events
4	Fault/event adaptive vehicle
5	Real-time multi-vehicle coordination
6	Real-time multi-vehicle cooperation
7	Battlespace knowledge
8	Battlespace cognizance
9	Battlespace swarm cognizance
10	Fully autonomous

Each ACL is based on three aspects characterizing autonomy, namely the level of independence from human involvement, the complexity of the mission and the complexity of the environment [27]. Of course, some of the distinctions between the ACL defined in Table 7.8 may not be of value for regulatory purposes and some of them are not applicable for civil UAS. A classification that takes into account only the level of human involvement and is compatible with the four operational modes proposed in [17] is provided below:

- Remotely piloted: A certified pilot remotely controls the system either within LOS or with feedback from the UA sensors.
- Remotely operated (semi-autonomous): The UA is given high-level commands (waypoints, objects to track, etc.) and its performance is monitored by a trained operator. In this case the flying is performed by the UA itself, but all the decision-making is delegated to a human.
- Fully autonomous: The UA is given general tasks and is capable of determining how to accomplish them, even at the face of unforeseen events. It can also monitor its health and take remedial action after the occurrence of faults.

Such a classification is more useful for regulatory purposes because it takes into account who is responsible for each system function and how the responsibility is shared between the pilot and entity that owns and operates the UAS. In addition to that, for multi-vehicle operations additional requirements may be imposed to ensure that appropriate separation is maintained at all times, in order to avoid mid-air collisions. It should be noted, that as autonomy increases new regulatory issues will arise. For example, highly autonomous systems may exhibit non-deterministic behavior [9, 39], something that is prohibited by current regulatory approaches.

7.3.4 Other Classifications

The military has defined its own taxonomy of UAS with different airworthiness requirements as shown in Table 7.9. For all categories, airworthiness and operator qualifications will need to be demonstrated. In addition to that, Cat I aircraft are limited to LOS operations.

Table 7.9 Military UAS categories and relevant UAS regulations. Source: [50]

Category	FAA Regulation	Airspace Usage	Airspeed Limit (KIAS)
Cat I - R/C Model Aircraft	None (AC 91-57)	Class G	100[1]
Cat II - Nonstandard Aircraft	FAR Parts 91, 101 and 103	Class E,G & non-joint-use D	250[1]
Cat III - Certified Aircraft	FAR Part 91	All	None

[1] proposed.

The NATO JCGUAV presented in September of 2009 a classification guide based on MTOW. All UAS are divided into three classes; Class I for those weighing less than 150 kg, Class II for those in the range 150–600 kg and Class III for those over 600 kg [44]. Class I is subdivided into small (20–150 kg), mini (2–20 kg) and micro. Class III is also subdivided, but based on the operational role of the UAS.

The JUAS CoE has defined its own categories that depend on the operational characteristics and other UAS attributes. These categories include Tactical, Operational and Strategic UAS that have different scope and operate under different commands [50]. Also defined are six levels of domestic use as shown in Table 7.11.

Finally UAS —like other aircraft— can be categorized based on their ownership as *public* or *state* when they are owned and operated by public entities like federal agencies or local law enforcement and *civil* when they are owned by industry or private parties [23].

Class	Category	Normal Operating Altitude	Normal Mission Radius	Example Platforms
CLASS I (less than 150 kg)	SMALL > 20 kg	Up to 5,000 ft AGL	50 km (LOS)	Luna, Hermes 90
	MINI 2–20 kg	Up to 3,000 ft AGL	25 km (LOS)	Scan Eagle, Skylark, Raven, DH3, Aladin, Strix
	MICRO < 2 kg	Up to 200 ft AGL	5 km (LOS)	Black Widow
CLASS II (150–600 kg)	TACTICAL	Up to 10,000 ft AGL	200 km (LOS)	Sperwer, Iview 250, Hermes 450, Aerostar, Ranger
CLASS III (> 600 kg)	STRIKE COMBAT	Up to 65,000 ft AGL	Unlimited (BLOS)	
	HALE	Up to 65,000 ft AGL	Unlimited (BLOS)	Global Hawk
	MALE	Up to 45,000 ft AGL	Unlimited (BLOS)	Predator B, Predator A, Heron, Heron TP, Hermes 900

Table 7.10 NATO UAS Classification Guide from the Sep. 2009 JCGUAV meeting. Source: [44]

Table 7.11 Domestic use UAS levels and corresponding system attributes as defined by the JUAS CoE. Source: [50]

Level	Airspeed (KIAS)	Weight (kg)	Operating Altitude (ft)
0	≤ 250	≤ 0.9	$\leq 1,200$
1	≤ 250	$0.9 - 9$	$\leq 3,000$
2	≤ 250	$10 - 594$	$\leq 18,000$
3	≤ 250	$595 - 5,625$	$\leq 18,000$
4	≥ 250	$\leq 5,625$	$\leq 18,000$
5	Any	$\geq 5,625$	$\geq 18,000$

7.4 Certification Paths

As mentioned in Sect. 4.3.4, currently there are only two avenues for UAS certification, either by applying for a COA in the case of public UAS or by applying for a special certificate in the experimental category for civil UAS. The latter presents prohibitive problems for the industry because it takes time and there are no clearly defined procedures for UAS. In addition to that, experimental certificates are quite restrictive and do not permit commercial applications.

Current certification paths are counter-productive for the FAA as well, because they force the FAA to allocate resources for thoroughly investigating each application instead of producing the required regulation [1]. Although the FAA is under pressure to present a UAS airworthiness certification roadmap, the document is still in development and not currently available. Regardless, it can be expected that because UAS technology is new and untested, the FAA will take a cautionary approach to regulation development and as a result the process of UAS integration in the NAS may take several years. During this time the required technology will first be developed, tested and verified, and then standards will be drafted before the FAA produces the required regulations.

To speed-up this process, a step-by-step integration of UAS in the NAS is proposed, starting with the small and simple designs and progress towards the larger and more complicated ones. This process has the advantage of allowing fast integration of the smaller and "safer" classes of UAS and aiding in developing technology, expertise and standards that can be used to regulate the larger classes. In addition to that, integration can be achieved incrementally. At first UAS will be restricted to low population/low air traffic areas but gradually this restriction will be relaxed as technology matures [12]. As a consequence the micro/mini and ultralight categories should be the main focus of current regulatory efforts.

A significant factor that will affect the length of the regulation development process and the nature of the resulting regulation is the perceived reliability of UAS platforms. Early reliability studies [48] performed on military UAS showed that those systems were more prone to significant mishaps by a factor of 100, compared with their manned counterparts. Nevertheless, a significant part of military UAS mishaps can be attributed to the rushed deployment of still experimental technology. Moreover, with accumulated flight hours this situation seems to be improving as demonstrated in Fig. 7.2. Currently several military systems are approaching or have reached the reliability of manned military aircraft. Although civil aviation requirements are stricter than those of the military, current prospectives seem encouraging.

7.5 Equipment Certification

A UAS, besides the aircraft itself, normally includes a number of other components such as a launch and recovery mechanism, a ground control station, a command and control link, different payloads and other elements that can be both on the ground or on-board the system [28]. It is therefore important for all components, especially those that are critical for flight safety, to be certified airworthy. Perhaps the simplest solution to this problem is through standardization. The use of standard communication frequencies and protocols or standard ground control user interfaces would mean that the corresponding compliant systems would be immediately certifiable. Nevertheless, they would still need to be certified for operating in combination with the particular UA under consideration. As a result, the current favored approach, especially in Europe, is to certify the UAS as a whole [14].

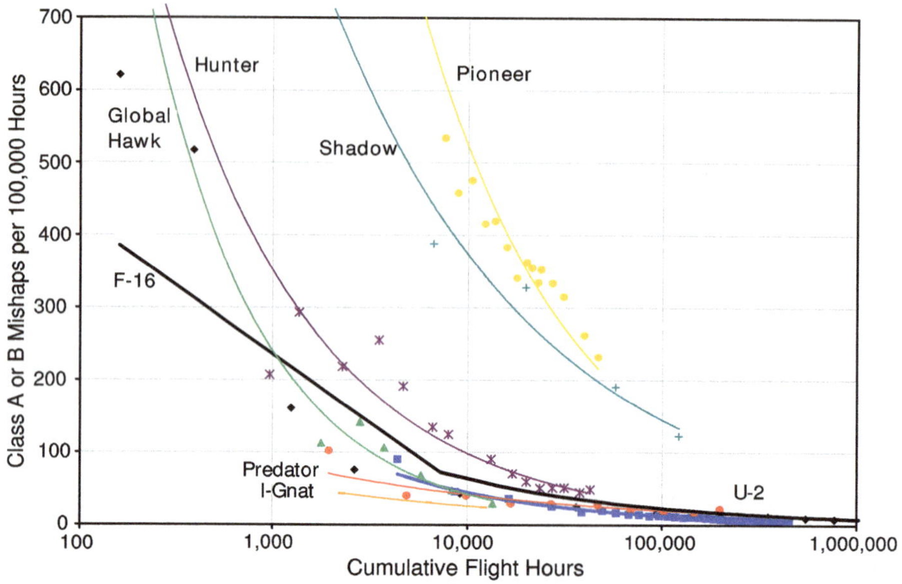

Fig. 7.2 Comparison of lifetime class A mishaps of the F-16 military aircraft and various military UAS (1986-2006) [50].

7.6 Operator Training and Certification

A significant percentage of UAS accidents has been attributed to human errors, errors which in several cases can be attributed to inexperience [54]. As a result operator training is a major concern for UAS and already some research centers are working on procedures and requirements for UAS operator and crew training. The military is also taking this matter seriously and is developing appropriate policies and procedures.

Up to now UAS operators were typically pilots that had been trained on conventional aircraft. It was soon realized that the differences in the control interface, the lack of certain information that would have been available if the pilot was onboard and other factors required specialized training for operators of UAS. Moreover, the lack of simulator training for certain UAS types and the use of ground control stations that were developed for engineers without taking into account human factors considerations [15], presented additional issues to be resolved.

There has been a broader need to recognize the human performance issues related to UAS operation, even before the UAS is flying so that appropriate design decisions are made as early as possible [4]. In the same framework, the procedures, skills and medical qualifications of future UAS pilots will need to be determined [46]. It is very likely that UAS operators will need to develop new skills compared to manned aviation pilots. For example, currently ground control stations come in different designs offering different means of controlling the aircraft. Operating more than one vehicle at the same time will also require new types of training, so that operators are fully aware of the situation of each controlled UAS, applying the correct action for each one. Despite interest for operating multiple UAS at the same time, no study has been done to determine under what conditions this would be safe [46]. In 2006 the USAF put together an experimental ground control station that allowed multi-UAS control, but flight crews are not comfortable with flying more than two UAS at the same time [5]. Similarly for semi-autonomous and fully autonomous UAS, it is possible that fail-safety is provided by manual pilot override. As a result in that case the operator will need to be able to switch from providing high level commands, to fully commanding the aircraft.

In addition to that, ground crew training is required to familiarize essential personnel with the new technologies and avoid accidents such as those occurring from unexpected movement of the platform.

7.7 Technology Issues

As mentioned in Sect. 4.1, a major concern for the FAA is that vital technology that would allow unrestricted access to the NAS is not mature yet. Moreover, up to now many UAS were designed for performance and low cost or as experimental vehicles, without performing rigorous engineering and safety studies [46]. In fact, in certain cases, not even the functional requirements of the UAS were determined before development begun [29].

It is noteworthy, that one third of the USAF fleet of Predator UAS has crashed, despite their record being similar to that of manned fighters like the F16 [34]. Naturally, one must also take into account that there are several factors that contribute to the large failure rate, such as: i) many UAS were still research/prototype systems at the time they were deployed [15], ii) in contrast to manned systems, most UAS are not equipped with redundant systems [34], iii) several parts were not aviation quality either because of cost concerns or due to the need for smaller, lighter components [14], iv) the personnel operating them did not have long experience with the system, v) the logistics and maintenance support was problematic due to the lack of reliability data [14, 15], and vi) they were deployed in harsh, battlefield environments. As a result, the high failure rates do not necessarily lead to the conclusion that UAS are inherently unsafe. In fact, several types of hazardous conditions such as pilot vertigo, cockpit smoke and other human factors related issues are not relevant in the case of UAS. On the other hand, UAS do introduce new hazards (e.g. lost command and control links) and compared to their manned counterparts they are more susceptible to certain issues like weather due to their lighter construction.

Results of a study related to reasons causing UAS accidents are presented in Table 7.12. It is noteworthy, that the accident rates due to human error constitute only around 20%, when for manned systems that percent rises to 70–80%. Nevertheless, it should be noted that other studies have shown that accidents directly attributable to human error or where human error was a significant contributing factor constitute a much larger part of total accidents, close to that of manned systems.

Table 7.12 Sources of Failures for US and Israeli Military UAS: [49].

	US	Israel
Power/Propulsion	38%	32%
Flight Control	19%	28%
Communications	14%	11%
Human/Ground	17%	22%
Misc	12%	7%

Regardless of the actual contribution of human error in the UAS accident total, the effect of human error is expected to decrease as the level of autonomy increases and operators gain more experience. Similarly communications is expected to be a smaller problem for civilian applications, since in the longterm an adequate and reliable communications infrastructure can be assumed to be available for most of such applications. Improvements on the power/propulsion systems are also expected, especially in smaller UAS, where simple and reliable electric propulsion is already available. As a result, mitigation measures for UAS should be primarily concentrated in the flight control system and secondarily on propulsion.

The following sections concern some of the technological issues that will need to be resolved so that UAS can be safely integrated in the NAS. The first and perhaps most important is that of collision avoidance.

7.7.1 Collision Avoidance

As already mentioned, for UAS to achieve an ELOS to that of manned aircraft, system reliability needs to be increased and the probability of mid-air collisions needs to be reduced. Collision avoidance has been a major area of research since most of the corresponding manned aviation procedures and technology rely primarily on pilots and secondarily on ATC instruction.

The FAA sUAS ARC in its recommendations stressed the importance of taking a multi-layered approach for ensuring safety with respect to mid-air collision accidents [42]. On the highest-level, procedures were recommended to reduce the probability of encounters by appropriate training, operating far from airspace with dense traffic and fly-away protection systems. Of course this requirement is not compatible with many types of the routine commercial operations envisioned. The next two levels are concerned with collision avoidance.

Mid-air collision avoidance can be divided into two parts. The first part is involved with ensuring appropriate separation of aircraft, something that is achieved via procedural rules and ATC instruction [1], but may not apply to all aircraft and airspace classes. The second part is involved with actually avoiding a collision in the case of inadequate separation. This entails systems like the TCAS-II and Automatic Dependent Surveillance-Broadcast (ADS-B) as well as pilot action based on the FAR-mandated "see and avoid" requirement.

Collision avoidance in manned aviation is achieved through various mechanisms that build additional layers of security to minimize the probability of collision, shown in Fig. 7.3. The first layer, cooperative collision avoidance, is currently realized through the ADS-B system meant to replace ground surveillance. This system operates by using the Global Positioning System (GPS) to determine the position of the aircraft and then broadcast it to other aircraft in the area as well as controllers on the ground. Although this system offers superior deconfliction, it may fail even if one aircraft in the area is not equipped with it. Since it is currently in the very early stages of adoption [1], its effectiveness is greatly reduced.

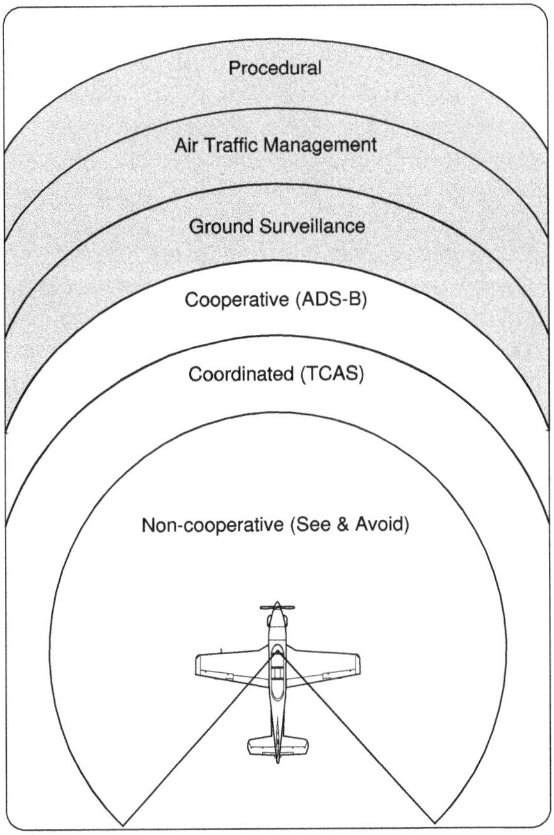

Fig. 7.3 Collision avoidance mechanisms available to civil aviation. In gray are the techniques to ensure separation. Adapted from: [3].

TCAS and its current implementation, TCAS-II, is an older collision avoidance system that has been already mandated for certain classes of passenger aircraft [1]. It operates by requesting information on a specific communications frequency and receiving responses from other aircraft equipped with TCAS. Using that information it is capable of evaluating the positions of all the TCAS-capable aircraft in the area and giving simple, auditory advice to alert the pilot of a possible collision as well as suggest a vertical maneuver to resolve the problem [10]. Significant modifications would be required to successfully use the system in UAS due to the differences in aircraft characteristics and the nature of possible collisions [10, 46, 55].

Based on current regulations, it can be expected that for the foreseeable future UAS collision avoidance cannot depend exclusively on either the ADS-B or the TCAS. This is because there will be airspace users that will not be equipped with any of these systems. Moreover, current regulations state that "*regardless of whether an operation is conducted under instrument flight rules or visual flight rules, vigilance shall be maintained by each person operating an aircraft so as to see and avoid other aircraft*". A change in current regulation is also unlikely, since UAS integration in the NAS should not incur any burden to current airspace users [51]. In any case, it is doubtful that systems like ADS-B or TCAS will be available in any but the larger UAS, due to the cost and size of the equipment involved [29]. Furthermore, these systems are not capable of terrain and other obstacle (e.g. birds and powerlines) avoidance [55] and cannot be used to demonstrate compliance with see and avoid requirements.

A Sense, Detect and Avoid (SDA) system or sense and avoid (S&A) system as is the most common use of the term, will be likely used as acceptable means of compliance with FAR Part 91 rules. A S&A system installed on a UAS should be capable of operating under various weather conditions and situations [55] and, as autonomy increases, with limited operator involvement. The latter is indeed important since different requirements can be expected depending on whether a pilot will be in the decision loop or not [56]. In any case a successful S&A system will likely entail information fusion from multiple sensors to avoid the "blind-spots" of a single sensor. It will also be necessary for the system to interface with ground-based sensors (e.g. radar) for supplementing its own information and with other sensors and on-board systems to be able to make successful decisions for collision avoidance and execute them with precision and timeliness.

A significant obstacle for the development of see-and-avoid-compliant systems is that FAR Part 91 does not quantify this requirement. Based on the principles of fairness and equivalence, it would be reasonable to demand that S&A systems achieve a performance comparable to that of human pilots. Nevertheless, both research and practice has shown that pilots are not very good at detecting potential collisions [14]. Moreover, currently manned aviation pilots can only monitor a limited part of the airspace around them, mostly in front of the aircraft. As a result, the development of S&A systems that are capable of exceeding pilot capabilities would be beneficial for the manned aviation sector, as well. In any case, it can be expected that some minimum requirements will be mandated that will include minimum detection distances and aircraft separation margins. The "reaction time" of the system should also be at least comparable to that of a pilot. Based on FAA AC 90-48C, it takes on average 6 s for a pilot to detect another obstacle and become aware of a possible collision [18]. An additional 6.4 s are required for deciding the necessary maneuver and for the aircraft to start executing the turn [18]. Also, although initially research has been focused on duplicating pilot vision [1], efficient deconfliction algorithms will also need to be developed to replace pilot action in the case of a possible collision.

Another approach that was employed in the VUSIL project (See Sect. 4.4.5) and is also being investigated by the US DoD [37], is to provide the pilots with ground-based radar data that have sufficient accuracy to allow them to perform the see and avoid function.

When a mid-air collision cannot be avoided, a fourth layer was proposed by the FAA sUAS ARC to reduce the effects of the accident [42]. This layer involves measures such as limiting the airspeed and mass of the UAS. This measure is also difficult to apply in practice, and is only useful until proper regulation is available.

It should be noted that the see-and-avoid requirement will put additional burden on both manned and unmanned aviation pilots that share the same airspace with other UAS [14]. This is because UAS are typically significantly smaller and often less maneuverable. As a result, distances could be misjudged while the necessary maneuvers to avoid a collision may need to be initiated sooner.

7.7.2 Sensors

From a regulation standpoint, sensor capabilities are an issue, since they will determine S&A performance which in turn is a major factor affecting safety of operations. S&A sensor research has investigated electro-optical, acoustic and microwave sensors [1]. Acoustic sensors are not very susceptible to problems from the environment and are quite robust, but have low resolution and exhibit problems tracking multiple targets. Electro-optical sensors have good resolution and allow multiple target tracking but are more susceptible to environmental interference (sun glare, temperature, vibration). Microwave sensors can give good range estimates but are typically too large, heavy and expensive especially for smaller UAS, although research is being done on ultra-light-weight radar systems for this class of UAS [33].

When combined, these sensors offer unique characteristics that enable a UAS to detect and in some cases track one or more targets in difficult conditions like fog, glare or darkness. In technology demonstrations S&A systems were able to surpass human pilots in detecting approaching aircraft from greater distances [1]. Besides developing sensors with higher resolutions and improved capabilities, efficient fusion algorithms will also be needed. Finally sensor miniaturization will play an important role in the integration of small UAS.

7.7.3 Communications

A major concern of UAS manufacturers and operators will be the security of the communications link and the availability of adequate bandwidth. The FAA with other international stakeholders is planning to bid for UAS-dedicated spectrum at the World Radiocommunication Conference in 2011. On the other hand, there is significant competition for spectrum allocation for other services and spectrum can be very expensive [38].

Several UAS functions that depend or may depend on communications such as UA control, ATC relay and sense and avoid, are also related to flight safety [38]. As a result both link capacity as well as reliability will be required.

Nevertheless protected spectrum for smaller UAS is unlikely to be available [43]. Security of the communications link is also of high importance to ensure that UAS are not susceptible to unintentional interference or malicious intent and hostile take-over [46]. Finally contingency planning and reliable procedures to ensure safety will be required for the instances where the link has been lost or the latency is such that remote operation is no longer possible. This will fall under a *continued safe flight and landing* requirement described in Sect. 7.1.5. A major issue in these cases is the predictability of the UAS performance to ensure separation with other air traffic [46]. For example, military systems assume a holding pattern until the link is re-established. In fact, during incidents where contact with multiple systems was lost, this allowed the recovery of all the UAS involved [34]. Different behaviors may also be acceptable, as long as it is clear to other airspace users what the UAS can be expected to do.

It is very likely that the architecture and communication protocols will be common among all UAS. This is both because of safety reasons relating to interference, as well as due to the cost of developing, maintaining and regulating different systems [38].

Within the regulatory framework covering communications, the issue of two-way communication with ATC will also need to be addressed. This presents a challenging technological problem as well, since it will likely involve relaying voice communication from the UA to its pilot [38]. Moreover, the communication and navigation equipment currently required for manned aircraft is often bulky and difficult to integrate into smaller UAS designs, while their operational characteristics (e.g. low level flying) may prevent some of those systems from operating as intended [14].

In addition to the aforementioned communication needs, the issue of accommodating the communication requirements of the payload must also be addressed [38]. It should be noted that the latter is currently out of the scope of the 2011 bid [38].

7.7.4 Power and Propulsion Systems

Power and propulsion systems have proved to be a major source of UAS accidents. As a result higher reliability will be required to comply with future regulations. This will have to be developed concurrently with higher endurance, power and lower cost and size demanded by the market.

7.7.5 Launch, Recovery and Flight Termination Systems

The most crucial phases of aircraft operations is take-off and landing. There are several launch systems available to UAS, ranging from hand-launch to regular runway take-off. Similarly retrieval of UAS may also impose some challenges, especially for operations away from dedicated facilities. It is expected that some sort of certification will be required for launch and retrieval systems, since they present a significant part of a UAS and can be crucial for the safety of operations.

Moreover, besides regular landing/retrieval, there is a number of different options for emergency flight termination systems that may need to be developed to mitigate the risk of certain UAS hazards. These systems are also important from a regulatory standpoint since they directly affect the safety performance of UAS. Nevertheless, it should be noted that certain types of flight termination systems may not be acceptable for demonstrating flight safety [20] or may not be acceptable because of secondary hazards that they may introduce. In any case, regardless of the system used, the behavior of a UAS under an emergency should be predicable so that other airspace users and ATC can take appropriate actions to isolate it [14].

7.8 Technology Testing and Evaluation

In parallel with standards and regulations development, other efforts are required to streamline the integration process of UAS in the NAS. Existing regulations severely limit access to airspace so that UAS developers can evaluate their systems. As a result, it is of foremost importance is to build test centers to evaluate UAS and their subsystems for both R&D as well as certification purposes. Such a test center recently opened in New Mexico State University under FAA oversight [25, 26]. The FAA is also using the William J. Hughes Technical Center in New Jersey it conduct its own experiments on UAS [25]. Similar centers are planned or have already been built in France, Canada, Australia and New Zealand. Such centers can be used to certify UAS hardware and software components as well as provide UAS crew training, as is the goal for the ARCAA in Australia [51].

Testing of S&A without access to the NAS has also been problematic, further underlining the importance of test centers which will allow airborne technology evaluation. Although successful demonstrations of various S&A systems have been made, extensive simulations and rigorous field testing is required to evaluate their performance under various conditions and collision scenarios, before they can be used in civilian applications.

The development of a database like the ASRS [36] to store flight logs and all incidents and accidents from UAS operations can facilitate UAS-related activities. This database will provide invaluable information for UAS developers and operators. It will also facilitate standardization and regulation efforts by providing insight on key technological issues and UAS failure modes.

Furthermore the UAS reliability data are also useful for insurance providers, since companies operating civil UAS will be liable for damages incurred due to UAS operations and will require indemnification [35]. Nevertheless, the lack of airworthiness standards will still be a matter of concern for insurance providers, especially in certain application scenarios [46].

References

1. Anand S (2007) Domestic use of unmanned aircraft systems: Evaluation of policy constraints and the role of industry consensus standards. Journal of Engineering and Public Policy 11
2. van Blyenburgh P (2006) UAV systems: Global review. presented at the Avionics '06 Conference, Amsterdam, the Netherlands
3. Bryner M (2007) Developing sense & avoid for all classes of UAS. Unmanned Aircraft Systems, The Global Perspective 2007/2008 pp 142–146
4. Burns J, Capriglione C, Duma KM, Hamilton BA (2010) Unmanned aircraft systems standards the role of standards and standardization in unmanned aircraft systems airspace integration. Bulletin of the Survivability/Vulnerability Information Analysis Center, Vol. 27, No. 1
5. Button K (2009) Multi-aircraft control of UAVs by a single pilot won't come easily, experts warn. C4ISR Journal
6. Cary L (2009) ICAO activities toward integration of UAS into the global ATM system. Presented at the ICAO Global Air Traffic Management Forum on Civil/Military Cooperation
7. Clothier R, Walker R (2006) Determination and evaluation of UAV safety objectives. In: Proc. 21st International Unmanned Air Vehicle Systems Conference, Irvine, CA, USA, pp 18.1–18.16
8. Clough B (2002) Metrics, schmetrics! how do you track a UAV's autonomy? In: Proc. AIAA 1st Technical Conference and Workshop on Unmanned Aerospace Vehicles, Portsmouth, VA
9. Clough BT (2002) Unmanned aerial vehicles: Autonomous control challenges, a researcher's perspective. In: Murphey R, Pardalos PM (eds) Cooperative Control and Optimization, Kluwer Academic Publishers, chap 3, pp 35–53
10. Coulter DM (2007) Airspace modeling for UAS sense and avoid. Unmanned Aircraft Systems, The Global Perspective 2007/2008 pp 147–148
11. Dalamagkidis K, Valavanis K, Piegl L (2008) Current status and future perspectives for unmanned aircraft system operations in the US. Journal of Intelligent and Robotic Systems 52(2):313–329

12. Davis B (2007) Students recommend a way ahead for UAVs in the NAS. Unmanned Systems 25(5):23
13. Davis KD (2008) Unmanned aircraft in the national airspace system - the certification path. Presented at the EASA Workshop on UAV
14. DeGarmo MT (2004) Issues concerning integration of unmanned aerial vehicles in civil airspace. Tech. rep., MITRE, Center for Advanced Aviation System Development, McLean, Virginia
15. Drew JG, Shaver R, Lynch KF, Amouzegar MA, Snyder D (2005) Unmanned aerial vehicles — end-to-end supoort considerations. Tech. rep., Rand Corporation
16. European Aviation Safety Agency (EASA) (2005) A-NPA, No. 16/2005, policy for unmanned aerial vehicle (UAV) certification
17. Federal Agencies Ad Hoc Autonomy Levels for Unmanned Systems (ALFUS) Working Group (WG) (2004) Autonomy levels for unmanned systems (ALFUS) framework - version 1.1. NIST Special Publication 1011
18. Federal Aviation Administration (1983) Pilot's role in collision avoidance. AC 90-48C
19. Federal Aviation Administration (1999) Equipment, systems and installations in part 23 airplanes. AC 23.1309-1C
20. Haddon DR, Whittaker CJ (2002) Aircraft airworthiness certification standards for civil UAVs. UK Civil Aviation Authority
21. Hayhurst KJ, Maddalon JM, Miner PS, Dewalt MP, Mccormick GF (2006) Unmanned aircraft hazards and their implications for regulation. In: Proc. 25th IEEE/AIAA Digital Avionics Systems Conference, Portland, OR, USA, pp 1–12
22. Hayhurst KJ, Maddalon JM, Miner PS, Szatkowski GN, Ulrey ML, Dewalt MP, Spitzer CR (2007) Preliminary considerations for classifying hazards of unmanned aircraft systems. Tech. Rep. NASA TM-2007-214539, National Aeronautics and Space Administration, Langley Research Center, Hampton, Virginia
23. Hempe D (2006) Unmanned aircraft systems in the United States. Presented to the US/Europe International Safety Conference
24. Hobbs A, Herwitz S (2005) Human factors in the maintenance of unmanned aircraft. In: Krebs WK (ed) Unmanned Aerial Vehicles Human Factors Program Review, Federal Aviation Administration, Office of the Chief Scientist for Human Factors, pp 3–8
25. Hoffmann T (2010) Eye in the sky: Assuring the safe operation of Unmanned Aircraft Systems. FAA Safety Briefing
26. Hottman SB (2009) NMSU-PSL-TAAC. UAS Yearbook 2009/2010, UVS International pp 61–62
27. Huang HM (2007) Autonomy levels for unmanned systems (ALFUS) framework: Safety and application issues. In: Proc. Performance Metrics for Intelligent Systems (PerMIS) Workshop, Gaithersburg, MD, USA, pp 48–53
28. ICAO/IMO Joint Working Group on harmonization of aeronautical and maritime search and rescue (ICAO/IMO JWG-SAR), International Civil Aviation Organization (ICAO) (2008) SAR operational principles, procedures and tech-

niques — development of procedural strategies for the practical provision of SAR services. Information Paper
29. Ingham LA (2008) Considerations for a roadmap for the operation of unmanned aerial vehicles (UAV) in South African airspace. PhD thesis, Department of Electrical and Electronic Engineering, Stellenbosch University
30. Joint Capability Group on Unmanned Aerial Vehicles (2007) STANAG 4671 - Unmanned Aerial Vehicle Systems Airworthiness Requirements (USAR). draft, NATO Naval Armaments Group
31. Joint JAA/Eurocontrol Initiative on UAVs (2004) A concept for european regulations for civil unmanned aerial vehicles (UAV). Final Report
32. Masutti A (2009) Proposals for the regulation of unmanned air vehicle use in common airspace. Air and Space Law 34(1):1–12
33. McGroggan L (2008) Raven project to develop small UAS S&A. UAS Yearbook 2008/2009, UVS International p 105
34. Mockenhaupt B (2009) We've seen the future, and it's unmanned. Esquire Magazine, [online] http://www.esquire.com/features/unmanned-aircraft-1109, accessed December 2009
35. Montgomery J (2006) Opening civil airspace to unmanned aerial systems. Unmanned Systems 24(4):35–38
36. NASA (2007) ASRS - Aviation Safety Reporting System. [online] http://asrs.arc.nasa.gov/
37. Pease F (2008) Four toolsets to achieve integration of UAS into the national airspace. In: UVS International Conference, Paris, France
38. Pelmoine C, Neale M, Tränapp N, Bittner JR, Klaeyle A (2008) Uas airspace integration – the communications challenges. In: UVS International Conference, Paris, France
39. Protti M, Barzan R (2007) UAV autonomy — which level is desirable? — which level is acceptable? Alenia Aeronautica viewpoint. In: Platform Inovvations and System Integration for Unmanned Air, Land and Sea Vehicles (AVT-SCI Joint Symposium), Florence, Italy, pp 12–1–12–12
40. Range Safety Group, Range Commanders Council (1999) Range safety criteria for unmanned air vehicles - rationale and methodology supplement. Supplement to document 323-99
41. Range Safety Group, Range Commanders Council (2007) Common risk criteria standards for national test ranges: Supplement. Supplement to document 321-07
42. Small Unmanned Aircraft System Aviation Rulemaking Committee (2009) Comprehensive set of recommendations for sUAS regulatory development
43. Tarbert B (2008) Gaining access to national airspace with UAS. Keynote Presentation - International Symposium on Unmanned Aerial Vehicles
44. The Joint Airforce Competence Centre (2010) Strategic concept of employment for unmanned aircraft systems in NATO
45. United States Air Force (2009) Unmanned aircraft systems flight plan 2009-2047

46. United States Government Accountability Office (2008) Unmanned Aircraft Systems: Federal actions needed to ensure safety and expand their potential uses within the national airspace system. Report to Congressional Requesters
47. US Department of Defense (2000) Standard practice for system safety. MIL-STD-882D
48. US Department of Defense Office of the Secretary of Defense (2003) Unmanned aerial vehicle reliability study. Report
49. US Department of Defense Office of the Secretary of Defense (2005) Unmanned aircraft systems roadmap 2005-2030. Report
50. US Department of Defense Office of the Secretary of Defense (2007) Unmanned systems roadmap 2007-2032. Report
51. Walker R, Gonzalez LF (2007) Australian research centre for aerospace automation. Unmanned Aircraft Systems, The Global Perspective 2007/2008 pp 17–18
52. Weibel RE (2005) Safety considerations for operation of different classes of unmanned aerial vehicles in the national airspace system. Master's thesis, Department of Aeronautics & Astronautics, Massachusetts Institute of Technology
53. Weibel RE, Hansman RJ (2004) Safety considerations for operation of different classes of UAVs in the NAS. In: Proc. AIAA 4th Aviation Tehcnology, Integration and Operations Forum and AIAA 3rd Unmanned Unlimited Technical Conference, Workshop and Exhibit, Chicago, IL, USA
54. Williams K (2004) A summary of unmanned accident/incident data: Human factors implications. DOT/FAA/AM 04-24
55. Zeitlin A (2007) Sense & avoid evaluations and standards for civil airspace access. Unmanned Aircraft Systems, The Global Perspective 2007/2008 pp 156–157
56. Zeitlin A (2009) UAS S&A standards: Challenges and progress. UAS Yearbook 2009/2010, UVS International pp 134–136

Chapter 8
Epilogue

Let's think the unthinkable, let's do the undoable, let's prepare to grapple with the ineffable itself, and see if we may not eff it after all.

Douglas Adams
(Dirk Gently's Holistic Detective Agency, 1987)

This Chapter is the epilogue that attempts to bring closure to this Second Edition of the book. This has been a difficult task. It should not be, but it is! After all, the reason is so simple that it was almost entirely ignored and overlooked during the course of revising and enhancing the book: Although several federal agencies and multiple governments (with perhaps competitive agendas) are involved in accomplishing the common goal of UAS integration in to civilian airspace, everything seems to move so fast, that when one considers a specific subject completed, new changes/modifications are in effect. It is true that there still exist major obstacles and that this process has been delayed considerably. However, it is common knowledge that 'something needs to be done' to at least integrate Class I UAVs in to the national airspace. The imposed deadlines seem to differ, but it has been argued that some final decisions need to be made by the end of 2012. Nevertheless, the ultimate goal remains intact and it is the same. Therefore, this Chapter presents recent available information published on the subject. Without loss of generality, this last Chapter may be also viewed as ... justification for publishing this revised manuscript.

8.1 Why UAS?

At first, one should start by trying to answer the question(s) *"Why UAS? What do they offer? Why are they useful if at all?"* The answers nowadays are more than straightforward and obvious. This seems hard to believe since the lack of a human pilot raises eye brows and makes the public suspicious, if not skeptical, against "those things that fly alone in the sky", to say the least.

However, the benefit of the doubt should be credited and granted to the general public. One should not forget what happened in the field of robotics about fifty five years ago; it took more than thirty years fighting for survival before robotics was accepted in the US as a viable, even better, alternative to completing a wide range of applications, which were tedious for humans. Needless to mention that when robotics was accepted in the US, auto makers in Japan had already automated their

K. Dalamagkidis et al., *On Integrating Unmanned Aircraft Systems into the National Airspace System*, Intelligent Systems, Control and Automation: Science and Engineering 54, DOI 10.1007/978-94-007-2479-2_8,

car assembly lines and research and development on humanoids had already began. Even European industries used robots before it happened in the US.

The rationale for making the above statement stems from the fact that *those robots were not pilot-less and they did not fly alone*. Therefore, it is natural to assume that it will be much more difficult for the public to accept UAS and their use in civilian applications. However, this is what the future holds, and UAS are on their way to entering everyday life.

8.2 UAS for Military Applications and Related Challenges

There is a very bright side when one focuses on military applications. The focus on and the perspective from the military point of view (overlooking civilian applications) is justified by the fact that today UAS are basically and mostly used in the battlefield. Further, the statement that UAS are best suited for 'dull, dirty and dangerous' missions has merit and it is supported because:

1. Dull operations that require more than 30- or 40- hour missions are best carried out using UAS, since crew members in manned aircraft are used to much shorter duty cycles. Before the 1990's crews were used to four-hour sorties and missions. Fatigue and mission duration compromise proficiency and functionality of crew members, thus, the UAS alternative prevails.
2. Dirty operations may require that UAS fly into nuclear clouds (as happened in 1946-1948) immediately after bomb detonation, a mission that is clearly dangerous to human crews and threatens human lives.
3. Dangerous operations like reconnaissance over enemy territory may result in loss of human lives, thus UAS are preferred.

As stated in [8] and [9], UAS are preferred over manned aircraft not only because of downsizing risk and increasing confidence in mission success avoiding at the same time the human cost of losing lives if the mission is unsuccessful, but also because unmanned vehicles have better and sustained alertness over humans during dull operations.

Future Combat System (FCS) initiatives and directives (now known under a different name), as well as the Unmanned Systems Roadmap 2007–2032 do envision an Air Force based mostly on unmanned planes [10]. This becomes more evident since future urban warfare, search and rescue, border patrol, homeland security forest fire detection, traffic monitoring, and other applications will utilize an unprecedented level of automation in which human-operated, autonomous, semi-autonomous air and ground platforms will be linked through a coordinated control system to perform complex missions.

The main challenge in future operations will relate to networked UAS with adaptable operational procedures, planning and asset deconfliction, increased situation awareness, coupled with cutting edge technologies to realize autonomous collaborative operations.

Technical challenges will stem from real-time sensing, computing and communication requirements, environmental and operational uncertainty, hostile threats and the emerging need for improved UAS and UAS team autonomy and reliability. Significant challenges will also relate to inter-UAS communications, links to command and control, contingency management, etc.

The requirement and emerging need for improved UAS autonomy is depicted in Fig. 8.1 that shows the tentative time table for autonomous control level trends [8, 9]. In essence, Fig. 8.1 tabulates unmanned aircraft sophistication levels from the US DOD perspective, which cover the whole spectrum from the teleoperated and preprogrammed flight of a single aircraft to self-actuating and fully autonomous group vehicle flights.

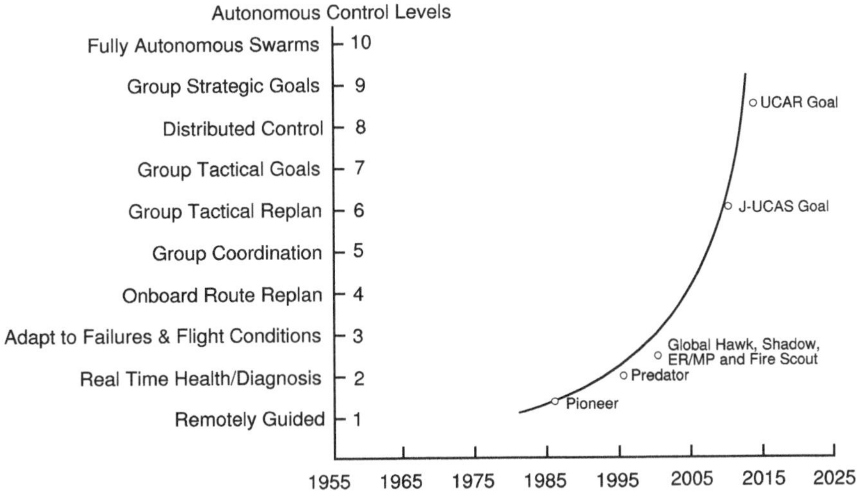

Fig. 8.1 Autonomous control level trend (taken from [8, 9]).

Challenges increase significantly as one moves up the hierarchy of the chart from single to multi-vehicle coordinated control. Only moderate success has been currently reported in meeting the lower echelon challenges, leaving open the whole field for subsequent developments.

Technically, to meet stated challenges, innovative coordinated planning and control technologies such as distributed artificial intelligence (DAI), multi agent System (MAS) theory, computational intelligence and soft computing, generalized system theory, as well as game theory and dynamic optimization, coupled with sophisticated hardware and software architectures will be needed. Even though related approaches and methodologies have been investigated intensively in recent years, most work has been focused on solving particular problems, such as formation control and autonomous search, while less attention has been paid to the overall 'system architecture' concept, especially from an implementation and integration point of view [11].

In addition to all of the above, NATO members, as part of the STANAG 4586, have been working towards 'plug and play' capabilities for UAS [7], attempting to improve the flow of information among shared systems and enhance battlefield efficacy. STANAG 4586 includes the 4586 standard and four volumes of supporting documentation towards establishing interoperability of present and future UAS in a NATO Combined/Joint Service Environment. The STANAG 4586 Annex B is composed of three appendices, which detail Data Link Interface, Command and Control Interface, and, Human Computer Interface requirements [7].

8.3 UAS for Civilian Applications: Challenges and Issues

It has been claimed that the future of UAS is on civilian and public applications. AUVSI Board Member Kitty Higgins, during an NTSB two day seminar on UAS in the NAS in Washington D.C., joked that one day *"..our pizzas will be delivered by UAS"* [10].

At the same time, Doug Davis who was at that time the head of the FAA's Unmanned Aircraft Program Office, made it clear that "no one should look to the sky for pizza just yet" [10]! However, the FAA, recognizing that for some special cases UAS may be very useful and effective, has partnered with police departments in Miami and Houston to evaluate the use of small UAS for surveillance [9]. Along the same lines, it was stated that special permits could be set for very restricted applications and small UAS, but he (Davis) also mentioned that a fully certified avionics suite for larger vehicles that could provide S&A capability will not be ready until 2020 or 2025.

Further, in AUVSI's Unmanned Systems Magazine, in a two-part interview with Margaret Gilligan, Associate Administrator for Aviation Safety with the FAA [5], she states that *"WE are working hard on the rulemaking for small UAS and 2013 continues to be our goal"*. However, she also states that larger systems are more of a challenge, although there are standards. And if UAS meet these standards, then they can fly in the NAS. It is clear from this interview that the major obstacle to be overcome is the level of safety, and the right metrics to measure safety.

Other concerns are raised by Davis [4], who reports:

> US government agencies and other enthusiasts are eager to get unmanned aerial vehicles into the National Airspace System (NAS) but many regulatory hurdles remain and other airspace users stress that safety should trump technology.

When surfing the web, an interesting article was found, entitled "Unmanned aircraft pose myriad problems to US airspace, GAO reports" [2]. It states what is going on nowadays in the US, despite progress made. Because of its importance, it is presented in its entirety next:

> A ton of work needs to be done by military, federal and civil aviation groups if the rapidly growing unmanned aircraft community is allowed routine access to public airspace.

In a wide-ranging report on the impact of unmanned aircraft on the country's commercial airspace, congressional watchdogs at the Government Accountability Office today called on Congress to create an overarching body within Federal Aviation Administration to coordinate unmanned aircraft development and integration efforts.

The GAO also called on the FAA to work with the Department of Defense, which has extensive unmanned aircraft experience to issue its program plan. In addition, the Department of Homeland Security (DHS) assesses the security implications of routine unmanned aircraft access to commercial airspace, the GAO said. Even if all issues are addressed, and there are a number of critical problems, unmanned aircraft may not receive routine access to the national airspace system until 2020, the GAO concluded.

But such access is certainly on the minds of the unmanned aircraft community. That's mainly because the market for government and commercial-use unmanned aircraft could explode in the coming years. Federal agencies such as the DHS, the Department of Commerce, and NASA alone use unmanned planes in many areas, such as border security, weather research, and forest fire monitoring. Researchers at the Teal Group said in their 2008 market study estimates that UAV spending will more than double over the next decade from current worldwide UAV spending of $3.4 billion annually to $7.3 billion, totaling close to $55 billion in the next ten years. The forecast also indicates that the US could account for 73% of the world's research and development investment unmanned flight in the next decade.

Still, routine unmanned aircraft access to the national airspace system poses technological, regulatory, workload, and coordination challenges, the GAO said. A key technological challenge is providing the capability for unmanned aircraft to meet the safety requirements of the national airspace system. For example, a person operating an aircraft must maintain vigilance so as to see and avoid other aircraft. However, because the airplanes have no person on board, on-board equipment, radar, or direct human observation must substitute for this capability. No technology has been identified as a suitable substitute for a person on board the aircraft in seeing and avoiding other aircraft, the GAO report stated.

Additionally, the aircraft's communications and control links are vulnerable to unintentional or intentional radio interference that can lead to loss of control of an aircraft and an accident, and in the future, ground control stations - the unmanned airplane equivalent to a manned aircraft cockpit — may need physical security protection to guard against hostile takeover, the GAO said.

There are other issues as well, the GAO report states, including:

- Many unmanned airplanes, particularly smaller models, will likely operate at altitudes below 18,000 feet, sharing airspace with other objects, such as gliders. Sensing and avoiding these other objects represents a particular challenge for unmanned aircraft, since the other objects normally do not transmit an electronic signal to identify themselves and FAA cannot mandate that all aircraft or objects possess this capability so that the aircraft can operate safely. Many small unmanned do not have equipment to detect such signals and, in some cases, are too small to carry such equipment. The Aircraft Owners and Pilots Association, in a 2006 survey of its membership, found that unmanned aircraft's inability to see and avoid manned aircraft is a priority concern.
- The effort to develop the Traffic Alert and Collision and Avoidance System (TCAS), used widely in manned aircraft to help prevent collisions, demonstrates the challenge of developing a detect, sense, and avoid capability for unmanned airplanes. Although FAA, airlines, and several private-sector companies developed TCAS over a 13-year period, at a cost of more than $500 million, FAA officials point out that the designers did not intend for TCAS to act as the sole means of avoiding collisions and that the on board pilot still has the responsibility for seeing and avoiding other aircraft. FAA officials also point out that TCAS computes collision avoidance solutions based on characteristics of manned aircraft, and does not incorporate unmanned aircraft's slower turn and climb rates in developing conflict solutions. Consequently, FAA officials believe that developing the detect, sense, and avoid technology that unmanned aircraft would need to

operate routinely in the national airspace system poses an even greater challenge than TCAS did. FAA officials believe that an acceptable detect, sense, and avoid system for airplanes could cost up to $2 billion to complete and is still many years away.
- The lack of protected radio frequency spectrum for unmanned operations heightens the possibility that an operator could lose command and control of the plane. Unlike manned aircraft, which use dedicated, protected radio frequencies, unmanned aircraft currently use unprotected radio spectrum and, like any other wireless technology, remain vulnerable to unintentional or intentional interference. This remains a key security vulnerability for unmanned aircraft, because in contrast to a manned aircraft where the pilot has direct, physical control of the aircraft, interruption of radio frequency, such as by jamming, can sever the plane's only means of control. One of the experts we surveyed listed providing security and protected spectrum among the critical airplane integration technologies.
- Unmanned aircraft have the capability to deliver nuclear, biological, or chemical payloads, and can be launched undetected from virtually any site. In response to the events of September 11, 2001, entry doors to passenger airplane cockpits were hardened to prevent unauthorized entry. However, no similar security requirements exist to prevent unauthorized access to unmanned aircraft ground control stations - the unmanned system equivalent of the cockpit. Security is a latent issue that could impede unmanned airplane developments even after all the other challenges have been addressed, according to one study.
- Although DOD has obtained benefits from its unmanned operations overseas, the agency notes in its Unmanned Systems Roadmap that unmanned aircraft reliability is a key factor in integrating unmanned systems into the national airspace system. Our analysis of information that DOD provided on 199 military unmanned airplane accidents, of varying degrees of severity, that occurred over 4.5 years during operations Enduring Freedom and Iraqi Freedom, indicates that reliability continues to be a challenge. About 65% of the accidents resulted from materiel issues, such as failures of aircraft components. FAA officials noted that unmanned aircraft today are at a similar stage as personal computers in their early years before newer, more user-friendly operating systems became standard.
- The variety of ground control station designs across unmanned aircraft is another human factors concern. For example, pilots of the Predator B control the aircraft by using a stick and pedals, similar to the actions of pilots of manned aircraft. In contrast, pilots of the Global Hawk use a keyboard and mouse to control the aircraft. Differences in unmanned system missions could require some variation among control station designs, but the extent to which regulations should require commonalities across all ground control stations awaits further research.
- Because unmanned aircraft have never routinely operated in the national airspace system, the level of public acceptance is unknown. One researcher observed that as unmanned aircraft expand into the non-defense sector, there will inevitably be public debate over the need for and motives behind such proliferation. One expert we surveyed commented that some individuals may raise privacy concerns about a small aircraft that is "spying" on them, whether operated by law enforcement officials or by private organizations, and raised the question of what federal agency would have the responsibility for addressing these privacy concerns.

While those issues are just a few outlined in the report, the GAO said a number of activities are also ongoing to address concerns. The GAO report states some of those activities include:

- The DoD plans to spend over $7 billion in research, development, test, and evaluation funds for unmanned aircraft between fiscal years 2007 and 2013. Data from these efforts could facilitate FAA's development of a regulatory framework to allow unmanned aircraft to have routine access to the national airspace.
- The FAA has budgeted $4.7 million for fiscal years 2007 through 2009 for further unmanned systems research on topics such as detect, sense, and avoid; command and con-

trol; and system safety management. NASA, FAA, and others have conducted tests to determine the capabilities of and potential improvements to detect, sense, and avoid technology. For example, in 2003, NASA installed radar on a manned aircraft that was equipped for optional control from the ground. The tests indicated that the radar detected intruding aircraft earlier than the onboard pilot, but also revealed the need for further work on the onboard sensing equipment to ensure adequate response time for the remote pilot. According to a summary of the lessons learned from these tests, the results showed some promise, but indicated that much work and technology maturation would need to occur before the tested system could be deemed ready for operational use.

- The FAA has established a 12,000 square mile unmanned system test center to provide airspace for testing and evaluating unmanned aircraft and to provide data for use in developing regulations. FAA expects to obtain additional data from increased coordination with the DoD. However, FAA has not yet analyzed the limited data that it has already accumulated on recent unmanned operations in the national airspace system, citing resource constraints. To address expected workload increases, FAA is introducing more automation into its work processes and has granted DoD authority to operate small unmanned systems weighing 20lbs or less, over its installations without receiving prior FAA approval.
- Addressing the challenge of radio frequency allocation for unmanned operations is moving forward, but may not be completed for several years. The International Telecommunication Union allocates radio frequency spectrum and deliberates such issues at periodic World Radiocommunication Conferences, the most recent of which was held in the fall of 2007. To obtain spectrum allocation for unmanned aircraft, FAA has participated with the Department of Commerce in a national preparation process to place spectrum allocation decisions on the conference's future agenda. At the 2007 conference, delegates agreed to discuss at the next conference, in 2011, the spectrum requirements and possible regulatory actions, including spectrum allocations, needed to support the safe operation of unmanned systems.
- The DoD is urging manufacturers to increase reliability while keeping costs low by using such practices as standard systems engineering, ensuring that replacement parts are readily available, and using redundant, fail-safe designs. The DoD also notes in its Unmanned Systems Roadmap that, although unmanned planes suffer accidents at one to two orders of magnitude greater than the rate incurred by manned military aircraft, accident rates have declined as operational experience increased. For some airplanes, the accident rates have become similar to or lower than that of the manned F-16 fighter jet, according to the roadmap. According to a study by The MITRE Corporation, General Atomics designed the Predator B with reliability in mind, and the Altair airplane, which is a modified version of the Predator, has, among other things, triple redundant avionics to increase reliability.
- FAA has established an unmanned system program office and is reviewing the body of manned aviation regulations to determine the modifications needed to address unmanned aircraft, but these modifications may not be completed until 2020. As an interim step, the FAA has begun an effort to provide increased access to the national airspace system for small unmanned aircraft. The FAA is taking steps to develop data to use in developing standards, but has been slow to analyze the data that it has already collected. FAA is also coordinating with other countries to harmonize regulations.

Feedback from European agencies is pretty much the same. D. Hughes' article in [6] states below the title that *"Unmanned aircraft need lots of flying hours to be 'civilized' "*. The article talks about the Eurocontrol plans to develop a roadmap by the end of 2009 outlining how unmanned aircraft will be integrated into the European ATM system. A key issue is harmonization with what the FAA, ICAO,

EUROCAE WG-73 and RTCA SC-203 are deciding. According to H. Matthiesen, Senior Specialist in ATM [6]:

> What the industry needs is flying hours and experience and lessons [learned] to build confidence in these machines. Airworthiness requires lots of flying hours, especially to support airworthiness on the civil side.

M. Bennett states in [1] that the European Defense Agency (EDA) awarded in January of 2008 a contract to a consortium of defense and aerospace companies to develop a detailed roadmap for integrating UAS into European airspace so that they can fly routinely with other air traffic by 2012 or 'the end of 2015 at the latest'.

It is essential to clarify that the thought process in US and European agencies is very similar if not identical; this is because of population density and the fact that UAS will need to fly mostly over metropolitan areas; hence the increased safety concerns.

Regardless, the potential is huge, and when the roadmap is complete, UAS will be utilized for a wide range of civilian and public applications like traffic monitoring, inspection (power lines, water dams, bridges, highways and pipelines), forest protection and fire detection, border patrol, to mention just a few applications. Discussion on the subject is endless; however, the next Section attempts documenting some thoughts of what lie ahead.

8.4 Challenges, Enabling Technologies

All of the above notwithstanding, there is consensus on a number of challenges that need be met before UAS fly routinely in civilian airspace. Even though smaller UAS will fly first [3, 5] the following are true:

- Safety, safety, and more safety, with all prerequisites and aftermath attached to it. The public will not tolerate accidents; it is as simple as that.
- Sense and Avoid technology. The NTSB members expressed particular interest in the ability of UAS to handle contingencies beyond close encounters in shared airspace [4]. This is perhaps the most important challenge to be overcome, as the FAA states that UAS should function as if there were a pilot on-board. In addition, redundancy is a must, and as such, it is obvious that sense and avoid technology should include both visual and non-visual sensor components.
- Bandwidth regulation.
- Lost Link Procedures: In all cases, the UAS must be provided with a means of automatic recovery in the event of a lost link. There are many acceptable approaches to satisfy the requirement. The intent is to ensure airborne operations are predictable in the event of lost link.
- Flight Termination System (FTS): It is highly desirable that all UAS have system redundancies and independent functionality to ensure the overall safety and predictability of the system. If a UAS is found to be lacking in system redundancies,

an independent flight termination system that can be activated manually by the UAS pilot in command, may be required to safeguard the public.
- Autonomous Operations: At first only those UAS that have the capability of pilot intervention, or pilot-in-the-loop, shall be allowed in the NAS outside of Restricted, Prohibited, or Warning areas. UAS that are designed to be completely autonomous, with no capability of pilot intervention, are going to be the last to be authorized for operations in the national airspace system.

New modeling techniques will be required to capture coupling between individual system/sensor dynamics, communications, etc., with 'system of systems' behaviors. Hybrid system approaches will play a key role in capturing complex behaviors and defining the means to represent and manage uncertainty, including spatial-temporal models of distributed sensors to integrate system and motion dependencies, contingency planning and situation awareness.

Intelligent and hierarchical/distributed control concepts must be developed and expanded to address 'system of systems' configurations. Game theoretic approaches and optimization algorithms running in real-time to assist in cooperative control and adversarial reasoning will be needed. Comprehensive approaches to control of networks of dynamic agents will be essential to tackle coordination and cooperation issues.

Networking and communications will deviate from traditional communication protocols and standards, requiring novel solutions to overcome jamming. Security metrics will need to be re-defined.

Sensors and sensing strategies will need innovative technologies and solutions that will include wireless communications. This will be coupled with building improved, reliable and cost-effective sensor suites, as well as 'smart' sensors, leading to better sensing strategies for massive data processing, data mining and sensor fusion.

The need for defining new 'system of systems' performance and effectiveness metrics for verification, validation and assessment of networked systems is more than obvious.

8.5 The Road Ahead

During the last almost twenty years, we have witnessed significant advances in the unmanned aircraft and unmanned systems state-of-the-art. Several advances have been presented in this book.

Major research and development projects, sponsored primarily by the military sector, have contributed towards development and deployment of aerial, undersea and ground unmanned systems with new designs surfacing at an increasing rate. Common objectives and goals of any such designs are: improved autonomy through innovative sensing and control strategies, enhanced reliability and endurance through fault-tolerant control, advanced materials and high efficiency power plants.

Yet, new and novel concepts and technologies are required for a more widespread use of such critical assets, not only for military but also for commercial and other applications such as homeland security, rescue operations, forest fire detection, delivery of goods, to name just a few applications.

Federated systems consisting of multiple unmanned aerial vehicles performing complex missions present new challenges to the control community. UAS must possess attributes of autonomy in order to function effectively in a 'system of systems' configuration.

Coordinated and collaborative control of UAS swarms demands new and novel technologies that integrate modeling, control, communications and computing concerns into a single architecture. Typical application domains include reconnaissance and surveillance missions in an urban environment, target tracking and evasive maneuvers, search and rescue operations, homeland security, etc.

Major technological challenges remain to be addressed for such UAS swarms, or similar federated system of systems configurations to perform efficiently and reliably. Excessive operator load, autonomy issues and reliability concerns have limited thus far their widespread utility. The systems and controls community is called upon to play a major role in the introduction of breakthrough technologies in this exciting area.

References

1. Bennett M (2008) Schedule for UAS flights in europe slips, but hope stays on track. Unmanned Systems 26(5):30–31
2. Cooney M (2008) Unmanned aircraft pose myriad problems to US airspace, GAO reports. Online, [online] http://www.networkworld.com/community/node/27876, accessed June 2008
3. Davis B (2008) Small UAS to take to the skies first, FAA says, but hurdles remain. Unmanned Systems 26(1):32–33
4. Davis B (2008) UAS in the national airspace: The NTSB takes a look. Unmanned Systems 26(6)
5. Gilligan P (2010) Unmanned aircraft in civil airspace: A chat with the FAA. Unmanned Systems 28(8):17–21
6. Hughes D (2008) UAV road map for europe. Aviation Week & Space Technology
7. Kirschbaum A (2008) 'plug and play' for UAS: STANAG 4586 shows the way. Unmanned Systems 26(5):28–29
8. US Department of Defense Office of the Secretary of Defense (2002) OSD UAV roadmap 2002-2027
9. US Department of Defense Office of the Secretary of Defense (2005) Unmanned aircraft systems roadmap 2005-2030. Report
10. US Department of Defense Office of the Secretary of Defense (2007) Unmanned systems roadmap 2007-2032. Report

11. Valavanis KP (ed) (2007) Advances in Unmanned Aerial Vehicles: State of the Art and the Road to Autonomy, Intelligent Systems, Control and Automation: Science and Enginee, vol 33. Springer

Glossary

Accident An unplanned event or series of events that results in damages [2]. The term mishap is often used to refer to an accident, as well. Accidents that do not occur directly, but rather as a result of other accidents, are referred to as secondary accidents. Note that the National Transportation Safety Board (NTSB) defines an accident as an occurrence incidental to flight in which, as a result of the operation of an aircraft, any person (occupant or non-occupant) receives fatal or serious injury or any aircraft receives substantial damage.

Accident Probability Levels An arbitrary categorization that provides a qualitative measure of the most reasonable likelihood of occurrence of an accident resulting from personnel error, environmental conditions, design inadequacies, procedural deficiencies, or system, subsystem, or component failure or malfunction [8].

Accident Risk Categories An arbitrary categorization of accident risk assessment values often used to generate specific actions such as mandatory reporting of certain hazards to management for action, or formal acceptance of the associated accident risk [8].

Accident Severity An assessment of the consequences of the most reasonable credible mishap that could be caused by a specific hazard [8].

Air Traffic Control (ATC) ATC is a service provided under appropriate authority to promote the safe, orderly and expeditious flow of air traffic [1].

Air Traffic Management (ATM) ATM refers to the dynamic, integrated management of air traffic and airspace —safely, economically and efficiently— through the provision of facilities and seamless services in collaboration with all involved parties [1].

Air Vehicle There are certain types of aircraft like moored balloons, unmanned balloons, unmanned rockets and ultralights that are considered "vehicles" and, thus, are allowed to fly without an airworthiness certificate. More specifically most requirements regarding pilot certification, operating and flight rules, vehicle registration and marking, maintenance certification that are normally applicable to aircraft, do

K. Dalamagkidis et al., *On Integrating Unmanned Aircraft Systems into the National Airspace System*, Intelligent Systems, Control and Automation: Science and Engineering 54, DOI 10.1007/978-94-007-2479-2,

not apply for this category [5], although operational restrictions are in place. For example the following pertain to the operation of ultralight vehicles (FAR Part 103): i) single occupant; ii) daylight operations; iii) recreation or sport purposes only, and, iv) no flight over congested areas in cities, towns or open areas when crowds are present.

Aircraft An aircraft is defined in the United States Code (USC) as *any contrivance invented, used, or designed to navigate, or fly in, the air* [49 USC §40102]. In general, aircraft exist in a variety of possible configurations: fixed-wing, rotary-wing or rotorcraft, helicopters, vertical take-off and landing (VTOL) vehicles, or short take-off and landing (STOL) ones. An aircraft may also be either heavier or lighter than air, with balloons and airships belonging to the latter category.

Airman Airman as defined in [49 USC 40102], is an individual:

1. In command, or as pilot, mechanic, or member of the crew, who navigates aircraft when under way;
2. Except to the extent the Administrator of the Federal Aviation Administration may provide otherwise for individuals employed outside the United States, who is directly in charge of inspecting, maintaining, overhauling, or repairing aircraft, aircraft engines, propellers, or appliances; or
3. Who serves as an aircraft dispatcher or air traffic control-tower operator.

Airport The airport is defined as a landing area used regularly by aircraft for receiving or discharging passengers or cargo [49 USC 40102].

Background Risk Risks voluntarily accepted in the course of normal activities [4].

Collision Avoidance Considered a last resort maneuver of an aircraft to avoid an imminent collision. Without the maneuver a collision might occur [6].

Compounding conservatism An analysis approach that results in extremely conservative results by making a series of conservative assumptions [4].

Conflict Avoidance Activity which seeks to ensure that aircraft remain safely separated and well clear of each other as to not present a collision hazard [6].

Conservatism As used in risk analysis modeling, conservatism is a set of modeling assumptions that exaggerates the risk by overstating event probabilities, hazard probabilities or consequences [4].

Damage An undesired outcome that may include injury, fatality as well as physical, functional and/or monetary loss.

Expected Fatalities The mean number of fatalities predicted to occur as a result of an operation if the operation were to be repeated many times [4].

Exposure The number of personnel or resources affected by a given event or, over time, by repeated events. This can be expressed in terms of time, proximity, volume, or repetition. This parameter may be included in the estimation of severity or probability, or considered separately [7].

Fail-safe A design feature that ensures the system remains safe, or in the event of a failure, causes the system to revert to a state that will not cause a mishap [8].

Fixed-wing Aircraft Aircraft with immovable wings that require a runway to take-off and land, or alternatively can be catapult launched. Nevertheless, fixed-wing configurations that require very short runways (STOL) or can take-off vertically (VTOL) are also available.

Flight Termination System (FTS) A flight termination system ends the flight of a vehicle and consists of the entire system on an airborne vehicle used to receive, decode, and execute a flight termination command [4].

Fragmentation The break up of an in-flight vehicle into fragments (components of the vehicle, pieces of the structure, chunks of solid propellant, miscellaneous hardware, etc.) due to explosive loads, aerodynamic and inertial loads, activation of a flight termination system, intercept with another vehicle, or impact on a surface [4].

General Aviation General aviation is a term used to describe all non-military and non-airline flying, encompassing everything from recreational aircraft to experimental aircraft to privately owned and operated business jets [1].

Hazard Any real or potential condition that can cause injury, illness, or death of personnel, or damage to or loss of equipment or property [4]. Hazards can be further distinguished as initiating, contributory and primary [2]. Initiating hazards include events and conditions that start an adverse chain of events that can lead to an accident. Primary hazards are events that directly and immediately cause an accident. Finally contributory hazards are the hazards that are not initiating or primary, although in [2] this term is defined equivalently with hazard.

Hazard threshold The lowest level at which adverse outcomes are expected to appear [4].

Hazardous operation Those activities which, by their nature, expose personnel or property to dangers not normally experienced in day-to-day actions [4].

Helicopter A helicopter refers to an aircraft that takes-off and lands vertically; it is also known as a rotary-wing aircraft with the ability to hover, to fly in very low altitudes, to rotate in the air and move backwards and sideways. It is capable of performing non-aggressive or aggressive flights. A helicopter may have different configurations, with a main and a tail rotor (most common), with only a main rotor, with tandem configuration, with coaxial but opposite rotary rotors, as well as with one, two or four rotors.

Individual Risk Individual risk is the risk that a person will suffer a consequence. Unless otherwise noted, individual risk is expressed as the probability that an individual will become a casualty due to all hazards from an operation at a specific location [4].

Involuntary Activity No choice was made by the person affected which placed them in a position of increased risk; or the activity participated in or the item used was one that is generally done or used by more than 99 percent of the population [4].

Model Aircraft A small UAS used by hobbyists and flown within visual line-of-sight under direct control from the pilot, which can navigate the airspace, and which is manufactured or assembled, and operated for the purposes of sport, recreation and/or competition [6].

Mishap See *Accident*.

Mission A flight test or operation. It may include multiple vehicles or all phases of the flight beginning with lift-off/launch [4].

National Airspace System (NAS) The National Airspace System refers to the common network of US airspace, air navigation facilities, equipment and services, airports or landing areas [1].

Outrage Factor The components of outrage regarding public perception of imposed risk [4]. Such components may include the source of the risk, the transparency/clarity of the risk presentation by the responsible party, the familiarity of the public with the risk and the level of the perceived risk.

Risk An expression of consequences in terms of the probability of an event occurring, the severity of the event and the exposure of personnel or resources to potential loss or harm [7].

Risk analysis A study of potential risk under a given set of conditions. Risk Analysis is an activity that includes the complete array of tasks from data gathering, identification of hazards, estimation of associated risks, and verification of results [4].

Risk Assessment The process of detecting hazards and their causes, and systematically assessing the associated risks [7].

Risk Control An action designed to reduce risk by lowering the probability of occurrence and/or decreasing the severity of an identified hazard [7].

Risk Management See *Risk Control*.

Risk Profile A plot that shows the probability of an accident causing a given number of casualties (vertical axis) vs. the number of casualties (horizontal axis). The area under the plot is a measure of the casualty expectation [4].

Safety Critical A term applied to any condition, event, operation, process, or item whose proper recognition, control, performance, or tolerance is essential to safe system operation and support [8].

Sensitivity Analysis The computation of the effect of changes in input values or assumptions (including boundaries and model function form) on the outputs. The study of how uncertainty in a model output can be systematically apportioned to different sources of uncertainty in the model input [4]. By investigating the "relative sensitivity" of model parameters, a user can become knowledgeable of the relative importance of parameters in the model.

Transponder A transponder is an electronic device that "responds" to interrogation by ground-based radar with a special four-digit code that air traffic control specifically assigns to the aircraft on which it is located. Certain transponders have the ability to transmit automatically the altitude of the aircraft in addition to the special code [1].

Uncertainty The absence of perfectly detailed knowledge [4]. Uncertainty includes incertitude (the exact value is unknown) and variability (the value is changing). Uncertainty may also include other forms such as vagueness, ambiguity, and fuzziness (in the sense of border-line cases).

Visual Line-of-Sight Unaided (corrective lenses and/or sunglasses exempted) visual contact with aircraft sufficient to be able to maintain operational control of the aircraft, know its location, and be able to scan the airspace in which it is operating to decisively see and avoid other air traffic or objects [6].

Visual Observer A sUAS flight crew member who assists the sUAS PIC in the duties associated with collision avoidance. This includes, but is not limited to, avoidance of other traffic, airborne objects, clouds, obstructions, and terrain [6].

Voluntary Activity The person affected made a choice, which placed them in an increased position of risk compared to the rest of the population [4]. This can be occupation-related (e.g. being a UAS crew member) or not (e.g. doing recreational paragliding).

Waiver Granted use or acceptance of an article that does not meet the specified requirement [3].

Worst-case A semi-quantitative term referring to the maximum possible exposure, dose, or risk, that can conceivably occur, whether or not this exposure, dose, or risk actually occurs in a specific population [4].

References

1. Air Transport Association (ATA) (2008) Learning center. [online] http://learningcenter.airlines.org/
2. Federal Aviation Administration (2000) FAA System Safety Handbook, FAA, chap Appendix A
3. Range Safety Group, Range Commanders Council (1999) Range safety criteria for unmanned air vehicles - rationale and methodology supplement. Supplement to document 323-99
4. Range Safety Group, Range Commanders Council (2007) Common risk criteria standards for national test ranges: Supplement. Supplement to document 321-07

5. Schultz R (2006) Ultralights, LSAs and kit airplanes – what's the difference? Florida Aviation and Business Journal [online] http://www.airportjournals.com/Display.cfm?varID=0609005, accessed April 2008
6. Small Unmanned Aircraft System Aviation Rulemaking Committee (2009) Comprehensive set of recommendations for sUAS regulatory development
7. United States Air Force (1998) Operational risk management (ORM) guidelines and tools. Air Force Pamphlet 91-215
8. US Department of Defense (2000) Standard practice for system safety. MIL-STD-882D

Appendix A
Human Vulnerability

When I look at you, I see a bag filled with blood, bones, viscera, and meat, held in by the most fragile and vulnerable covering, so easily punctured, so facilely ripped or torn.

Douglas Preston and Lincoln Child
(Dance of Death, 2005)

As discussed in Sect. 5.3, the modeling of ground impact fatalities depends on estimating the $P(\text{fatality}|\text{exposure})$ term among others. In turn, this term depends on several factors, including the shape of the UAS and/or its fragments, the vulnerability of the human body to injury and the presence of other objects that may provide sheltering such as buildings, trees and vehicles.

Human vulnerability has been the subject of extensive studies in the past, especially in the area of automotive safety. In the military domain, the capacity of fragments to seriously injure or kill personnel has been investigated, to determine areas that need to be controlled around firing ranges, rocket tests, explosions, etc. As a consequence, several models have been proposed to assess the risks of certain injury scenarios. These models can be taken into account in determining a better estimate of the actual risk posed by a UAS, which can lead to more realistic and cost-effective safety measures.

This Appendix aims at providing some additional information on the issue of human vulnerability modeling and when possible how it can be applied to assess UAS safety. The definition of the injury types and their severity classification is provided first. This is followed by a section on different issues that need to be considered in vulnerability modeling. Finally selected vulnerability models are presented, spanning different injury mechanisms.

A.1 Injury Types and Severity

Impact of an object on the human body can cause stress to human tissue that in turn can cause tearing or deformation beyond recoverable limits, a process known as trauma [17]. There are several ways a UAS may mechanically cause trauma. If an aircraft or a fragment hits a person, it can cause what is known as blunt trauma. This occurs by imparting a high acceleration and/or loading to a body part or tissue, which may suffer functional changes as a result [4, 14, 17]. If movement of the body is constrained by a hard surface such as a wall, then the person will suffer crushing

K. Dalamagkidis et al., *On Integrating Unmanned Aircraft Systems into the National Airspace System*, Intelligent Systems, Control and Automation: Science and Engineering 54, DOI 10.1007/978-94-007-2479-2,

injuries [4]. Small sized fragments with sufficient velocity will cause penetrating injuries, while, on the other hand, sharp edges can cause lacerations [4, 14]. Besides mechanical injuries, in certain cases, other types need also be evaluated such as injuries from possible explosions either direct (burn, overpressure) or indirect (window breakage). For UAS that carry flammable, toxic or radioactive payloads, the effects of possible leaks after a crash will also need to be taken into account in the safety evaluation.

To consistently estimate the level of injury, the Abbreviated Injury Scale (AIS), presented in Table A.1, is widely used [4]. In the AIS scale fatalities correspond to level 6 injuries. It is noteworthy, that the Range Commanders Council in [14] proposes limits based on *casualties*. The latter were defined as people that have suffered injuries with AIS level 3 or higher [14].

Table A.1 The Abbreviated Injury Scale. Source [4]

AIS	Severity	Type of Injury
0	None	None
1	Minor	Superficial
2	Moderate	Reversible injury: medical attention required
3	Serious	Reversible injury; hospitalization required
4	Severe	Life threatening; not fully recoverable without care
5	Critical	Non-reversible injury; not fully recoverable even with medical care
6	Virtually unsurvivable	Fatal

To facilitate mapping between an incident and the AIS level of injuries, different criteria have been developed for trauma to different body regions. For an impact to the head, a very common criterion used is the Head Injury Criterion (HIC), proposed by the National Highway Traffic Safety Administration in 1972 [5, 10]. The HIC is given by:

$$\mathrm{HIC} = \max\left[(t_2 - t_1)\left(\frac{1}{t_2 - t_1}\int_{t_1}^{t_2} \alpha(t)dt\right)^{2.5}\right] \tag{A.1}$$

where α is the head acceleration within a time interval defined by t_1 and t_2. Very small or very large intervals have been found not to increase risk of injury. As a result, only accelerations lasting between 3 ms and 35 ms need be considered [5]. A common approach known as HIC-15 is to fix the time interval to 15 ms around the maximum acceleration [6]. There is a direct mapping between the HIC value and the expected level of injury. For example, a maximum HIC-15 value of 700 is currently allowed in the US for motor vehicle crashes, which corresponds to a 5% probability of AIS level 4 injuries [6].

Other head injury criteria include the Head Impact Power (HIP) [12] that considers linear as well as angular accelerations, the Peak Linear Acceleration (PLA) [15] and a number of criteria that also employ a model of the head, which is simulated using finite element methods [10].

Rapid accelerations to the head can also cause neck injuries. The associated level of injury is a function of the compression and tension forces to the neck [6]. Nevertheless, and especially in the case of debris impact, it can be assumed that if the level of head injuries is controlled, neck injuries will be within acceptable limits as well.

For thorax and abdominal injuries, two major criteria in use are the blunt criterion and the viscous criterion, that will be discussed in the sections that follow. In addition to those, several other criteria have been proposed based on compression, force, velocity, acceleration and functions of the aforementioned parameters [8].

A.2 Vulnerability Modeling Considerations

Human vulnerability to impact of objects is subject to the influence of a large number of factors. Specifically with respect to the impacting object, its size, shape, mass, density, frangibility, deformability, velocity and angle of impact among other factors, can affect the severity of the injuries caused [4, 7, 14]. Additionally, the position, posture, clothing, weight, height, age, health and sex of the affected person, must also be taken into consideration [4, 7, 14]. The location where the object will strike the body is also significant, since body regions exhibit different sensitivity and injuries may have different outcomes [14]. A complete vulnerability study should also take into consideration all possible injury mechanisms; blunt trauma, penetration, lacerations and crashing injuries, as well as, the possibility of multiple strikes [14]. Moreover, a UAS mid-air collision accident may cause a number of fragments that can strike a person as they fall to the ground, after bouncing off of other objects or by generating new fragments after the initial impact [14]. As a result, accurate modeling of the potential effects on the human population of an explosion, UAS impact or other similar accident, relies on information that is typically not known a priori. It is, therefore, important to determine the key factors involved and, when possible, determine average or conservative values for certain parameters to reduce the complexity of the problem.

Historically, vulnerability modeling in the area of flight safety was based exclusively on the characteristics of the impacting fragment, specifically its kinetic energy on impact [4]. Nevertheless, studies found that depending on the injury mechanism, kinetic energy may not be accurate. This led to the development of generalized models of the form mv^x [4]. Specifically, in 1968 Feinstein et al reported that while crushing injuries were best modeled using kinetic energy, impulse loading was better modeled using momentum and penetrating or lacerating injuries with mv^4 [3]. Other researchers followed, investigating different injury types and mechanics, but mostly maintaining the relationship between mv^x and injury severity.

The sensitivity to trauma and the potential for lethal injury varies throughout the human body. On the other hand, taking into consideration each possible impact location separately, would impose a very high burden on vulnerability modeling. To overcome this problem, the body is often divided into regions that are assumed to be of the same sensitivity [7]. Such regions typically involve the head, thorax, abdomen and limbs, as shown in Fig. A.1. Using the dimension of the standard man, it is possible to determine the expected exposed area of each body region as a function of the angle of impact and body position. The latter can take several forms, but typically three body positions are considered; standing, sitting and prone. From these three the prone position exposes the largest area for vertical and near-vertical impacts. Depending on the characteristics of the area in consideration, certain assumptions may be made on the mix of body positions within the population. For example, on a beach it can be expected that the prone position will be prevalent. On the other hand, it would be over-conservative to consider all persons in that position [2].

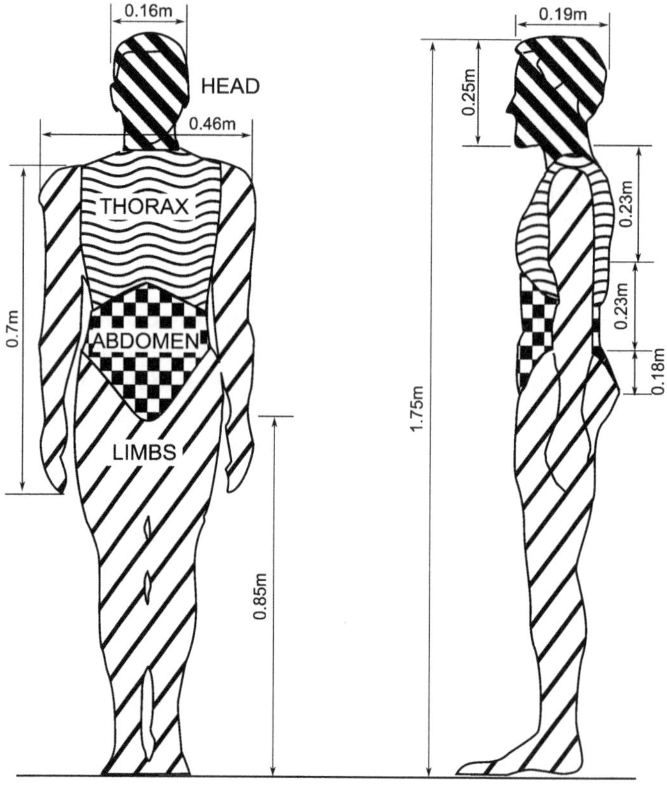

Fig. A.1 The human body divided into four regions of similar sensitivity to trauma. The dimensions shown are those for the standard man. Source [7]

To improve accuracy, models that are based on the biological response of the body have also been developed [4]. Naturally, the increased precision comes at the cost of computational complexity and the need to define incident conditions in detail. An intermediate approach in complexity as well as accuracy, is to consider the trauma mechanisms to each body region [4]. For example, when injury to the thorax is considered, and depending on the characteristics of the impact, different types of injuries may occur, each of different severity. Even low levels of chest compression may stress internal organs and vessels [17]. If the compression exceeds the tolerance of the rib cage, rib fractures and possible organ puncture may occur [17]. Similarly, abdominal injuries can occur by blunt trauma, penetration or acceleration [8]. The latter may result in internal organ or blood vessel rupture [8]. This type of modeling allows adjustments to incorporate higher or lower precision models depending on requirements and capabilities [4]. Depending on the characteristics of the impact, more than two injury mechanisms may be in effect. For instance, high velocity/high energy fragments will cause penetrating injuries. For lower energies, blunt force trauma may need to be considered as the predominant injury mechanism [4].

Another major consideration when modeling human vulnerability is that of uncertainty [4]. In most cases, the population is modeled using an "average person" and in effect an average response to injury is used. Nevertheless, actual populations include a wide range of people that usually include groups that are more vulnerable to trauma, such as children. One solution to this problem is to assume that every individual member of the population belongs to the most vulnerable group [4]. Of course, such an assumption would inevitably lead to an over-conservative model. An alternative approach is to incorporate a probabilistic approach that is capable of considering the variations in vulnerability within the population [4].

A.3 Vulnerability Models

This Section briefly presents a selection of some of the vulnerability models that have been proposed and/or have been used during the last 50 years. It is important to note that a significant part of the research in this area involves the risk to personnel from projectiles and fragmentation following explosions. As such, the conclusions drawn may not be suitable for impacts of larger objects. Additionally, results from motor vehicle crashes that represent the other major part of current research should also be used with care since the injury mechanisms may differ.

For the subsequent presentation, a chronological order was preferred when possible, to afford the reader with a glimpse on the evolution of vulnerability modeling. Comparisons between models and other popular fatality/casualty criteria are also provided. Although there is strong connection between injury level and fragment mass and velocity, there is no single function of those variables that can model it with accuracy [14]. Nevertheless, criteria based on kinetic energy and other mv^x type functions have been proposed and have been used extensively in the past. As a result, all the criteria shown are energy based criteria.

A.3.1 Vulnerability Thresholds

A basic approach to modeling vulnerability is to assign a threshold to separate impacts that may cause a fatality or casualty and those that cannot. In reality, the goal is not to achieve zero probability of fatality or casualty, but rather an arbitrarily low propability [14]. The latter is defined based on the target level of safety that needs to be met and the confidence required. For several decades a threshold in fragment kinetic energy of 79 J was used extensively to limit personnel exposure to explosions [11]. This limit was created to correspond to an AIS of 2 or 3 [1]. Although its origins are not known exactly, this criterion was probably derived in the late 19th century as a rough rule of thumb and there have since been concerns, regarding its accuracy [1, 11].

The Range Commanders Council in the latest version of their risk criteria standards have advocated towards using vulnerability thresholds as a more conservative modeling approach and employing other vulnerability models only when needed [14]. The latter refers to situations where the thresholds would be over-conservative or would impose unreasonable restrictions on range operations. The proposed thresholds are summarized in Table A.2. They include both casualty, as well as fatality thresholds.

According to [14] the 15 J casualty threshold was derived as the lower bound of all previously used criteria. In fact, according to the same document, this criterion has been used in certain cases as a threshold for any type of injury. As a result, it may be over-conservative, although on the other hand it offers significant protection to children, as well. It should be noted, that this criterion has also been adapted by the FAA Office for Commercial Space Transportation for commercial launches [14].

The lethality threshold was derived from Feinstein's curves [14], presented in the following Section, and represents a probability of fatality of 1%. Since those curves were derived for adult males, an adjustment of approximately 14% was made to account for the entire population.

Table A.2 Vulnerability thresholds for different injury mechanisms. Source [14]

Hazard	Injury severity	Threshold
Blunt trauma / Crashing injuries	Casualty	15 J
Blunt trauma / Crashing injuries	Fatality	34 J
Chunky penetration	Casualty	$71.5\,\mathrm{kJ\,m^{-2}}$
Overpressure	Casualty	13.8 kPa

A.3.2 Log-normal Vulnerability Models

Use of a threshold criterion is not representative of reality since there is no single value where the probability of fatality jumps from zero to one. In fact, this probability has been found to increase gradually as the energy absorbed is increased. A popular model of this increase is through the use of a log-normal function. One such work, is the one undertaken by Feinstein et al in 1968 [3].

Feinstein et al used injury data from US Department of Defense experiments to determine a relationship between fragment weight and velocity with injury severity. Since different body parts have different response to impact, three relationships were determined for impact to the head, the thorax and to the abdomen and limbs, respectively. Figure A.2 shows the threshold for serious injury and 10%, 50% and 90% probability of fatality, derived for an impact to the head. Based on Fig. A.2 and using the slope of the curve derived, it can be shown that the probability of fatality is constant with respect to kinetic energy for fragment masses up to about 2 kg. It should be noted, that in the case of impact to the thorax, the slope yielded a probability of fatality that was constant with respect to $mv^{1.5}$. Due to the lack of data regarding lethality of larger fragments, conservative estimates were used, resulting in the abrupt change to zero slope in the right part of Fig. A.2 [2].

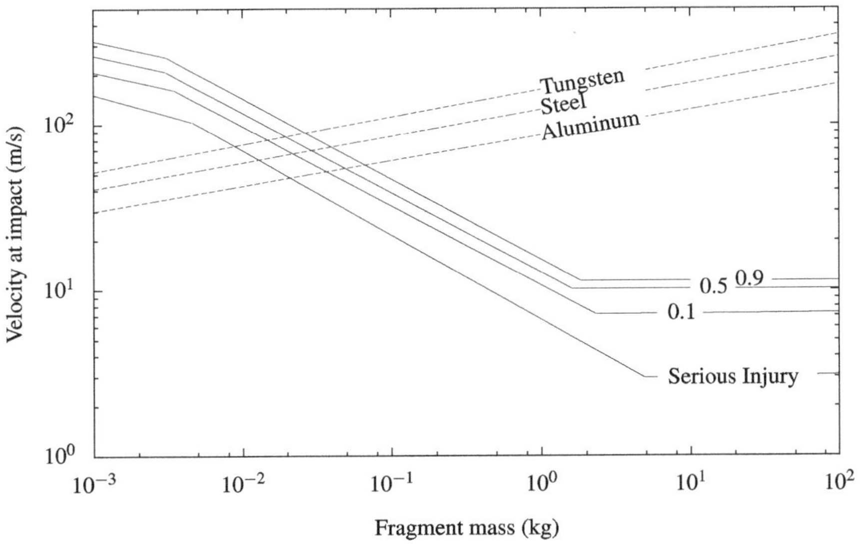

Fig. A.2 Feinstein's probability of fatality after an impact to the head as a function of fragment velocity and weight. For comparison purposes the terminal velocity of a sphere made from different materials and dropped from 10,000 ft is shown with dashed lines. Source [2]

To relate the probability of fatality with the severity of the impact, Feinstein et al used the following log-normal distribution:

$$P(\text{fatality}|K) = \int_0^K \frac{1}{x\beta_i\sqrt{2\pi}} e^{-\frac{(\ln x - \ln \alpha_i)^2}{2\beta_i^2}} dx \tag{A.2}$$

where K is the fragment kinetic energy and α_i, β_i are the scale and shape parameters of the log-normal. The scale and shape parameters depend on the body part that was impacted.

Having different curves for different body parts presents a significant problem, since in general it is very difficult to know a priori which body part is going to be struck by the fragment [2]. To overcome this limitation, an "average" curve was constructed. This curve was derived by taking into account the average area presented by each body part with respect to the total area exposed to an impact [2]. As shown in Table A.3, this area depends on the position of the body (prone, sitting or standing), resulting in three different curves for each body position.

Table A.3 The percent of area presented by the head, thorax, abdomen and limbs for different body positions.

Body Position	Total Area Presented	Head	Thorax	Abdomen and Limbs
Standing	$0.093\,m^2$	0.29	0.43	0.28
Sitting	$0.269\,m^2$	0.10	0.14	0.76
Prone	$0.474\,m^2$	0.09	0.23	0.68

The log-normal curves relating kinetic energy to probability of fatality with respect to body position can also be averaged to produce a single curve. The averaging is based on the assumption of equal probabilities of people being in any of the aforementioned positions [2]. This average curve is also proposed by the Range Commanders Council as a possible measure for estimating the probability of fatality following a UAS ground impact [13]. The aforementioned log-normal curves, as well as the averaged one, are shown in Fig. A.3.

Similarly to Feinstein, Janser presented in 1982 [7] what he named basic lethality. This lethality was a function of the kinetic energy of the fragment. This basic lethality was then aggregated for a debris shower following a ground detonation. The analysis also took into account soil and fragment characteristics, as well as, the body region each fragment impacted. The latter was based on the calculation of the probability of impacting a body region as a function of impact angle and the ratio of the exposed area of that region to that of the whole body.

Specifically, he presented four curves of basic lethality, for the thorax, abdomen, head and limbs body regions respectively. These curves are shown in Fig. A.4. It should be noted, that based on the curves presented in [7], the model used is not exactly log-normal, but neither the model nor the way its parameters were derived, were provided.

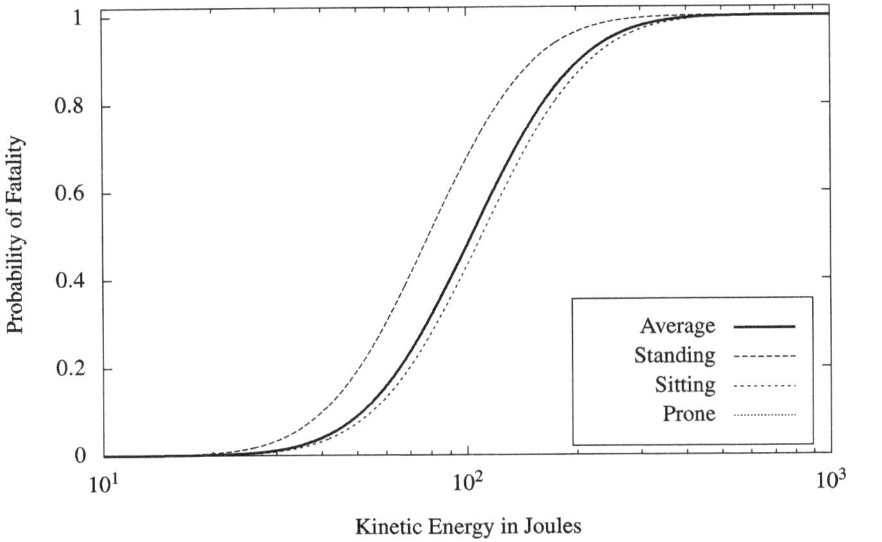

Fig. A.3 The log-normal curves of probability of fatality with respect to kinetic energy for different body positions as derived by Feinstein et al. Also provided is the averaged curve. It should be noted, that the prone and the averaged curve are almost identical. Source [2]

Figure A.4 also provides a comparison between Janser's and Feinstein's curves. It is evident that, although there is rough agreement on the head and abdomen regions, Feinstein is significantly more conservative regarding trauma to the thorax and limbs. Nevertheless, since the data based on which Janser's curves were derived are unknown, it is not possible to determine the reasons for this discrepancy.

A.3.3 The Viscous Criterion (VC) and the Blunt Criterion (BC)

Many experiments in the past have attempted to correlate injury to the thorax and abdomen with force. Nevertheless, this approach presented problems, since impact force is not sufficient to model all the injury mechanisms adequately [17]. During an impact to the thorax, the body is compressed. This puts pressure on internal organs and may even result in the collapse of the rib cage and possible organ puncture [17]. Additionally, especially for high velocity impacts, one must also consider how fast or slow a body part is loaded during an impact, since each organ has different viscous tolerance [17]. Viscous tolerance refers to how fast a body part may absorb energy without failing [17]. The viscous response or Viscous Criterion (VC) of a body is calculated as the product of velocity (V) and compression (C):

$$VC = V \cdot C \qquad (A.3)$$

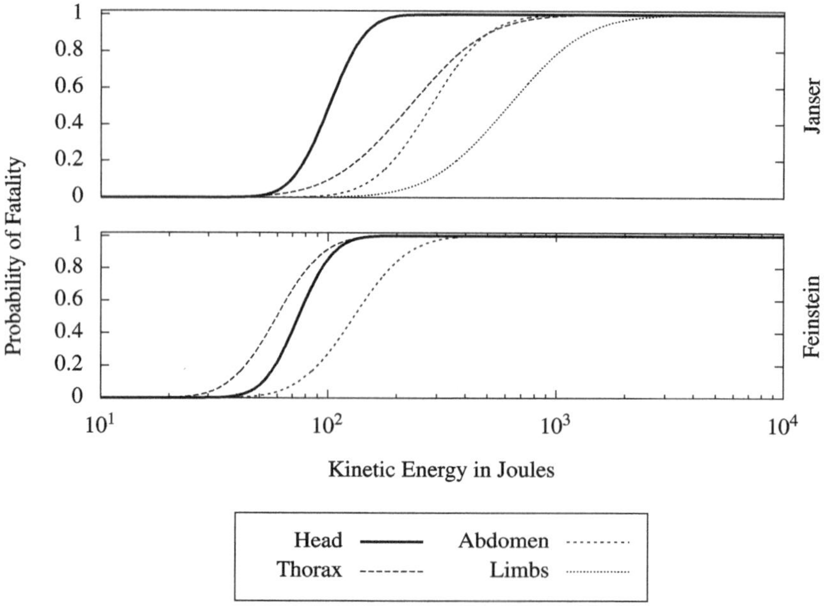

Fig. A.4 The log-normal curves of probability of fatality with respect to kinetic energy for different body positions as derived by Janser in [7]. Feinstein's curves also provided for comparison purposes. It should be noted that Feinstein considered injuries to the abdomen and limbs together.

The Viscous Criterion has been used particularly in the automotive industry, since a number of studies have found good correlation between VC and injury severity. Selected results from some of these studies are presented in [8]. In addition to the Viscous Criterion, the Blunt Criterion (BC) is a newer criterion proposed for use by the US Department of Defense. Its purpose is to assess the risk of blunt force trauma from projectiles and blunt weapons [16]. Similarly to VC, the BC is also based on the energy absorbed during impact and is defined as [1, 16]:

$$\mathrm{BC} = \ln\left(\frac{E}{W^{1/4}TD}\right) \tag{A.4}$$

where E is the kinetic energy at impact, D is the effective diameter of the projectile, W is the mass of the person impacted and T is the thickness of the body wall at the point of impact. The D parameter refers to the surface of the projectile that is in contact with the body. To estimate it, the curvature of the body wall when compressed may also need to be considered. Nevertheless, the effect of curvature on D and as a consequence BC has been found to be relatively small [16].

In the case of larger objects, the impacting projectile does not expend all its kinetic energy during impact. As a result, instead of using its kinetic energy prior to impact, an effective energy may be used instead. The latter is calculated by [16]:

$$E = \frac{1}{2} m_1 v_1^2 \left(1 - \frac{m_1}{m_1 + m_2}\right) \tag{A.5}$$

where m_1 and v_1 correspond to the mass and velocity of the projectile, while m_2 refers to the effective mass of the body part that was struck. The value of m_2 can be estimated as a percentage of the mass of the entire body.

Based on a number of experiments from previous researchers, Sturdivan et al [16] performed logistic fits that mapped the BC to the probability of sustaining injuries at different AIS levels. These fits are shown in Fig. A.5 for level 3 and 6 injuries, which correspond to the casualty and fatality characterizations of [14]. Statistical analysis has shown that BC is a better predictor of injury severity than kinetic energy, despite the fact that the former depends on the latter [1]. A significant correlation between BC and VC has also been found [16].

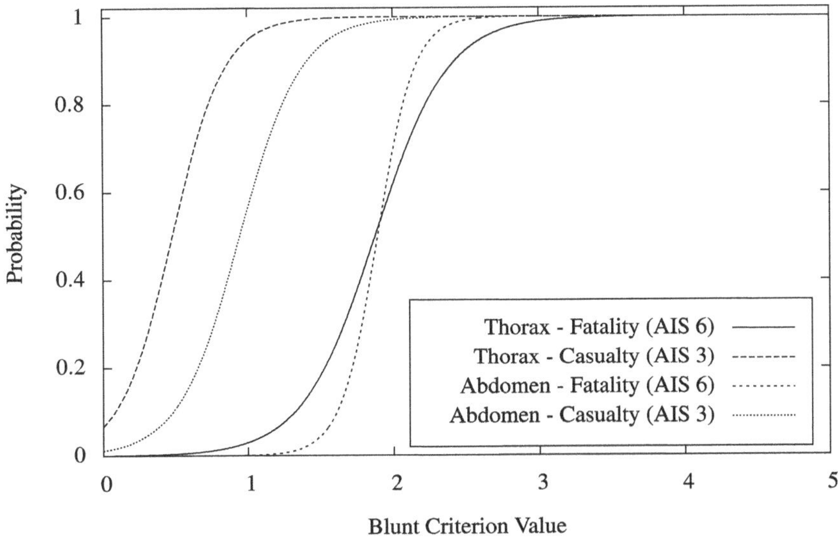

Fig. A.5 The log-normal curves of AIS level 3 and 6 injuries to the thorax and abdomen with respect to the blunt criterion [16]. In [16] the abdomen was divided into five regions, only the most sensitive is shown here.

Both VC and BC are good predictors of the severity of blunt injuries. Nevertheless, BC's dependence on impact characteristics that are easier to estimate, make it more attractive for use in a number of scenarios [16]. Finally, it should be noted that for very slow objects, neither BC nor VC are applicable. This is because they don't take into account the rate at which energy is absorbed by the body [16]. In this case, criteria designed for crushing injuries may be more suitable.

A.3.4 Penetrating Injuries

Penetrating injuries typically occur with fragments of small contact area and high velocity, so that they can penetrate the clothing and skin layers [14]. In the case of UAS, this injury mechanism would require significant fragmentation. This could occur from fragmentation after a UAS ground impact at high velocity, from helicopter rotor strike, an explosion or other similar scenario.

In 1984, Neades and Rudolph presented a survey of the then state of the art on wounding mechanisms and vulnerability modeling [11]. Due to the nature of the problem they were interested in, the focus was on low mass/high velocity projectiles. Although they did consider non-penetrating injuries, they concluded that due to the dearth of experimental data at the time, only provisional models of questionable accuracy could be derived [11].

With respect to penetrating injuries, Neades and Rudolph presented an earlier study of Lewis et al [9]. This study related the probability of full skin penetration with the ratio of the fragment's kinetic energy to it's area using the Walker-Duncan criterion [18]:

$$P(\text{penetration}) = \frac{1}{1+e^{-\alpha-\beta\ln\left(mv^2/A\right)}} \tag{A.6}$$

where α, β are fitting constants and A is the presented area of the fragment. The curves for different levels of clothing protection are presented in Fig. A.6.

Use of kinetic energy may seem counter-intuitive since the fragment typically retains a significant percent of its kinetic energy during penetration. Nevertheless, results have shown that it remains a good predictor of significant injury [14].

References

1. Bir C, Viano DC (2004) Design and injury assessment criteria for blunt ballistic impacts. Journal of Trauma 57(6):1218–1224
2. Cole JK, Young LW, Jordan-Culler T (1997) Hazards of falling debris to people, aircraft, and watercraft. Sandia Report SAND97-0805, Sandia National Laboratories
3. Feinstein DI, Haugel WF, Kardatzke ML, Weinstock A (1968) Personnel casualty study. Tech. Rep. Project No. J 6067, Illinois Institute of Technology Research Institute
4. Haber JM, Linn AM (2005) Practical models of human vulnerability to impacting debris. In: Proc. First IAASS Conference: "Space Safety, a New Beginning", Nice, France, ESA SP-599, pp 543–548
5. Henn HW (1998) Crash tests and the head injury criterion. Teaching Mathematics and its Applications 17(4):162–170
6. Insurance Institute for Highway Safety (2009) Frontal offset crashworthiness evaluation: Guidelines for rating injury measures

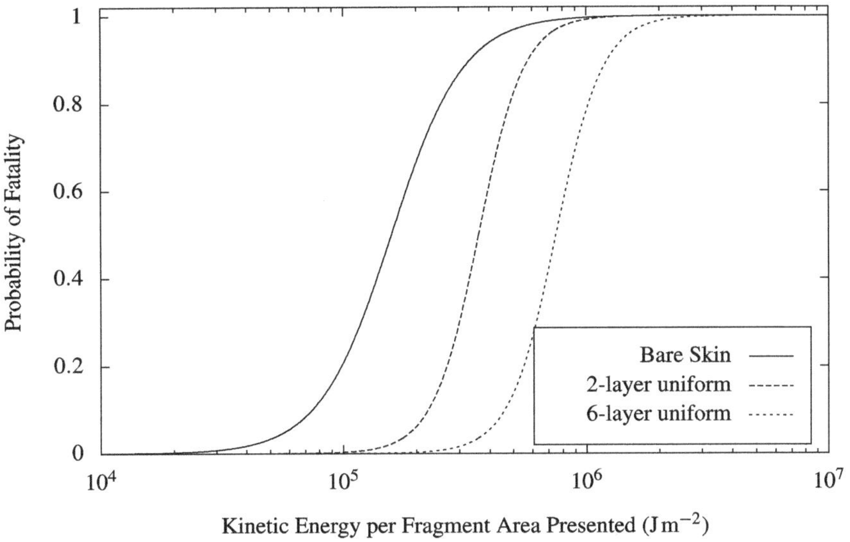

Fig. A.6 The probability of skin penetration with respect to the ratio of fragment kinetic energy over it's presented area as determined by Lewis et al [9]. Source [11]

7. Janser PW (1982) Lethality of unprotected persons due to debris and fragments. In: Proc. 20th Explosives Safety Seminar, Norfolk, VA, USA, pp 1513–1529
8. Johannsen H, Schindler V (2006) Review of the abdomen injury criteria. Integrated Project on Advanced Protection Systems (FP6-PLT-506503) AP-SP51-0039-B, Institute National de Reserche sur les Transports et leur Securité (INRETS)
9. Lewis JH, Coon PA, R CV, Sturdivan LM (1978) An empirical/mathematical model to estimate the probability of skin penetration by various projectiles. ARCSL-TR-78004, U.S. Army Armament Research & Development Command, Chemical Systems Laboratory, Aberdeen Proving Ground, MD
10. Marjoux D, Baumgartner D, Deck C, Willinger R (2008) Head injury prediction capability of the hic, hip, simon and ulp criteria. Accident Analysis and Prevention 40(3):1135–1148
11. Neades DN, Rudolph RR (1984) An examination of injury criteria for potential application to explosive safety studies. In: Minutes of the 21st Explosives Safety Seminar, Houston, TX, USA, vol 2, pp 1343–1354
12. Newman JA, Shewchenko N, Welbourne E (2000) A proposed new biomechanical head injury assessment function - the maximum power index. Stapp Car Crash Journal 44:215–247
13. Range Safety Group, Range Commanders Council (1999) Range safety criteria for unmanned air vehicles - rationale and methodology supplement. Supplement to document 323-99

14. Range Safety Group, Range Commanders Council (2007) Common risk criteria standards for national test ranges: Supplement. Supplement to document 321-07
15. Shuaeiba FM, Hamoudaa AMS, Umara RSR, Hamdana MM, Hashmi MSJ (2002) Motorcycle helmet: Part I. biomechanics and computational issues. Journal of Materials Processing Technology 123(3):406–421
16. Sturdivan LM, Viano DC, Champion HR (2004) Analysis of injury criteria to assess chest and abdominal injury risks in blunt and ballistic impacts. Journal of Trauma 56(3):651–663
17. Viano DC (1988) Cause and control of automotive trauma. Bulletin of the New York Academy of Medicine 64(5):376–421
18. Walker SH, Duncan DB (1967) Estimation of the probability of an event as a function of several independent variables. Biometrica 54:167–179

Appendix B
Ground fatality probability model sensitivity analysis

"That is what we meant by science. That both question and answer are tied up with uncertainty, and that they are painful. But that there is no way around them. And that you hide nothing; instead, everything is brought out into the open."

Peter Høeg
(Borderliners, 1995)

B.1 Background

In Sect. 5.3.1, the following model was proposed to calculate the probability of fatality given a ground impact:

$$P(\text{fatality}|\text{exposure}) = \frac{1-k}{1-2k+\sqrt{\frac{\alpha}{\beta}}\left[\frac{\beta}{E_{imp}}\right]^{\frac{3}{p_s}}} \tag{B.1}$$

where $k = \min\left[1, \left(\frac{\beta}{E_{imp}}\right)^{\frac{3}{p_s}}\right]$

The model presented in (B.1) is an improvement of the model presented in the first edition of this book, which is given by the following equation:

$$P(\text{fatality}|\text{exposure}) = \frac{1}{1+\sqrt{\frac{\alpha}{\beta}}\left[\frac{\beta}{E_{imp}}\right]^{\frac{3}{p_s}}} \tag{B.2}$$

In this Appendix a sensitivity analysis will be carried out to evaluate the effect of perturbations in the inputs and parameters of both models. Typically, such an analysis assumes that model inputs and parameters are random variables. It is then possible to determine the joint distribution of these random variables and consequently the distribution of the model as a function of the characteristics (i.e. the parameters of the probability distribution function) of the random variables. Alternatively, each parameter and input can be evaluated individually, considering the values of the rest fixed and known.

Assuming a function $g(x)$ of a random variable x with a distribution function $f(x)$, then the expected value of $g(x)$ is given by [2]:

$$E\left[g(x)\right] = \int_{-\infty}^{+\infty} g(x)f(x)\mathrm{d}x \tag{B.3}$$

K. Dalamagkidis et al., *On Integrating Unmanned Aircraft Systems into the National Airspace System*, Intelligent Systems, Control and Automation: Science and Engineering 54, DOI 10.1007/978-94-007-2479-2,

Using the expectation of $g(x)$, it is possible to calculate its bias as follows:

$$Bias\left[g(x)\right] = g(\mu) - E\left[g(x)\right] \tag{B.4}$$

where μ is the true value of x. Similarly the variance of $g(x)$ is given by:

$$Var\left[g(x)\right] = E\left[g^2(x)\right] - E\left[g(x)\right]^2 \tag{B.5}$$

Nevertheless, the functional form of the models of (B.1) and (B.2) makes an analytical derivation of the corresponding expectation $E[P(\text{fatality}|\text{exposure})]$ difficult. This is true regardless of the assumed probability distribution functions for the parameters and input and whether they are considered separately or together. An alternative approach is to consider small perturbations of each parameter and input and calculate their impact on the model. This method, known as infinitesimal perturbation analysis is typically carried out by taking a Taylor series approximation of the function $g(x)$. Assuming that $f(x)$ is negligible outside an interval $(\eta - \varepsilon, \eta + \varepsilon)$, $E\left[g(x)\right]$ can be approximated as [2]:

$$E\left[g(x)\right] \simeq g(\eta) + g''(\eta)\frac{\sigma^2}{2} \tag{B.6}$$

with the variance being:

$$Var\left[g(x)\right] \simeq \left[g'(\eta)\right]^2 \sigma^2 \tag{B.7}$$

Holtzman [1] proposed the use of differences instead of derivatives. Although differences provide less accuracy for approximation purposes, Holtzman asserts that the use of differences is advantageous for sensitivity analysis, since derivatives only allow for very small changes without allowing one to define how small. As a result, he was able to prove that using differences for perturbation analysis affords higher accuracy, even for high values of standard deviation.

Using a second order central difference approximation instead of the Taylor expansion, the expected value of $g(x)$ can be approximated from [1], as:

$$E\left[g(x)\right] \simeq g(\mu) + \frac{1}{2}\frac{g(\mu+h) - 2g(\mu) + g(\mu-h)}{h^2}\sigma^2 \tag{B.8}$$

where μ and σ are the mean and variance of the x random variable. The bias and variance of $g(x)$ can then be calculated from (B.6) and (B.7).

B.2 Analysis

In (B.8) a value for h needs to be chosen. For this analysis the value of $\sqrt{3}\sigma_i$ used in [1] is selected. For each random variable, (B.8) will be given by:

$$E\left[g(x_i|x_j, j \neq i)\right] \simeq g(\mu_i) + \frac{g(\mu_i + \sqrt{3}\sigma_i) - 2g(\mu_i) + g(\mu_i - \sqrt{3}\sigma_i)}{6} \tag{B.9}$$

The perturbation analysis will be carried out for the following three random variables: $x_1 = \log_{10} E_{imp}$, $x_2 = \log_{10} \alpha$ and $x_3 = p_s$. Each of these variables is considered to be normally distributed with mean μ_i and variance σ_i. Parameter β will be considered fixed, with a value of 34 J. After substituting the parameters in (B.1), the function to be analyzed is given by:

$$P(\text{fatality}|x_1, x_2, x_3) = \frac{1}{1 - 2 \cdot 10^{\frac{3\log 34 - 3x_1}{x_3}} + 10^{\frac{x_2 - \log 34}{2} + \frac{3\log 34 - 3x_1}{x_3}}} \tag{B.10}$$

Substituting (B.10) in (B.6) and taking into account (B.9), the model bias is obtained by:

$$Bias\left[P(\text{fatality})\right] = \simeq \frac{2g(\mu_i) - g(\mu_i + \sqrt{3}\sigma_i) - g(\mu_i - \sqrt{3}\sigma_i)}{6} \tag{B.11}$$

For the results presented in the following section, it is assumed that each random variable is $x_i \sim \mathcal{N}(\mu_1, 0.152)$. This corresponds to an 80% certainty that $\mu_1 - 0.5 < x_1 < \mu_1 + 0.5$. In the case of the estimated kinetic energy at impact and the α parameter, this means that we are only 80% confident that the value used is within an order of magnitude from the actual. Similarly for the p_s, it means that there is 80% confidence that the actual value is within ± 0.5 of the actual.

The resulting $Bias\left[P(\text{fatality})\right]$ corresponds to the bias of the probability of fatality considering that one of the parameters/inputs is a normally distributed random variable. As a result, a bias of 0.01 means that on average the estimated probability of fatality will be 1% higher. To assist the reader in evaluating the variance results, Table B.1 is provided, which gives the confidence intervals of the probability of fatality for different values of variance and required confidence levels.

	Confidence Level			
Variance	80%	90%	95%	99%
10^{-1}	±0.405	±0.520	±0.620	±0.815
10^{-2}	±0.128	±0.164	±0.196	±0.258
10^{-3}	±0.041	±0.052	±0.062	±0.081
10^{-4}	±0.013	±0.016	±0.020	±0.026
10^{-5}	±0.004	±0.005	±0.006	±0.008

Table B.1 The confidence interval of the probability of fatality given its variance and a required confidence level.

B.2.1 Kinetic Energy at Impact Results

When the kinetic energy at impact is considered to be a random variable, the bias of the probability of fatality is presented in Fig. B.1 and the corresponding variance in Fig. B.2.

It is obvious that there is significant bias for low kinetic energies, low α and low sheltering factors, especially for $p_s < 3$ and $E_{imp} < 1$ kJ. This occurs due to the fact that under those conditions, the model behaves like a threshold function and a small variation can lead to a big change in the output. This region is also where the more significant difference between the two models can be found, for the same reason. This effect quickly disappears for higher sheltering factors and kinetic energies. The variance can be significant for low values of p_s, but only when combined with low kinetic energies $E_{imp} < 1$ kJ. The parameter α has little effect away from the 34 J limit on both the bias and the variance magnitudes.

In general the model is behaves well with respect to perturbations of the E_{imp}, considering that the input variance was considerable.

B.2.2 Parameter α Results

For α, the bias is presented in Fig. B.3 and the variance in Fig. B.4.

Some bias is exhibited for above average sheltering factors, and especially for low kinetic energies. Nevertheless, in all cases the bias can be considered negligible since it didn't exceed 0.01. Under the same conditions, some variance was also exhibited, although it only approached values up to 0.015.

B.2.3 Sheltering Factor Results

The fatality probability bias attributed to the sheltering factor being assumed to be a random variable is presented in Fig. B.5. The corresponding variance is presented in Fig. B.6.

High α in combination with low sheltering factor values gave the worst results in terms of bias, reaching up to 0.05. The effect was however tempered as kinetic energy increased. The variance peeked for certain combinations of p_s and E_{imp}. It is considerable for low kinetic energies and high α, exceeding in certain cases the value of 0.02. Nevertheless, considering the input variance it can be considered acceptable.

B.3 Discussion

Although the variance chosen for each random variable was significant, the model was in general well behaved, with perturbations generally resulting in small biases and variance. An exception to this was the combination of small kinetic energy and low value of α and/or p_s. This is expected because in this case the model curve is steep and parameter and input perturbations can result in significant changes in the model output.

Nevertheless, it should be noted that this model is proposed for the calculation of an order of magnitude for UAS T_{GI} and not an exact value. As a result the biases exhibited will only have a minimal effect on the final outcome sought.

The reader is also cautioned that this does not mean that the model is accurate with respect to reality. A thorough model validation would require a wealth of experimental results, some of which are difficult if not impossible to obtain.

References

1. Holtzman JM (1989) On using perturbation analysis to do sensitivity analysis: derivatives vs differences. In: Proc. 28th IEEE Conference on Decision and Control (CDC'89), Tampa, FL, USA
2. Papoulis A, Pillai SU (2002) Probability, Random Variables, and Stochastic Processes, 4th edn. Mc-Graw Hill

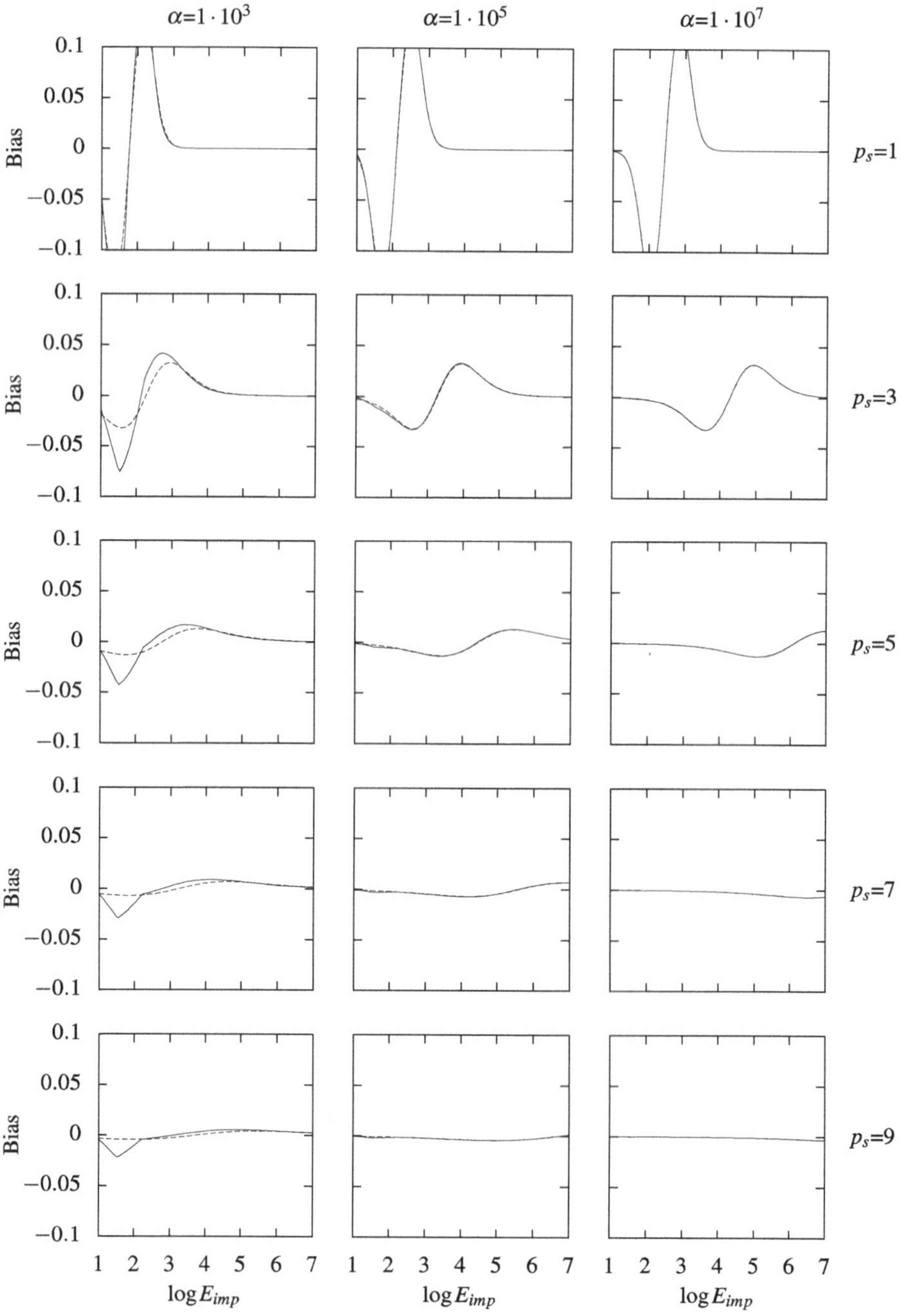

Fig. B.1 The bias calculated for perturbations of the x_1 random variable, corresponding to the $log_{10}E_{imp}$ as a function of the actual value of x_1. The results are given for different values of α and p_s. The dashed lines correspond to the model without the correction factor k.

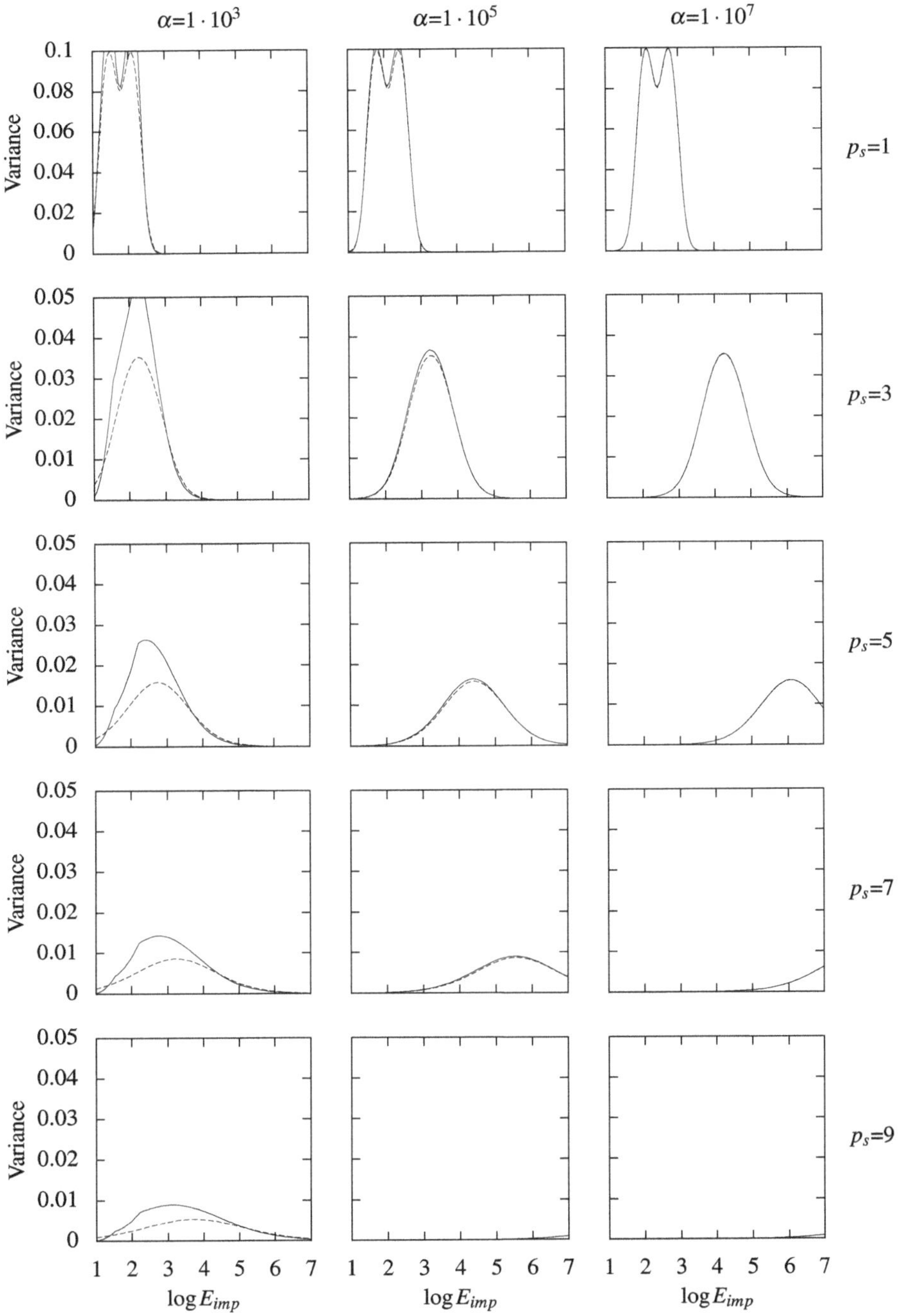

Fig. B.2 The variance calculated for perturbations of the x_1 random variable, corresponding to the $log_{10}E_{imp}$ as a function of the actual value of x_1. The results are given for different values of α and p_s. The dashed lines correspond to the model without the correction factor k.

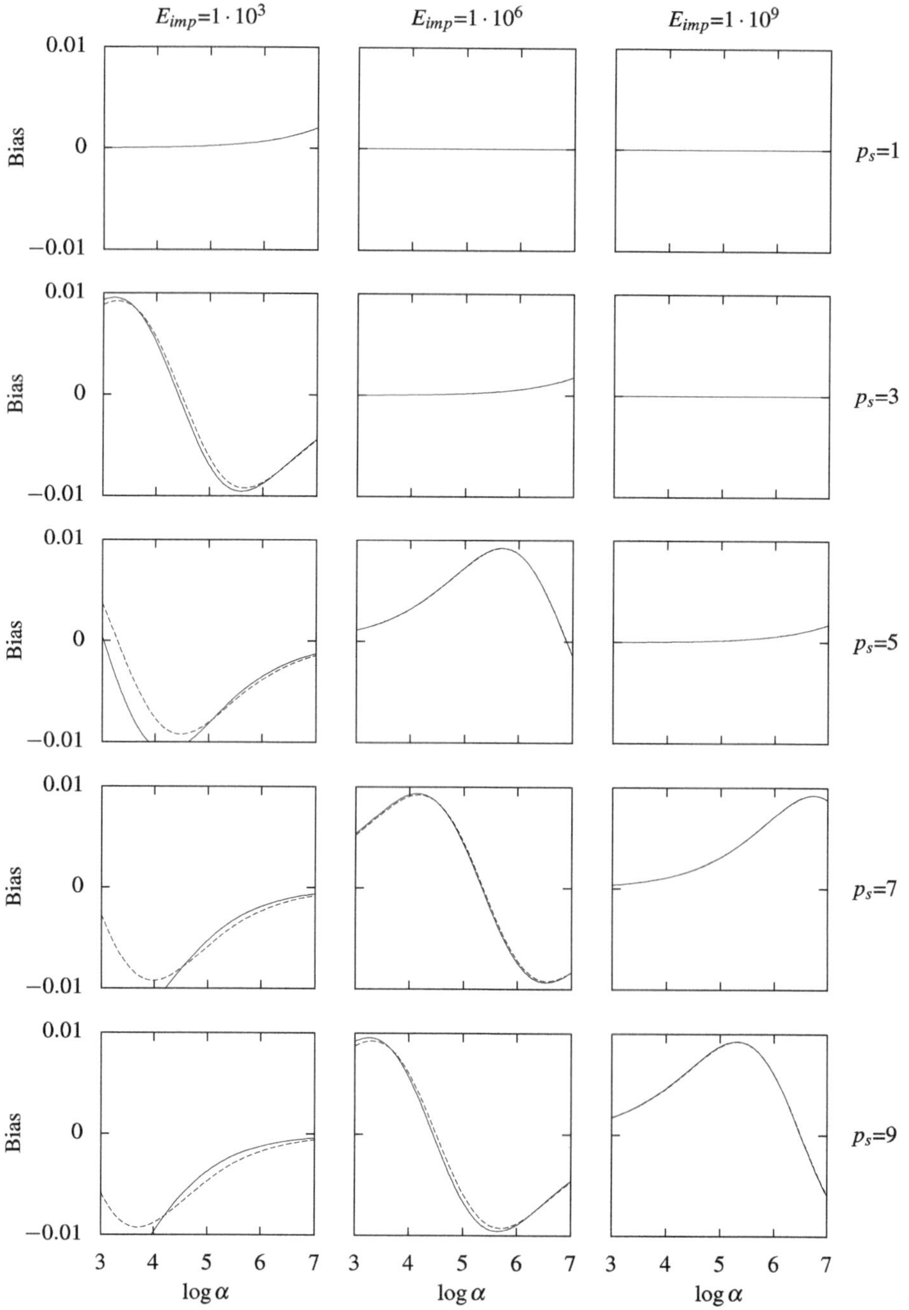

Fig. B.3 The bias calculated for perturbations of the x_2 random variable, corresponding to the $log_{10}\alpha$ as a function of the actual value of x_2. The results are given for different values of E_{imp} and p_s. The dashed lines correspond to the model without the correction factor k.

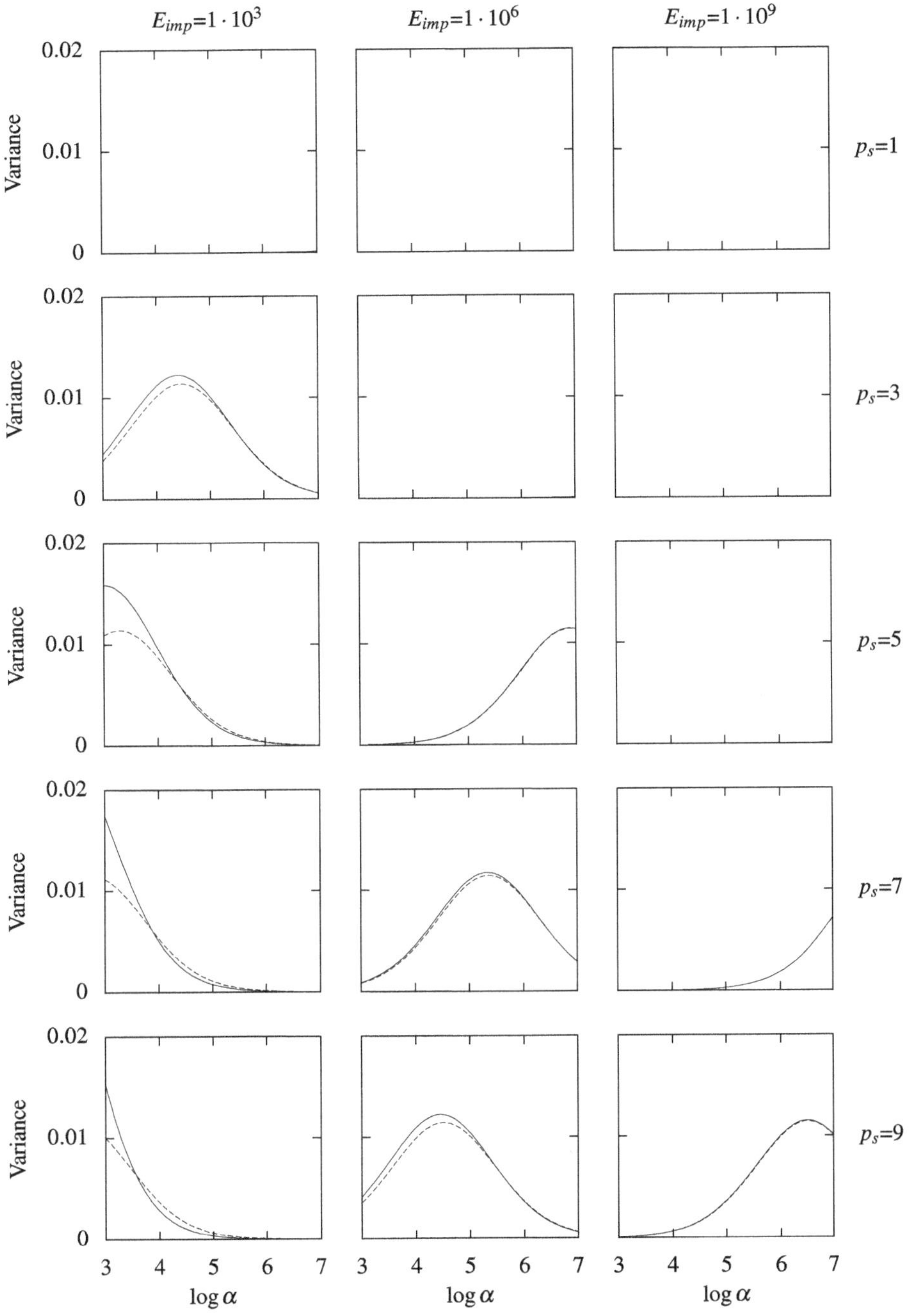

Fig. B.4 The variance calculated for perturbations of the x_2 random variable, corresponding to the $log_{10}\alpha$ as a function of the actual value of x_2. The results are given for different values of E_{imp} and p_s. The dashed lines correspond to the model without the correction factor k.

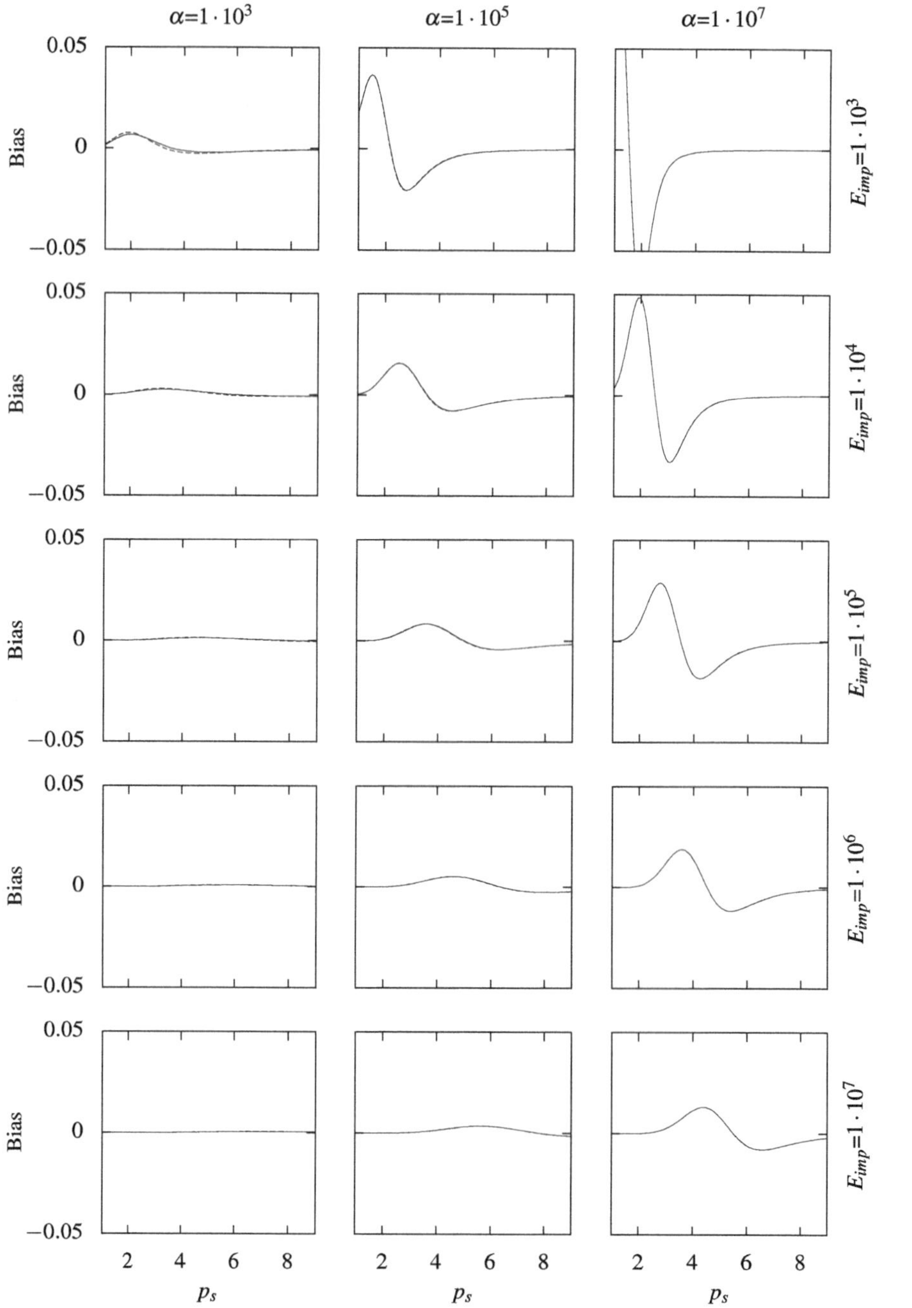

Fig. B.5 The bias calculated for perturbations of the x_3 random variable, corresponding to the sheltering factor p_s as a function of the actual value of x_3. The results are given for different values of α and E_{imp}. The dashed lines correspond to the model without the correction factor k.

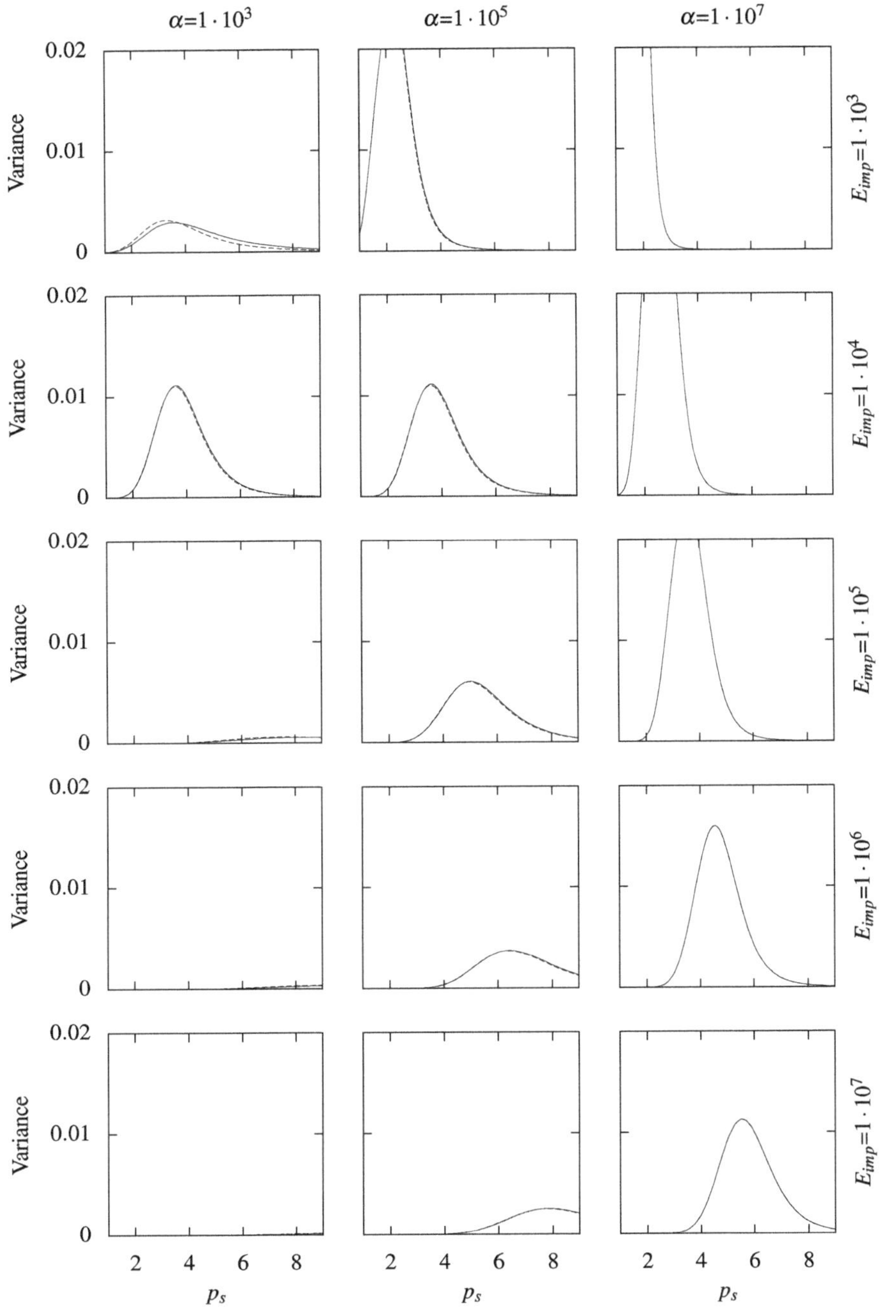

Fig. B.6 The variance calculated for perturbations of the x_3 random variable, corresponding to the sheltering factor p_s as a function of the actual value of x_3. The results are given for different values of α and E_{imp}. The dashed lines correspond to the model without the correction factor k.

Appendix C
UAS Reference

This Appendix contains a reference table of Unmanned Aircraft Systems in development, production or currently decommissioned. The data were kindly provided by UVS-International and are available in the UVS-International UAS Yearbook 2008/2009.

In the maximum speed column, numbers in parentheses indicate cruising speed. In the payload column, wherever the weight is provided in parentheses, it also includes fuel weight. Additionally, the (i) and (e) indicators refer to internal and external storage, respectively.

Key for the class column:

CC	Civil Commercial
DP	Dual Purpose (civil/military)
DV	Developmental Vehicle
M	Military
RV	Research Vehicle

Key for the status column:

DC	Development continuing
DS	Ordered as test/demo system
ES	Ordered/Entering service
IS	In inventory and/or in service
MR	Developed & market ready
ND	No longer in production or development
POC	Proof-of-concept, demonstrator

Key for the airframe column:

FLW	Flapping wing
FW	Fixed wing
LtA	Lighter than air
Prf	Parafoil
RW	Rotary wing
SRW	Shrouded rotary wing
TB	Tilt body
TR	Tilt rotor
TW	Tilt wing

K. Dalamagkidis et al., *On Integrating Unmanned Aircraft Systems into the National Airspace System*, Intelligent Systems, Control and Automation: Science and Engineering 54, DOI 10.1007/978-94-007-2479-2,

Country	Producer(s)/Developer(s)	System Designation	Class	Airframe	Status	Max. Speed (km/h)	Endurance (hrs)	Endurance (km)	MTOW (kg)	Payload Capacity (kg)
Argentina	AeroDreams UAV	ADS-101 Strix	CC	FW	DC		4			5
Argentina	AeroDreams UAV	ADS-301 Nancu	CC	FW	DC		1	15		
Argentina	AeroDreams UAV	ADS-401	CC	FW	DC					
Argentina	Air Force	Guardian	M	FW	DC	(120)		50	80	30
Argentina	Air Force	Proyecto UAV Etapa 1	M	FW	DC	(213)	8		300	50
Argentina	Air Force	Proyecto UAV Etapa 2	M	FW	POC		15			30
Argentina	Air Force	Proyecto UAV Etapa 3	M	FW	POC		3 months			100
Argentina	Argentine Army	Lipan M3	M	FW	DC	80	5	40	60	20
Argentina	Nostromo Defensa	Centinela	M	RW	DC	40	1		8	1.4
Argentina	Nostromo Defensa	Cabure	M	FW	DS		1		3.5	
Argentina	Nostromo Defensa	Cabure II	M	FW	POC	90	1	10	3.5	
Argentina	Nostromo Defensa	Yagua E	M	FW	DS,DC		1.5		12	
Argentina	Nostromo Defensa	Yagua C	M	FW	DS,DC		16		12	
Argentina	Nostromo Defensa	Yarara B	M	FW	POC,DC	150	4	20	22	5
Argentina	Nostromo Defensa	Yarara C	M	FW	DC	147	6	50	30	7
Australia	AAI Corp - Aerosonde	Aerosonde Mk III & IV	CC	FW	IS	150+	24+	3000+	15	5
Australia	ADI (Thales subsidiary)	Cybird-2	M	FW	DS	420	1.5			
Australia	ADI (Thales subsidiary)	Jandu	M	FW	DS	350	4+			35+
Australia	ADRO	Pelican Observer	CC	FW	DC	220	1-12		27.2	13.6
Australia	AeroCam Australia	Trainer	CC	FW	IS				2.5	

(Continued)

Country	Producer(s)/Developer(s)	System Designation	Class	Airframe	Status	Max. Speed (km/h)	Endurance (hrs)	Endurance (km)	MTOW (kg)	Payload Capacity (kg)
Australia	AeroCam Australia	Shadow UAV	CC	FW	DC			90	25	
Australia	BAE Systems	Nulka	M, DV	RW	IS				67.5	
Australia	BAE Systems & University of Sydney	Brumby Mk3	RV	FW	POC,DC	185	2		45	7
Australia	BAE Systems & University of Sydney	Kingfisher Mk1	RV	FW	POC,DC	185	3	64-185	60	12
Australia	BAE Systems & University of Sydney	Kingfisher Mk2	RV	FW	POC,DC	185	14		115	30
Australia	Codarra Advanced Systems	Avatar	DP	FW	IS	50	1	10	6	1
Australia	CSIRO	Mantis	DP	RW	DC		0.3		8	0.08
Australia	Entecho	Demipod	DV	SRW	DC	(120)	6		200	100
Australia	Entecho	Mupod	DV	SRW	DC	(120)	2		10	5
Australia	Silverstone, Australia & AUVA, USA	Flamingo	DV	FW	DC				20	
Australia	Sonacom & University of Sydney	Mirli	DP	TB	POC,DC	370	5	1000	300	100
Australia	University of South Australia & Aerospace Sciences Corp.	Tandem Wing	CC	FW	POC,DC					
Australia	UAV Vision	G18 Aeolus	CC	RW	DC	90		1	14	5
Australia	UAV Vision	T21	CC	RW	IS,DC	90		1	35	8
Australia	UAV Vision	T26	CC	RW	DC	90		0.8	45	12
Australia	V-TOL Aerospace	Hammerhead	DV	TR	DC		0.3		9	4
Australia	V-TOL Aerospace	i-copter Phantom	CC	RW	DC	180	2-12		250	135
Australia	V-TOL Aerospace	i-copter Seeker	CC	RW	DC	130	1.5-2		35-40	5-10

(Continued)

Country	Producer(s)/Developer(s)	System Designation	Class	Airframe	Status	Max. Speed (km/h)	Endurance (hrs)	Endurance (km)	MTOW (kg)	Payload Capacity (kg)
Australia	V-TOL Aerospace	Warrigal	CC	FW	DC	110	1	6	5	
Austria	AeroSpy	Quadrocopter	CC	RW	DC		0.3		0.9	0.3
Austria	AeroSpy	unnamed	M, RV	FW	DC		0.75		1	150
Austria	FH Joanneum (Univ. Graz)	JXP-S	CC, DV	FW	DC	(54)			25	3
Austria	KameraDrohne	KamerDrohne HD35	CC	RW	DC				40	
Austria	Schiebel Elektronische Geräte	Camcopter	DP	RW	IS	(90)	6	10	68	25
Austria	Schiebel Elektronische Geräte	S-100	M	RW	IS	220	6	130	200	50
Austria	University of Salzburg	Javiator	CC, RV	RW	DC				5.6	
Belgium	Flying-Cam	FlyingCam	CC	RW	IS	120	0.25	0.35	15	
Belgium	Gatewing	K120	CC	FW	MR	85	0.75	5	1.2	250
Brazil	Incubaero	Embravant	DP	FW	DC		3		40	
Brazil	Gyrofly Innovations	Gyro 500	CC	RW	DC	50	0.3	1.5	1.5	0.2
Brazil	Gyron Systemas Autonomas	Helix	DV	RW	MR,ND					
Brazil	Flight Solutions	FS-01 Watchdog	DV	FW	DC	(190)		70	65	25
Brazil	Flight Solutions	FS-02 AvantVison	DV	FW	DC		1	12	3	1.5
Brazil	Flight Solutions	FS-03 Starcopter	DV	RW	DC	90		250	240	113
Bulgaria	Aviotechnica	Yastreb	M	FW	POC?	180	1.5	50	66	4.5
Bulgaria	Kintex	Niti	DP	FW	MR	160	20		60	10
Canada	Advanced Subsonics	Grasshopper	DP	FW	DC	83	2		8.2	2.3
Canada	Aerial Insight	Ai-extended	CC	FW	MR		1.5		36.8	
Canada	Aerial Insight	Ai-multi	CC	FW	DC		1		3.1	
Canada	Aerial Insight	Ai-solo	CC	FW	MR		0.75		2.7	

(Continued)

Country	Producer(s)/Developer(s)	System Designation	Class	Airframe	Status	Max. Speed (km/h)	Endurance (hrs)	Endurance (km)	MTOW (kg)	Payload Capacity (kg)
Canada	Aerion Labs	Scout	CC	RW	MR	36	0.3	3	1	
Canada	CropCam	CropCam	CC	FW	IS		0.3	3	1.5	
Canada	Draganfly Innovations	DF-SAVS	CC	RW	IS					
Canada	Draganfly Innovations	DF-TSU	CC	RW	IS					
Canada	Draganfly Innovations	DX-PRO	CC	RW	IS					
Canada	Draganfly Innovations	Tango	CC	FW	IS	100	1		4	1.14
Canada	Draganfly Innovations	X6	DP	RW	IS	50			1.5	0.5
Canada	MicroPilot	MP-Trainer	DP	FW	MR					0.9
Canada	MicroPilot	MP-Vision	DP	FW	MR	60	0.9	4	2.7	
Canada	MMIST	SnowGoose	M	Prf	IS	60		450	609	250
Chili	Chilean Air Force Polytechnical Academy	Vantapa X-02	DP, DV	FW	DC	(150)	7		150	
Chili	RMS SA	Mantarraya	M	FW	MR	240	4	100	60	15
China (PR)	Beijng Black Buzzard Aviation Technology	HFT-40A	DV	FW	DC	500		80	45	
China (PR)	Beijng Black Buzzard Aviation Technology	HFT-60A	DV	FW	DC	602		150	90	
China (PR)	Beijing Strong Science & Technology Development	AW 12	M	FW	MR					
China (PR)	Beijing Strong Science & Technology Development	AW 2 Sun Ying	M	FW	MR	40-120	1.3	5	10	
China (PR)	Beijing Strong Science & Technology Development	AW 4 Shark	M	FW	MR					
China (PR)	Beijing University of Aeronautics & Astronautics	Chong Hong	M, DV	FW	POC,DC	800	3	2500	1700	70

(Continued)

Country	Producer(s)/Developer(s)	System Designation	Class	Airframe	Status	Max. Speed (km/h)	Endurance (hrs)	Endurance (km)	MTOW (kg)	Payload Capacity (kg)
China (PR)	Beijing University of Aeronautics & Astronautics	M-22	M, DV	RW	DC?		1.5		50	
China (PR)	Beijing University of Aeronautics & Astronautics	VT-UAV Seagull	M	RW	POC,DC?					
China (PR)	Beijing University of Aeronautics & Astronautics	WZ-5	M	FW	IS	800	3	2500	1700	
China (PR)	CASIC	LT series	M	FW	DC					
China (PR)	Chengdu Aircraft Industry Co. (CAC)	HALE UAV	M, DV	FW	POC,DC					
China (PR)	Chengdu Aircraft Industry Co. (CAC)	Tianyi	M, DV	FW	POC,DC	180	3	100	80	20
China (PR)	China Aviation Industry Co. (AVIC I)	AVIC I Sunshine	M	FW	DC	240	12	230	750	70
China (PR)	Guizhou Aircraft Ind. Corp.	Soar Dragon	M, DV	FW	DC	750		7000	7500	650
China (PR)	Guizhou Aircraft Ind. Corp.	WZ-2000 (formerly WZ-9)	M	FW	IS	800			1700	80
China (PR)	NRIST	I-Z	M	RW	IS	100	1		9	
China (PR)	NRIST	PW-1	?	FW	?	180	6	100	210	30
China (PR)	NRIST	PW-2	?	FW	?	180	7	200	210	30
China (PR)	NRIST	W-30	M	FW	POC?	150	4	10	18	5
China (PR)	NRIST	W-50	M	FW	POC?	180	4-6	100	95	20
China (PR)	NRIST	Z-2	M	RW	MR	108	1		35	10
China (PR)	NRIST	Z-3	M	RW	DC?		4	100	130	30
China (PR)	NUAA	Soar Bird	M	RW	MR	(150)			310	
China (PR)	Shenyang	Aircraft Design	M	FW	DC					

(Continued)

Country	Producer(s)/Developer(s)	System Designation	Class	Airframe	Status	Max. Speed (km/h)	Endurance (hrs)	Endurance (km)	MTOW (kg)	Payload Capacity (kg)
China (PR)	Xi'An ASN Technology Group	ASN-104/105	M	FW	IS	250	2	60/100	140	30/40
China (PR)	Xi'An ASN Technology Group	ASN-105B	M	FW	DC	200	7	150	170	40
China (PR)	Xi'An ASN Technology Group	ASN-15	M	FW	IS	80	1	10	6.5	
China (PR)	Xi'An ASN Technology Group	ASN-206	M	FW	IS	210	4-8	150	222	50
China (PR)	Xi'An ASN Technology Group	ASN-207	M	FW	MR	180	16	600	410-480	30-100
Colombia	EAFIT University	Colibri Project	RV	FW	DC					
Colombia	Efigenia Aerospace Robotics	EJ-2A Fatima	DV	FW	DC					
Colombia	Efigenia Aerospace Robotics	EJ-1B Mozart	DV	RW	DC		1		8	
Colombia	FromSky	Terraco	CC	FW	DC					
Croatia	Defense Research Est.	BL-50	M	FW	DC	110	5		53	
Croatia	Soko	B3	DP, DV	FW	DC	(106)	9		220	
Croatia	Soko	B4	DP, DV	FW	DC	(120)	12		220	
Czech Rep.	VTUL a PVO	Sojka III	M	FW	IS	207	2	100	145	25
Finland	Patria	Mini-UAV	M	FW	MR	120	1	20	3.5	0.5
France	ABS Aerolight	Maxi	DP, DV	Prf	DC	18-65	1		25	15-30
France	ABS Aerolight	Pixy	DP	Prf	IS	60	1		5.6-10.5	
France	Aeroart & Mercury Computer	Aelius 0	CC	FW	POC,DC	160	0.5		75	17
France	Aeroart & Mercury Computer	Aelius 1	CC	FW	POC,DC	180		250	20	30
France	AeroDrones	Aerodrone	DV	RW	DC					
France	Aeroart	A100	DV	FW	DC	160		6	44	30
France	Aeroart	Aves	DV	FW	DC	110	2		24	12
France	Aeroart	H250	DV	FW	DC		8		125	
France	Aeroart	Seagos	DV	FW	DC					
France	Alcore Technologies	Azimut 2001	M, DV	FW	DC	120	1.5	50	9	2.5

(Continued)

Country	Producer(s)/Developer(s)	System Designation	Class	Airframe	Status	Max. Speed (km/h)	Endurance (hrs)	Endurance (km)	MTOW (kg)	Payload Capacity (kg)
France	Alcore Technologies	Biodrone	M, DV	FW	DC	120	1.5	50	12	3.5
France	Alcore Technologies	Chacal 2	M, DV	FW	DC	280	3	700	85	10
France	Alcore Technologies	Drone Futura	M, DV	FW	DC	360	1	300	70	10
France	Alcore Technologies	Easycopter	M, DV	RW	DC		0.15	1	1.6	
France	Alcore Technologies	Epsilon 1	M, DV	FW	DC	60	0.15	1	0.45	0.1
France	Alcore Technologies	Maya	M, DV	FW	DC	108	0.5	50	2.5	0.5
France	Bertin Technologies	Flying Ball	M, DV	SRW	POC,DC	30	0.5		1.5	0.2
France	Bertin Technologies	HoverEye	M, DV	RW	DC		0.15	0.3	3.5	
France	Dassault Aviation	AVE C (Petit Duc C)	DV	FW	POC,ND				60	
France	Dassault Aviation	AVE D (Petit Duc)	DV	FW	POC,ND	0.5		150	60	
France	DSTU (Dassault Aviation & Sagem)	SlowFast UAV	DV	FW	POC,ND				60	
France	EADS Military Aircraft	EuroMALE	M	FW	ND		24	1500		450
France	EADS Military Aircraft & Dyn'Aéro (airframe)	Surveyor 2500	DP	FW	MR	360	12	185	450-750	100
France	EADS Military Aircraft & Eurocopter (airframe)	Orka 1200	M, DV	RW	DC	195	8	185	680	180
France	EADS Military Aircraft & IAI-Malat, Israel	Eagle 1	M	FW	ES,DS	220	30	1700	1200	250
France	EADS Military Aircraft & IAI-Malat, Israel	Eagle 2	M	FW	DC	460	24	2900	3600	450
France	EADS Military Aircraft & SurveyCopter	Scorpio 30	DP	RW	MR	(50)	2	10	38	15

(Continued)

Country	Producer(s)/Developer(s)	System Designation	Class	Airframe	Status	Max. Speed (km/h)	Endurance (hrs)	Endurance (km)	MTOW (kg)	Payload Capacity (kg)
France	EADS Military Aircraft & SurveyCopter	Scorpio 6	DP	RW	IS	(35)	1	10	13	6
France	EADS Military Aircraft & SurveyCopter	Tracker	M	FW	ES	(60)	2+	10	8.2	1
France	EADS Military Aircraft (former CAC Systems; ceased trading 2003)	Fox MLCS AT2	M	FW	POC,ND	180	3	50	90	15
France	ECT Industries & ISNAV	Hetel M	M, DV	RW	POC,DC	200	5	200	550	150
France	EuroMC	Aero-Drone 50	DP, DV	SRW	DC	(36)	0.3		1.5	
France	EuroMC	Aero-Drone 70	DP, DV	SRW	DC	(36)	0.3		1.85	
France	EuroMC	Aero-Drone 120	DP, DV	SRW	DC	(36)	0.3		6	
France	Flying Robots	FR 101	DP	Prf	MR		12	150	600	250
France	Flying Robots	FR A2	DP	FW	DC	120	1.15	10	15	4
France	Flying Robots	FR E1	DP	FW	DC	60	1	10		
France	Gates Technology	GT Aircraft	CC	LtA	POC,DC					
France	Infotron	IT 180-5 TH	DP	RW	DC	90	1.5	5+	15	5
France	Infotron	IT 180-5 EL	DP	RW	POC,DC	90	0.5	5	15	5
France	Lehmann Aviation	LP960	DP	FW	IS	75	0.5	1.5-2	0.9	
France	Lehmann Aviation	LV580	DP	FW	IS	75	0.5	1.5-2	0.75	
France	Novadem	NX110	DP	RW	MR					
France	Novadem	NX110m	M	RW	MR					
France	Novadem	U130	CC	RW	MR					
France	Onéra	Remanta	M, DV	FLW	POC,DC				<1	
France	Pix-Air & AirStar	Soulcam	CC	LtA	MR	20	3	1.5	90	30

(Continued)

Country	Producer(s)/Developer(s)	System Designation	Class	Airframe	Status	Max. Speed (km/h)	Endurance (hrs)	Endurance (km)	MTOW (kg)	Payload Capacity (kg)
France	Polyvionics	Vulcas	DV	FW	POC,DC				20	5
France	PR Automation (based on Vario, Germany)	PY Copter	CC	RW	MR		0.45-1		15.4	4
France	PY Design	O.V.O	DV	LtA	POC,DC	36	10		2	
France	Sagem Défense Sécurité (Safran)	Merlin	DP	FW	DC	75	1+	7+	6	0.8
France	Sagem Défense Sécurité (Safran)	Sperwer	M	FW	IS	240	6+	200	350	50
France	Sagem Défense Sécurité (Safran)	Sperwer B	M	FW	DC	150	12	200	350	100
France	Sagem Défense Sécurité (Safran)	Ugglan	M	FW	IS	220	6		330	
France	Sagem Défense Sécurité (Safran) & Bertin Technologies (airframe)	Odin	M	SRW	DC?	100	0.6	1	3.2	0.2
France	Sagem Défense Sécurité (Safran) & Meggitt Defence Systems, UK (airframe)	Crecerelle	M	FW	POC,ND	240	5	200	145	35
France	Sagem Défense Sécurité (Safran) & Meggitt Defence Systems, UK (airframe)	Crecerelle EW	M	FW	POC,ND	240	5		145	
France	Sagem Défense Sécurité (Safran) & Onéra, France & Stemme, Germany (airframe)	Busard	CC, RV	FW	DC		20	180+	1100	

(Continued)

Country	Producer(s)/Developer(s)	System Designation	Class	Airframe	Status	Max. Speed (km/h)	Endurance (hrs)	Endurance (km)	MTOW (kg)	Payload Capacity (kg)
France	Sirehna & KYU Microdrones, France & PY Automation, France	Cybird	CC	FW	MR	70	1	3	1.5	0.5
France	Sirehna & KYU Microdrones, France & PY Automation, France	Elytre (Elsa)	s CC	FW	DS	70	1	3	1.5	0.5
France	Survey-Copter	Copter 1	DP	RW	IS		0.5			5-7
France	Survey-Copter	Copter 1b	DP	RW	IS	70	0.75	1-8	12-15	5.5
France	Survey-Copter	Copter 4	DP	RW	DC	70	1	1-8	25	10
France	Survey-Copter	DVF-2000	DP	FW	MR	90	1.5	6	7	1
France	Survey-Copter	unnamed	DP	LtA	MR					
France	Tecknisolar-Seni	Bourdon	M, DV	FW	MR	20-60	1		7	
France	Tecknisolar-Seni	Buteo	M, DV	SRW	POC,DC					
France	Tecknisolar-Seni	Coccinelle	M, DV	FW	POC,DC					
France	Tecknisolar-Seni	D.E.R.E.	M	FW	POC,DC	(60)	0.5	1.5	2.5	
France	Tecknisolar-Seni	Eclaireur	M, DV	FW	POC,DC	(120)	1	10	25	
France	Tecknisolar-Seni	Libellule	M, DV	FW	POC,DC				4	
France	Thales	Spy Arrow LW	M, DV	FW	POC		0.5		0.5	
France	Thales	Spy Arrow HC	M, DV	FW	POC		0.5		1.2	
France	Vision du ciel	Cyclope 4.0	CC	FW	IS	25	0.25	1	2.5	1
France	Vision du ciel	I.Z.I 1.0	CC	FW	IS	35	0.25	0.8	1.7	0.4
France	Vision du ciel	Pixy 26-40	CC	Prf	IS	15	1	1		4
France	Vision du ciel	Pixy 29-40	CC	Prf	IS	35	1	1		6
Germany	AirRobot	AR70	DP, DV	RW	POC,DC		0.3	0.5	1	0.2

(Continued)

Country	Producer(s)/Developer(s)	System Designation	Class	Airframe	Status	Max. Speed (km/h)	Endurance (hrs)	Endurance (km)	MTOW (kg)	Payload Capacity (kg)
Germany	AirRobot	AR100	DP, DV	RW	POC,DC		0.3	0.5	1	0.2
Germany	AirRobot	Mikado	DP, DV	RW	POC,DC		0.3+	0.5	1	
Germany	Borjet	CoRex	CC	FW	DC		0.6		5	2.5
Germany	Borjet	FlyEye	CC	FW	DC				3	0.5
Germany	Diehl (see Microdrones, Germany)	SensoCopter	DP,DV	RW	POC,DC		0.3	3	0.9	150
Germany	EADS Military Aircraft Systems	Barrakuda	M, DV	FW	POC,DC				3250	300
Germany	EADS Military Aircraft Systems	D0-MAV	M	FW	MR	50	0.5	1.5	0.5	
Germany	EADS Military Aircraft Systems	Midas	DV	FW	POC,DC	40	0.5		0.2	
Germany	EADS Military Aircraft Systems	QuattroCopter	M, DV	RW	POC,DC	40	0.5		0.5	
Germany	EADS Military Aircraft Systems	Shark	M, DV	RW	POC,DC	160	4		190	60
Germany	EMT	Aladin	M	FW	IS	90	0.5+	5	3.2	
Germany	EMT	FanCopter	M	SRW	POC,DC	50	0.3+	0.5+	1.3	
Germany	EMT	Luna	M	FW	IS	(70)	3+	65	40	3
Germany	EMT	Mikado	M, DV	FW	DC		0.25+	0.5	0.5	
Germany	EMT	X13	M, DV	FW	DC	180	6	200	130	
Germany	EMT (based on NEO S-300, Swiss UAV)	Museco	M	RW	DC		2	100	75	
Germany	Imar Navigation	IFF-4.5	M, DV	RW	DC	300	10		220	
Germany	Mavionics	Carolo C40	M, DV	FW	MR	36	0.25	0.5	0.45	0.03

(Continued)

Country	Producer(s)/Developer(s)	System Designation	Class	Airframe	Status	Max. Speed (km/h)	Endurance (hrs)	Endurance (km)	MTOW (kg)	Payload Capacity (kg)
Germany	Mavionics	Carolo P200	M, DV	FW	MR	(55)	1		4	
Germany	Mavionics	Carolo P330	M, DV	FW	MR	108	1		5	0.4
Germany	Mavionics	Carolo P50	M, DV	FW	MR	90	0.33		0.53	0.05
Germany	Mavionics	Carolo P70	DP	FW	MR	80	0.33		0.55	0.1
Germany	Mavionics	Carolo T140	DP	FW	MR	70	0.75		1.85	0.3
Germany	Mavionics	Carolo T200	DP	FW	MR	90	0.75		4.5	1
Germany	Microdrones	Md4-1000	DP	RW	DC				4.8	1.2
Germany	Microdrones (see Diehl)	Md4-200	DP	RW	DC		0.3	0.5	1.1	0.2
Germany	Rheinmetall Defense Electronics	Fledermaus	M	FW	POC,ND	120	3.5	100	161	35
Germany	Rheinmetall Defense Electronics	Kolibri (Hummingbird)	M	RW	DC				1.6	
Germany	Rheinmetall Defense Electronics	KZO	M	FW	IS	220	3.5+	100	161	35
Germany	Rheinmetall Defense Electronics	Mücke	M	FW	POC,ND	180	5	100	168	
Germany	Rheinmetall Defense Electronics	Opale	DP	FW	MR	360	12	200	1785	190
Germany	Rheinmetall Defense Electronics	Tares (ex-Taifun)	M, DV	FW	DC		4	200	160	50
Germany	Rotrob	Rotrob	CC	RW	IS		2	125		18
Germany	SIM Security & Electronic Systems	SkyEye	M	RW	MR			0.5	0.9	0.2
Germany	Scalecopter	CamClone	DP	RW	DC	160	1.1	20	41.6	10
Germany	TFH Wildau	Bully	CC	FW	DC	160	0.5		17	7

(Continued)

Country	Producer(s)/Developer(s)	System Designation	Class	Airframe	Status	Max. Speed (km/h)	Endurance (hrs)	Endurance (km)	MTOW (kg)	Payload Capacity (kg)
Germany	UAV Services & Systems	MX-Sight	DP	FW	IS,DC	130	0.5	35	6.9	2.2
Germany	UAV Services & Systems	X-Sight	DP	FW	IS,DC	180	3	45	16	6
Greece	BSK Defense	Ideon		FW	DC	120		25		
Greece	BSK Defence	Phaeton		FW	DC	160				
Greece	EADS - 3 Sigma	Nearchos	DV, DP	FW	POC,DC	220	8-12	240	190	51-92
Hungary	HI Aero	Gabbiano	DP	FW	IS		2	15	4.5	
India	ADE Bangalore	Kapothaka	M, DV	FW	DC	180	1.5		125	
India	ADE Bangalore	Nishant	M, DV	FW	IS	185	4.5	100	375	60
India	Kadet Aero Systems	Trogon	M, DV	FW	DC	130	0.75	5	2.25	0.5
India	MKU	Terp	M	FW	DS	138	1.5	15		
India	MKU	Erasmus	M	FW	DC	184	5	64		
India	Speck Systems	BAAZ	M, DV	FW	DC	60	1	10	7	
International	ADE Bangalore, India & IAI, Israel	Gagan	M	?	POC,DS,DC			250		
International	ADE Bangalore, India & IAI, Israel	Pawan	M	?	POC,DS,DC		5	150	120	
International	ADE Bangalore, India & IAI, Israel	Rustam	M	?	POC,DS,DC		24+	300	1100	
International	ADE Bangalore, India & IAI, Israel	Rustom	M	FW	DC	225	24+	250	1800	350
International	Advanced Subsonics, Canada & Spartan Electronics, USA	Sonobuoy Flight Kit	M	FW	DC			185		
International	Airscan Consortium (EC funded)	Airscan	CC	LtA	POC,DS,DC					

(Continued)

Country	Producer(s)/Developer(s)	System Designation	Class	Airframe	Status	Max. Speed (km/h)	Endurance (hrs)	Endurance (km)	MTOW (kg)	Payload Capacity (kg)
International	Alcatel, Belgium & Vito, Belgium & Verhaert, Belgium & QinetiQ, UK	Pegasus	CC	FW	POC,DC				27	
International	Composites Technology, Malaysia & BAE Systems, USA	Eagle 150	DP	FW	IS	246	10	250	648	60
International	Dassault Aviation, France & Alenia Aeronautica, Italy & SAAB, Sweden & EADS Casa, Spain & Hellenic Aerospace Industry, Greece & RUAG Aerospace, Switzerland	Neuron	M, DV	FW	POC,DS,DC				2270	
International	EADS MAS, France - Germany	Advanced UAV	M	FW	DC					
International	EADS MA, France - Germany & Bombardier, Canada	CL-289	M	FW	IS,ND	(740)	0.5	180-200	240	30
International	EADS MAS, France & Selex Galileo (Galileo Avionica), Italy	Carapas (Surveyor 600)	M	FW	POC,DS,DC	972			330	
International	EADS MAS, Germany & Northrop Grumman, USA	EuroHawk	M, DV	FW	DC	555	30	3000	14000	
International	European Consortium	MAVDEM Project	M, DV	RW	POC,DS					
International	Selex Galileo, Italy & General Atomics-AS, USA	Predator-IT	M	FW	IS		24		1020	

(Continued)

Country	Producer(s)/Developer(s)	System Designation	Class	Airframe	Status	Max. Speed (km/h)	Endurance (hrs)	Endurance (km)	MTOW (kg)	Payload Capacity (kg)
International	IAI-Malat, Israel & HAL, India	Chetak	M	RW	DC	180	6		2200	220
International	IAI-Malat, Israel & Sonaca, Belgium	Hunter B	M	FW	IS	200	8	200	727	113
International	IMI, Israel & Brunswick Defence, USA	Delilah	M	FW	IS	800			185	30
International	IMI, Israel & Brunswick Defence, USA	Delilah AR	M	FW	IS	796			185	30
International	IMI, Israel & Brunswick Defence, USA	ITALD	M	FW	IS	925	0.6		172	36.3
International	IMI, Israel & Brunswick Defence, USA	TALD	M	FW	IS	926	0.6		181	36.3
International	Kawada, Japan & Schweizer, USA	RoboCopter 300	CC	RW	MR		1.6		794	(294)
International	Korea Aerospace Ind., South Korea & AAI Corp, USA	Bejo	M	FW	IS,ND	150	3		130	
International	Northrop Grumman, USA & IAI-Malat, Israel	E-Hunter	M	FW	MR	222	30+		998	(306)
International	Northrop Grumman, USA & IAI-Malat, Israel	Hunter II (= Heron)	M	FW	DC	300	30		1497	450
International	Northrop Grumman, USA & IAI-Malat, Israel	MQ-5B Hunter	M	FW	IS	222	21	100-200	816	(226)
International	Northrop Grumman, USA & IAI-Malat, Israel	RQ-5A Hunter	M	FW	IS	204	12	200	726	113
International	NRL, USA & CybAero, Sweden	Apid Vantage	M	RW	POC,DC	90	6		160	

(Continued)

Country	Producer(s)/Developer(s)	System Designation	Class	Airframe	Status	Max. Speed (km/h)	Endurance (hrs)	Endurance (km)	MTOW (kg)	Payload Capacity (kg)
International	Onéra, France & Royal Military Academy, Belgium	Mirador	M, DV	FW	POC,DC				<1	
International	Politecnico Torino, Italy & Euro Consortium	Heliplat	CC	FW	POC,DC	(71)	weeks		7200	
International	SmartFish, Swizterland & DLR, Germany	HyFish	RV	FW	POC,DC	200				
International	Raytheon Systems, USA & IAI-MBT, Israel	Cutlass	M, DV	FW	MR	(185)	6		135	
International	Sagem, France & Bell, USA & Rheinmetall, Germany	Euro Eagle Eye	M	TR	DC	400	5	1480	1350	90-135
International	Teledyne Brown, USA & Rheinmetall DE, Germany	Prospector	M	FW	DC		6		161	35
International	Teledyne Brown, USA & Rheinmetall DE, Germany	Thunder	M	FW	DC					
International	Thales, UK & Elbit Systems, Israel	WatchKeeper 450	M	FW	ES	176	20	200	450	150
International	AF UAV Research & Tech Centre, UAE & CybAero, Sweden	APID 55 - UAE	M	RW	DC					
International	AF UAV Research & Tech Centre, UAE & Schiebel, Austria	Al-Sber	M	RW	DC					
Iran	Aero Pars	R3	DP	FW	DC			5		
Iran	Aero Pars	R4	DP	FW	DC	(50)	0.5-1	10	3.8	
Iran	Aero Pars	R5	DP	FW	DC					

(Continued)

Country	Producer(s)/Developer(s)	System Designation	Class	Airframe	Status	Max. Speed (km/h)	Endurance (hrs)	Endurance (km)	MTOW (kg)	Payload Capacity (kg)
Iran	Aero Pars	R6	DP	FW	DC			10	3.5	
Iran	Akhondzadeh	Rotary Wing	DV	RW	DC					
Iran	Akhondzadeh	Shahin-1	DV	RW	DC	110				
Iran	Amirkabir University of Technology	Electric UAV	DP, DV	FW	POC	(120)				
Iran	Asr-e-Talaie	Alamdar	DP	FW	MR	90	0.75		4	1
Iran	Balzan Daneshgahe Tehran	Feathering Wing MAV	DV	FLW	DC					
Iran	Faraz Asia Technologies	Faraze-2	M	FW	MR	90	0.5		3	
Iran	Farnas	Black Eagle	M	FW	MR	60	0.5		2.75	0.5
Iran	Farnas	Ghasedak	RV	FW	IS				0.35	0.1
Iran	Farnas	Nazer	RV	FW	IS	(65)	0.66			
Iran	Farnas	Orooj-4	RV	FW	POC		8		4.2	0.5
Iran	Farnas	unnamed	M	FW	MR					
Iran	Farnas	Shahin-1	M	RW	IS	50	2		5.8	
Iran	Farnas	Shahin-2	M	RW	IS	50	3		12	
Iran	Farnas	Shahin-8	M	RW	IS	60	3		19	10
Iran	Farnas	Sabokal	DV	FW	POC	55	2	5	1.3	0.35
Iran	Iran Aircraft Manufacturing Industries (HESA)	Ababil A	DP, DV	FW	DC	60	0.5	10	6.5	1
Iran	Iran Aircraft Manufacturing Industries (HESA)	Ababil B	DP, DV	FW	DC	60	0.5	15	6.5	1.5
Iran	Iran Aircraft Manufacturing Industries (HESA)	Ababil C	DP, DV	FW	DC	60	0.5	15	6.5	1.5

(Continued)

Country	Producer(s)/Developer(s)	System Designation	Class	Airframe	Status	Max. Speed (km/h)	Endurance (hrs)	Endurance (km)	MTOW (kg)	Payload Capacity (kg)
Iran	Iran Aircraft Manufacturing Industries (HESA)	Ababil Jet(Hadaf-1)	M	FW	DC	700				
Iran	Iran Aircraft Manufacturing Industries (HESA)	Ababil-II	M	FW	IS	370		100	83	
Iran	Iran Aircraft Manufacturing Industries (HESA)	Ababil-S	M	FW	IS	300		100		
Iran	Iran Aircraft Manufacturing Industries (HESA)	Ababil-T	M	FW	IS			100		
Iran	Mechanics College of Isfahan University	Aria	DP, DV	FW	POC	(120)				
Iran	Qods Aviation Industries	Mohajer-2	M	FW	IS	200	1.5	50	85	16.5
Iran	Qods Aviation Industries	Mohajer-3	M	FW	IS	220	3			
Iran	Qods Aviation Industries	Mohajer-4	M	FW	IS	200	5		143.5	16.5
Iran	Qods Aviation Industries	Mohajer-5	M	FW	IS					
Iran	Qods Aviation Industries	Saeghe-1	M	FW	ND	250	1		60	5
Iran	Qods Aviation Industries	Saeghe-2	M	FW	IS	230	2		75	16.5
Iran	Qods Aviation Industries	Shekarchi	M	FW	IS					
Iran	Qods Aviation Industries	Talash-1	M	FW	ND	90	0.5		0.5	12
Iran	Qods Aviation Industries	Talash-2	M	FW	IS	120	0.5		0.5	11
Iran	Tadbirgaran Samane Energy	Tadbir-1	RV	FLW	POC					
Iran	Zanjan Copter (Tahsini)	SABA Z-2	DV	RW	POC					10
Israel	Aero Design & Development	Heron-ADD	M, DV	FW	ND					
Israel	Aero Design & Development	Hornet	M, DV	RW	ND	80	2	100	260	50
Israel	Aero Design & Development	Skylark-ADD	M, DV	FW	ND	80	1.5	10	4.6	

(Continued)

Country	Producer(s)/Developer(s)	System Designation	Class	Airframe	Status	Max. Speed (km/h)	Endurance (hrs)	Endurance (km)	MTOW (kg)	Payload Capacity (kg)
Israel	Aeronautics Defense Systems	Aerolight	M	FW	MR	180	4	150	40	8
Israel	Aeronautics Defense Systems	Aerosky	DP	FW	IS		5	100	70	18
Israel	Aeronautics Defense Systems	Aerosky 2	DP	FW	DC		5		80	
Israel	Aeronautics Defense Systems	Aerostar	DP	FW	IS	200	14+	200	200	50
Israel	Aeronautics Defense Systems	Dominator	M	FW	DC	278	24		800	400
Israel	Aeronautics Defense Systems	Dominator 2	M	FW	DC					400
Israel	Aeronautics Defense Systems	Orbiter 1	M	FW	MR	138	1.5	15	6.5	1.2
Israel	Aeronautics Defense Systems	Orbiter 2 (2008)	M	FW	MR		2-3	40		
Israel	Aeronautics Defense Systems	Orbiter 3 (2008)	M	FW	MR		3-8			
Israel	Aeronautics Defense Systems	unmanned	M	RW	DC				700	
Israel	BlueBird Aero Systems	Blueye	M	Prf	MR	65	9	50	55	18
Israel	BlueBird Aero Systems	Boomerang	M	FW	MR	100	2	10	5	0.5
Israel	BlueBird Aero Systems	MicroB	DP	FW	MR	45-90	1	10	1	0.450
Israel	BlueBird Aero Systems	Spyeye	M	Prf	MR	30-70	3	30	25	5
Israel	BTA	Canard	M, DV	FW	MR	148	6	50-100	40	
Israel	BTA	Mini Sheddon	M	FW	MR	130	2.5	10-50	27	
Israel	BTA	Sheddon	M	FW	MR	130	6	50-100	40	
Israel	Elbit Systems - Silver Arrow Division	Colibri	M	FW	IS	161	2		36	14
Israel	Elbit Systems - Silver Arrow Division	Hermes 1500	M	FW	MR	240	24+	200	1500	350
Israel	Elbit Systems - Silver Arrow Division	Hermes 180	M	FW	MR	194	10+	150	195	32
Israel	Elbit Systems - Silver Arrow Division	Hermes 450	M	FW	IS	176	20	200	450	150

(Continued)

Country	Producer(s)/Developer(s)	System Designation	Class	Airframe	Status	Max. Speed (km/h)	Endurance (hrs)	Endurance (km)	MTOW (kg)	Payload Capacity (kg)
Israel	Elbit Systems - Silver Arrow Division	Hermes 700	M, DV	FW	POC				700	
Israel	Elbit Systems - Silver Arrow Division	Hermes 900	M	FW	MR				970	300
Israel	Elbit Systems - Silver Arrow Division	Micro-Vee	M	FW	MR	204	5	50	55	8.2
Israel	Elbit Systems - Silver Arrow Division	Seagull	M	FW	POC	74	4	5-10	5.5	
Israel	Elbit Systems - Silver Arrow Division	Skylark I	M	FW	IS	72	2	10	5.5	
Israel	Elbit Systems - Silver Arrow Division	Skylark I LE	M	FW	ES		3	15	6.3	
Israel	Elbit Systems - Silver Arrow Division	Skylark II	M	FW	ES	129	6	50-60	43	10
Israel	Elbit Systems - Silver Arrow Division	Sniper	M	FW	MR	120	6		155	
Israel	Elbit Systems + IAI + Urban Aerospace	Mule	DP	SRW	POC	185	3-5		1725	
Israel	Elbit Systems + IAI + Urban Aerospace	Panda	DP	SRW	POC				14	1.5
Israel	EMIT Aviation	Blue Horizon II	M	FW	IS	220	10	150	180	37
Israel	EMIT Aviation	Butterfly	M	Prf	DC	55	4		450	230
Israel	EMIT Aviation	DragonFly 2000	M	FW	DC	200	14		140	16
Israel	EMIT Aviation	Mercury 3	M	FW	DC	260	30		550	150
Israel	EMIT Aviation	Sting II	M	FW	DC	(144)			130	35

(Continued)

Country	Producer(s)/Developer(s)	System Designation	Class	Airframe	Status	Max. Speed (km/h)	Endurance (hrs)	Endurance (km)	MTOW (kg)	Payload Capacity (kg)
Israel	EMIT Aviation, Israel & Cradance, Singapore	Sparrow N	M	FW	MR	184	4-6	20-120	45	12
Israel	IAI-Malat	Bird Eye 100	M	FW	MR	150	1	5	1.3	0.3
Israel	IAI-Malat	Bird Eye 400	M	FW	MR		1.3	15	4.1	1.2
Israel	IAI-Malat	Bird Eye 500	M	FW	MR	110	1	10	5	0.85
Israel	IAI-Malat	Bird Eye600	M	FW	MR		2	7.5	7.5	1.3
Israel	IAI-Malat	E-Hunter	M	FW	MR	200	25	200	954	114
Israel	IAI-Malat	Eitan	M	FW	POC					
Israel	IAI-Malat	EyeView B	DP	FW	MR	(111)	6		154	
Israel	IAI-Malat	Firebird	DP	FW	MR	(111)	6		154	
Israel	IAI-Malat	Harpy	M	FW	IS	250	2	500	135	
Israel	IAI-Malat	Harop (Harpy-2)	M	FW	DC	(185)		150	125	
Israel	IAI-Malat	Heron (Mahatz)	M	FW	IS	231	40	1000	1100	250
Israel	IAI-Malat	Heron TP (Heron II)	M	FW	ES,DC	450	30		5080	1800
Israel	IAI-Malat	Hunter	M	FW	IS	(184)	12	100/200	727	114
Israel	IAI-Malat	I-See	M	FW	MR		0.75-1	5-10	7.5	0.8
Israel	IAI-Malat	I-View	M	FW	ES		4-6	50-80	165	20-30
Israel	IAI-Malat	I-View 50	M	FW	MR		6	50	65	10
Israel	IAI-Malat	I-View 150	M	FW	MR		7	100	160	20
Israel	IAI-Malat	I-View 250	M	FW	MR		8	150	250	60
Israel	IAI-Malat	Mastiff 3	M	FW	POC,ND	185	7.5	135	138	37
Israel	IAI-Malat	Mosquito 1	M	FW	DC		1	1	0.5	
Israel	IAI-Malat	Mosquito 1.5	M	FW	DC		1	1.6	0.5	0.02

(Continued)

Country	Producer(s)/Developer(s)	System Designation	Class	Airframe	Status	Max. Speed (km/h)	Endurance (hrs)	Endurance (km)	MTOW (kg)	Payload Capacity (kg)
Israel	IAI-Malat	NRUAV	M	RW	DC	180	6		2200	220
Israel	IAI-Malat	Scout	M	FW	POC,ND	176	6	100	159	38
Israel	IAI-Malat	Searcher I	M	FW	POC,ND	194	14	120	372	63
Israel	IAI-Malat	Searcher II	M	FW	IS,ES	230	15	200	426	100
Israel	IAI-Malat	Searcher III	M	FW	DC		20	300	436	120
Israel	IAI-Malat & Technion University	Sun Sailor	DV	FW	POC					
Israel	Innocon	ASIS	DV	FW	POC	235			550	150
Israel	Innocon	Micro Falcon I	M	FW	DC		1		6	
Israel	Innocon	Mini Falcon I	M	FW	MR	185	5	100	100	17
Israel	Innocon	Mini Falcon II	M	FW	DC	200	10	200	160	25
Israel	Israel Military Industries (IMI)	Rainbow	M	FW	MR		2		6	
Israel	Israel Military Industries (IMI)	Samson	M	FW	IS	900		70	181.5	35
Israel	ITL Optronics	Lightener	M	FW	POC	(74)	2-3		5.5	
Israel	Rafael	Skylite	M	FW	MR	126	1	10	6	
Israel	Rafael	SkyLite B	M	FW	MR	100	1.5	10	6	1.2
Israel	Rafael	SkyLite B-LR	M	FW	DC			35	8	
Israel	Steadicopter	STD-5	DP	RW	MR		1-2	5	13.6	5.4
Israel	Topi-Vision	Casper - 200	M	FW	DC	80	1.5	10	2.3	0.24
Israel	Topi-Vision	Casper - 250	M	FW	DC		1.5			
Israel	Topi-Vision	Casper - 350	M	FW	DC	100	4	50	32	5.5
Israel	Topi-Vision	Casper - 420	DP	FW	DC	(110)	4	50	12	
Israel	Topi-Vision (sold to WB Electronis Po)	Sofar	M	FW	MR	70	1	10		3

(Continued)

Country	Producer(s)/Developer(s)	System Designation	Class	Airframe	Status	Max. Speed (km/h)	Endurance (hrs)	Endurance (km)	MTOW (kg)	Payload Capacity (kg)
Italy	A2Tech	RV-02	CC	FW	DC		0.3-0.6	2-10	2	
Italy	A2Tech	RV-160TD	CC	FW	DC			10		10
Italy	Aermatica	Anteos	CC	RW	DC					
Italy	Alenia Aeronautica	Sky-X	M, DV	FW	POC,DC	810	1	185	1200	200
Italy	Alenia Aeronautica	Sky-Y	M	FW	DC	(260)	14	93	1200	150
Italy	Alenia Aeronautica	BlackLynx	M	FW	DC		36		3500	800
Italy	Alenia Aeronautica & Selex Galileo & Thales Alenia Space	Molynx	CC	FW	DC	400	25	3700	3000	600
Italy	Alpi Aviation	Strix-A	M	FW	MR		2.25	12.5	6	1.5
Italy	Alpi Aviation	Strix-B	M	FW	MR					
Italy	CIRA	Castore	RV	FW	POC	> 1000				
Italy	International Aviation Supply	Archimede	DP	FW	MR	(160)	20	320	650	
Italy	International Aviation Supply	Corvo	M	FW	DC		4-8			2.3-6.8
Italy	International Aviation Supply	Gabbiano	M	FW	DC		1		4.5	0.5
Italy	International Aviation Supply	Pico		FW	DC	50	1.5	30	5	
Italy	International Aviation Supply	Pitagora	DV	FW	DC	130	5	150	25	2
Italy	International Aviation Supply	Raffaello	DP	FW	DC	160	16	150	180	35
Italy	International Aviation Supply	Sky Arrow U	M	FW	DC		8		450	100
Italy	MAVTech	MH600-AP	CC	FW	DC	75	1	1.5		0.10
Italy	MAVTech	MH2000	CC	FW	DC	75	1	0.75	11	1
Italy	Nautilus	NRC-Class D	CC	LtA	DC		0.25		9	
Italy	Nautilus	NRC-Class E	CC	LtA	DC		0.25		7	
Italy	Nimbus	NBS 20	CC	LtA	DC	50	2.5		20	4
Italy	Nimbus	NBS 35	CC	LtA	DC	50	6		35	6

(Continued)

Country	Producer(s)/Developer(s)	System Designation	Class	Airframe	Status	Max. Speed (km/h)	Endurance (hrs)	Endurance (km)	MTOW (kg)	Payload Capacity (kg)
Italy	Selex Galileo	Falco	M	FW	IS,ES	216	8-14	150	240-350	70+
Italy	Selex Galileo	Falco Evo	M	FW	MR		18-20		750	120
Italy	Selex Galileo	Mirach 150	M	FW	POC,ND	700	1	250	340	50
Italy	Selex Galileo	Mirach 26	M	FW	POC,ND	220	8		230	35
Italy	Selex Galileo	Nibbio	M, DV	FW	ND	970	1.5	380	330	70
Italy	Selex Galileo & U.T.R.I	Asio	M	SRW	POC,DC	46	0.8	10	4	0.5
Italy	Selex Galileo & U.T.R.I	Otus	M	FW	POC,DC	82	1	10	2.3	1
Italy	Selex Galileo & U.T.R.I	Strix	M	FW	POC,DC	82	1.5	12.5	20	1
Italy	Siralab	SR-H3	CC	FW	POC,IS	(36)	0.5			1
Italy	U.T.R.I	Crex	M	FW	DC				0.6	
Italy	U.T.R.I	Profalk	M	TR	DC		4		18	3
Italy	U.T.R.I	MHELI	DV	RW	POC,DC	80		15	4.8	1
Italy	U.T.R.I	Spyball	M	SRW	DC				1	
Italy	U.T.R.I	TSO-401	DV	SRW	DC	65	1	14	4.2	0.9
Japan	Epson & Sony	Micro VTOL	DV	RW	POC					
Japan	Fuji Heavy Industries	FFOS	M	RW	IS	120	3	150	275	
Japan	Fuji Heavy Industries	HSFD	RV	RW	DS				735	
Japan	Fuji Heavy Industries	RPH-1	DV	RW	DS	(36)	1		330	
Japan	Fuji Heavy Industries	RPH-2A	CC	RW	ES	120	1	150	305	60
Japan	Hirobo	Sky Surveyor	CC	RW	MR		1		48	
Japan	Kawada Industries & Hitachi	Colugo	DP	FW	DC	54	0.5		0.4	
Japan	Ministry of Defence-Techn. R&D Institute	Unnamed	M	FW	DC					
Japan	Nara Institute of Science + Technology	XB-2	DP	SRW	DC	435	24			

(Continued)

Country	Producer(s)/Developer(s)	System Designation	Class	Airframe	Status	Max. Speed (km/h)	Endurance (hrs)	Endurance (km)	MTOW (kg)	Payload Capacity (kg)
Japan	Nara Inst. of Science + Technology & TAO	Skyblade	CC, DV	FW	ND					
Japan	Yamaha Motors	Aerial RMAX	CC	RW	IS					
Japan	Yamaha Motors	Agricultural RMAX	CC	RW	IS					
Japan	Yamaha Motors	Autonomous RMAX II	DP	RW	IS	72	2.5	LOS 150m	94	10
Japan	Yamaha Motors	Autonomous RMAX IIG	DP	RW	IS	72	2.5	LOS 150m	94	10
Japan	Yanmar Agricultural Equipment Co.	YH-300SL	CC	RW	IS					
Japan	Yanmar Heli Service & Kobe Giken	KG-135	CC	RW	POC,ND					
Jordan	Jordan Aerospace Industries	Falcon	M	FW	DC	180	4	30-50	60	6
Jordan	Jordan Aerospace Industries	I-Wing	M	FW	DC			10		
Jordan	Jordan Aerospace Industries	Silent Eye	M	FW	DC	110	1	10	3.5	0.5
Malaysia	Composites Technology Research (CTRM)	Eagle ARV	DP	FW	IS	246	10	250	648	60
Malaysia	CTRM & System Consultancy Services & Ikramatic Systems	Alundra	M	FW	POC,DC	148	2	50		25
Malaysia	Sapura	Cyber Eye	M, DV	RW	DC					
Malaysia	Sapura	Cyber Shark	M, DV	FW	DC					
Malaysia	System Consultancy Services	UAV	DP	FW	DC	120	8	220	35	15
Mexico	Hydra Technologies	S3	M	FW	DC	300	2		40	3.2
Mexico	Hydra Technologies	S3E	M	FW	DC	200	5		45	10

(Continued)

Country	Producer(s)/Developer(s)	System Designation	Class	Airframe	Status	Max. Speed (km/h)	Endurance (hrs)	Endurance (km)	MTOW (kg)	Payload Capacity (kg)
Mexico	Hydra Technologies	S4	DP	FW	ES	170	8	80	55	9
Mexico	Hydra Technologies	S5	DP	FW	DC					
Netherlands	ASTI & TNO	DelFly I	DV	FLW	DC					
Netherlands	ASTI & TNO	DelFly II	DV	FLW	DC		0.13 +		16	0.0025
Netherlands	Delft Dynamics	Robot Helicopter	CC	RW	IS,DC				15	2
Netherlands	Dutch Space	MATE	M, DV	FW	DS,DC		1	5	6	
Netherlands	Geocopter (former E-Producties)	EKH-001	DP	RW	POC,DC				35	
Netherlands	Geocopter	Furore	CC	RW	MR		1		80	30
Netherlands	HighEye	HE 26	CC	RW	IS	110		10	15	7.5
Netherlands	HighEye	HE 26 C	CC	RW	IS		1	1.5		7
Netherlands	HighEye	HE 3.6 t	CC	RW	IS	110		10	24	14
Netherlands	HighEye	HE 60	CC	RW	IS	110		12	25	14
Netherlands	HighEye	HE 80	CC	RW	IS	110		11	34	20
Netherlands	UAV-Europe	MH 23	DP	RW	DC	100	2.5	11-30	40	15
New Zealand	Skycam UAV	Kahu -2E-B	DP	FW	MR	90				
New Zealand	Skycam UAV	Kahu Hawk	DP	FW	MR	55			25	
New Zealand	Skycam UAV	Kahu silver eye	DP	FW	MR	55			25	
New Zealand	TGR Helicorp Ltd.	Snark	M	RW	DC	289	24+	5500	1136	686
New Zealand	TGR Helicorp Ltd.	Wasp	CC	RW	DC					
Norway	ET-Air	Cruiser	DV	FW	POC,DC				20	
Norway	Norut IT	CryoWing	CC	FW	POC,DC			1500		
Norway	Odin Aero	Recce D6	DP	FW	POC,DC	(100)	0.55	10	2.8	
Norway	Prox Dynamics	Black Hornet PD-100	CC	RW	DC				0.20	

(Continued)

Country	Producer(s)/Developer(s)	System Designation	Class	Airframe	Status	Max. Speed (km/h)	Endurance (hrs)	Endurance (km)	MTOW (kg)	Payload Capacity (kg)
Norway	Prox Dynamics	Hornet-3	CC	RW	DC					
Norway	Proxflyer	BladeRunner	CC	RW	POC,IS				0.05	
Norway	Proxflyer	MicroFlyer	CC	RW	POC,MR		0.04		0.0078	
Norway	Proxflyer	Mosquito	CC	RW	POC,MR				0.11	
Norway	Proxflyer	Nanoflyer	CC	RW	POC,MR		0.02		0.0027	
Norway	Robot Aviation	unnamed	CC	FW	DC					
Norway	Scandicraft & CybAero, Sweden	Apid 55	DP	RW	MR	90	3-6	50	150	55
Norway	SiMiCon	SRC	M, DV	FW	POC,DS,DC	575	4		150	15
Pakistan	Air Weapons Complex	AWC Mk I	M	FW	DC?	175	2	30	30	14
Pakistan	Air Weapons Complex	AWC Mk II	M, DV	FW	DC?	175	3	50	60	34
Pakistan	Air Weapons Complex	Bravo	M	FW	POC,IS,ND	160	4+	80	110	15-20
Pakistan	Air Weapons Complex	Shaspar	M	FW	DC					
Pakistan	Air Weapons Complex	Vision I	M, DV	FW	DC	160	5+	100	120	25
Pakistan	Air Weapons Complex	Vision II	M, DV	FW	DC	160	5	150		30
Pakistan	Directorate General of Munitions Prod.	HudHud I	M, DV	FW	DC	165	2.5	50	35	20
Pakistan	Directorate General of Munitions Prod.	HudHud II	M, DV	FW	DC	165	3.5	80	70	40
Pakistan	Integrated Defence Systems	Hornet Mk 2	M	FW	POC?	277	1.5	10	46	46
Pakistan	Integrated Defence Systems	Hornet Mk 5	M	FW	DC	384	2.5	40-50	60	20
Pakistan	Integrated Defence Systems	Huma-1	M, DV	FW	DC	180	5	150	130	20
Pakistan	Integrated Dynamics	Border Eagle II	M	FW	DC	160	3	30	15	4
Pakistan	Integrated Dynamics	Desert Hawk	M	FW	POC,DS,DC		2			

(Continued)

Country	Producer(s)/Developer(s)	System Designation	Class	Airframe	Status	Max. Speed (km/h)	Endurance (hrs)	Endurance (km)	MTOW (kg)	Payload Capacity (kg)
Pakistan	Integrated Dynamics	Explorer	CC	FW	DC	160	4	20	20	8
Pakistan	Integrated Dynamics	Firefly	M, DV	FW	POC,DC		8sec			
Pakistan	Integrated Dynamics	Hawk Mk-1	M	FW	DC	240	3	50	46	12
Pakistan	Integrated Dynamics	Hawk Mk-2	M	FW	DC		4	50	50	15-20
Pakistan	Integrated Dynamics	Hawk Mk-5	M	FW	DC		4	80	60	15
Pakistan	Integrated Dynamics	Hornet	M	FW	DC	240	4	80	50	12
Pakistan	Integrated Dynamics	Rover	CC	FW	DC	100	1	5	5	1
Pakistan	Integrated Dynamics	Shadow-1	M	FW	DC	130	6	160	90	25
Pakistan	Integrated Dynamics	Vision Mk-1	M	FW	DC	207	3	50	35	10
Pakistan	Integrated Dynamics	Vision Mk-2	M	FW	DC	208	4	50	80	15
Pakistan	Integrated Dynamics	Vision X-1	M	FW	DC		8	250	140	30
Pakistan	National Development Complex	Vector Mk 1	M	FW	POC?	205	4.5	10	105	
Pakistan	National Development Complex	Vector Mk 2	M	FW	POC,DS	(120)	5		120	
Pakistan	Satuma Research	Flamingo	M	FW	MR	130	6-8	200	245	30
Pakistan	Satuma Research	Jassos HST	M	FW	DC	140	1.5	LOS	40	5
Pakistan	Satuma Research	Jassos II	M	FW	POC?	160	4-5	100	125	20
Pakistan	Satuma Research	Mukhbar	M	FW	DC	220	1.5	30	40	5
Pakistan	Satuma Research	NB-X2	M, DV	FW	DC					
Pakistan	Satuma Research	Parwaz	M	FW	POC?				40	5
Pakistan	Satuma Research	Stingray	M	FW	DC	90	1	15	7.5	1.5
Philippine	Roboteknik	Custos	DP	FW	DC		2	10		
Poland	Air Force Institute of Technology	HOB-bit	DP, DV	FW	POC,DC	90	1.5	20	3.5	

(Continued)

Country	Producer(s)/Developer(s)	System Designation	Class	Airframe	Status	Max. Speed (km/h)	Endurance (hrs)	Endurance (km)	MTOW (kg)	Payload Capacity (kg)
Poland	Air Force Institute of Technology	unnamed	DP, DV	FW	POC,DC	90		20	18	
Poland	Research & Development Centre for Mechanical Appliances	Bee	M, DV	FW	DC	50	0.15		0.07	
Poland	Research & Development Centre for Mechanical Appliances	CamBat	M, DV	FW	DC	40	0.4		1.7	
Poland	WB Electronics	Sofar	M	FW	MR	90	2	10	4.9	
Portugal	Faculty of Engineering -University of Porto	ASASF	RV	FW	POC,DC		0.4			
Portugal	IST & OGMA	Armor X7	M, DV	FW	POC,DC	100	12-15		200	
Portugal	School of Engineering, Minho University	AIVA	DV	FW	POC,DC				15	
Romania	Centrul de Invetica	Octogon	DV	RW	POC,ND					
Romania	Electromecanica Ploiesti & Politehnica Bucharest & INCAS & Romarm	Vigilent	DV	FW	POC,DC					
Romania	Inav Aviation Institute	Argus S	M, DV	FW	DC	135		250	15.5	3
Romania	Inav Aviation Institute	Argus XL	M, DV	FW	DC	230		300	140	40
Russian Fed.	A-Level Aerosystems	ZALA 421-01	DP	FW	IS	120	0.75	10	2.5	0.25
Russian Fed.	A-Level Aerosystems	ZALA 421-02		RW	DC	150	6	50	95	
Russian Fed.	A-Level Aerosystems	ZALA 421-03		FW	DC	95	1	25	12	2
Russian Fed.	A-Level Aerosystems	ZALA 421-04 electric		FW	DC	130	1	40	7	1

(Continued)

Country	Producer(s)/Developer(s)	System Designation	Class	Airframe	Status	Max. Speed (km/h)	Endurance (hrs)	Endurance (km)	MTOW (kg)	Payload Capacity (kg)
Russian Fed.	A-Level Aerosystems	ZALA 421-04 thermic		FW	DC	130	3	40	8	1
Russian Fed.	A-Level Aerosystems	ZALA 421-06		RW	DC	80	3	40		3.5
Russian Fed.	A-Level Aerosystems	ZALA 421-07	DP	FW	MR					
Russian Fed.	A-Level Aerosystems	ZALA 421-08		FW	DC	130	1.5	15	1.7	
Russian Fed.	A-Level Aerosystems	ZALA 421-09		FW	DC	130	10	50	70	
Russian Fed.	A-Level Aerosystems	ZALA 421-12		FW	DC	120	2		4	1
Russian Fed.	Enics Research Center	E90 (R90 system)	M, DV	FW	MR	180	0.5		45	
Russian Fed.	Enics Research Center	Eleron	DV	FW	POC,DC	(64)	1	10	2.9	
Russian Fed.	Irkut	Irkut-10	DP	FW	DC	120	2.5	70	8.5	1.5
Russian Fed.	Irkut	Irkut-20	DP	FW	DC	180	3	70	20	3
Russian Fed.	Irkut	Irkut-2F	DP	FW	DC	110	1	10	2.8	0.3
Russian Fed.	Irkut	Irkut-2T	DP	FW	DC	110	1	10	2.8	0.3
Russian Fed.	Irkut	Irkut-2M	DP	FW	DC	110	1	15	3	0.3
Russian Fed.	Irkut & Aeronautics, Israel (airframe)	Irkut-60	DP	FW	DC	180	6	70	65	15
Russian Fed.	Irkut & Aeronautics, Israel (airframe)	Irkut-200	DP	FW	DC	210	5	80	200	30
Russian Fed.	Irkut & Stemme, Germany (airframe)	Irkut-850	DP	FW	DC	270	12	200	860	200
Russian Fed.	Kamov	Ka-117	DV	RW	POC,DC			200		
Russian Fed.	Kamov	Ka-137	M	RW	MR	175	4	50	280	50
Russian Fed.	Kamov	Ka-226	DV	RW	POC,DC		16		3600	
Russian Fed.	Kamov	Ka-37	M	RW	DC?					

(Continued)

Country	Producer(s)/Developer(s)	System Designation	Class	Airframe	Status	Max. Speed (km/h)	Endurance (hrs)	Endurance (km)	MTOW (kg)	Payload Capacity (kg)
Russian Fed.	Kamov	Ka-37C	M	RW	DC?					
Russian Fed.	KB Lutch	Tipchak	M	FW	POC,DC	200	2	40	50	
Russian Fed.	KB Lutch	X01	M	FW	POC,DC	125	4.5		50	7
Russian Fed.	KB Lutch	X02	M	FW	POC,DC	180	1		35	4
Russian Fed.	MiG	Skat	M	FW	DC	800			10000	2000
Russian Fed.	Mil	Mi-34BP1	DV	RW	POC,DC	195	3.5	620?	920	360
Russian Fed.	Mil	Mi-34BP2	DV	RW	POC,DC	220	5.5	980?	790	520
Russian Fed.	NII Kulon	BLA-06 Aist	M	FW	POC,DC	250	12	250	500	
Russian Fed.	NII Kulon	BLA-07	M	FW	POC,DC	190	3		35	
Russian Fed.	NII Kulon	Filin	M	FW	DC	960	1		3000	
Russian Fed.	NII Kulon	Mokit (Yula)	M	FW	DC	200	1.5		290	
Russian Fed.	Radio MMS		DP	RW	DC					
Russian Fed.	Sokol Experimental Design Bureau	Dan-Baruk	M	FW	POC,DC					
Russian Fed.	Sukhoi	Zond-1	CC	FW	DC	0.5	18	12000	12000	1500
Russian Fed.	Sukhoi	Zond-2	DP	FW	DC	0.5	24	12000	12000	1500
Russian Fed.	Sukhoi	Zond-3	DP	FW	DC	250	12	2500	2000	500
Russian Fed.	Tranzas (Kronshtadt)	Dozor 2	DP	FW	MR	150	6		50	8
Russian Fed.	Tranzas (Kronshtadt)	Dozor 3	DP	FW	DC	150	16		500	115
Russian Fed.	Tranzas (Kronshtadt)	Dozor 4	DP	FW	ES	150	8		85	12
Russian Fed.	Tranzas (Kronshtadt)	Dozor 5	DP	FW	POC	150	10		95	15
Russian Fed.	Teknol	Mini UAV System	DP	FW	MR	75	0.6		3.75	
Russian Fed.	Teknol	Tactical UAV System	DP	FW	MR	80	1.5		7	

(Continued)

Country	Producer(s)/Developer(s)	System Designation	Class	Airframe	Status	Max. Speed (km/h)	Endurance (hrs)	Endurance (km)	MTOW (kg)	Payload Capacity (kg)
Russian Fed.	Tupolev	Berkut	DP	FW	POC,DC	180	8		180	
Russian Fed.	Tupolev	Tu-143	M	FW	ND	875	0.2	90	1400	
Russian Fed.	Tupolev	Tu-243 Reys	M	FW	ND	850-940	0.2	180	1400	
Russian Fed.	Tupolev	Tu-300 Korshun	M	FW	DC	950	1		3000	
Russian Fed.	Vega Radio Construction Company	Tipchak	M	FW	DC					
Russian Fed.	Yakovlev	Albatros	M	TR	DC?	300	7	100	450	
Russian Fed.	Yakovlev	Expert	M	FW	DC?	110	6	100	40	
Russian Fed.	Yakovlev	Klest	M	FW	POC,DC				130	
Russian Fed.	Yakovlev	Pchela-1T	M	FW	IS	180	2	50	138	70
Russian Fed.	Yakovlev	Proryv	M	FW	POC,DC				10.000	3.000
Russian Fed.	Yakovlev	Proryv-R	M	FW	POC,DC		20		9.800	1.200
Russian Fed.	Yakovlev	Strekoza	M	FW	DC?	110	6	50-10	0 40	
Russian Fed.	Yakovlev	Voron	M	FW	POC,DC		2		500	140
Serbia	EMA-UAV	Nikola Tesla 150	RV	FW	POC,DC	70		10	40	
Serbia	Utva Aircraft Industry	Gavran I	M	FW	POC,DC	(120)	0.75	10	16	4
Serbia	Utva Aircraft Industry	Gavran II	M	FW	POC,DC	120	0.75	10	30	4
Serbia	Utva Aircraft Industry	IBL-2004	M	FW	POC,DC					
Singapore	Cradance	Golden Eagle	DP	FW	MR	72	2	10	0.85	0.08
Singapore	Singapore Technologies Aerospace	Blue Horizon	M, DV	FW	DC	130	16		180	
Singapore	Singapore Technologies Aerospace	Extender	M	FW	POC	100	0.5	5	1.6	0.25

(Continued)

Country	Producer(s)/Developer(s)	System Designation	Class	Airframe	Status	Max. Speed (km/h)	Endurance (hrs)	Endurance (km)	MTOW (kg)	Payload Capacity (kg)
Singapore	Singapore Technologies Aerospace	Fantail	M, DV	SRW	MR	111	0.5+	5	5.5	0.4
Singapore	Singapore Technologies Aerospace	LALEE	M, DV	FW	POC,DC				5000	
Singapore	Singapore Technologies Aerospace	MAV-1	M, DV	FW	POC,DC				80	20
Singapore	Singapore Technologies Aerospace	Skyblade II	DP	FW	MR	130	1-2	8		
Singapore	Singapore Technologies Aerospace	Skyblade IV	DP	FW	MR	150	6-12	100	50	12
Singapore	Singapore Technologies Dynamics	Phantom Eye	M, DV	FW	DC	74	1		2	
Slovakia	Advanced Unmanned Systems International	AirSniper	DP	TR	DC	200	1-4		12	
Slovenia	Aviotech	RVM04	M, DV	FW	DC	145	4	80	36	11
Slovenia	C-Astral	Bramor	DP	FW	MR	104	2	20	4	0.6
Slovenia	C-Astral	Spectral System	DP	FW	MR	140	4		9.2	
Slovenia	University of Ljubliana	Karantania	M	FW	POC,DC	150	1	16	7.4	4.7
Slovenia	University of Ljubliana	Maister	M	FW	DC		6	20	50	20
Slovenia	University of Ljubliana	Vision - UPP	M	FW	DC		1	5	3.7	0.6
South Africa	ABAT	Posduif	DP	FW	ND			60		0.7
South Africa	Advanced Technologies & Engineering	Civil Vulture	CC	FW	DC	(120)	8-9	200	150	35
South Africa	Advanced Technologies & Engineering	Endurance Vulture	M	FW	DC	(120)	8-9	200	150	35

(Continued)

Country	Producer(s)/Developer(s)	System Designation	Class	Airframe	Status	Max. Speed (km/h)	Endurance (hrs)	Endurance (km)	MTOW (kg)	Payload Capacity (kg)
South Africa	Advanced Technologies & Engineering	Kiwit	DP	FW	MR	(120)	1	5	3	
South Africa	Advanced Technologies & Engineering	Night Vulture	M	FW	DC	(120)	3-4	200	135	35
South Africa	Advanced Technologies & Engineering	Roadrunner	DP	FW	MR		1		6	1
South Africa	Advanced Technologies & Engineering	Sentinel 500M	DP	FW	DC		5		150	
South Africa	Advanced Technologies & Engineering	Sentinel 620	DP	FW	DC		12		200	35
South Africa	Advanced Technologies & Engineering	Sentinel 800	DP	FW	DC		20		250	35
South Africa	Advanced Technologies & Engineering	Vigil EE	CC	FW	DC		8		80	20
South Africa	Advanced Technologies & Engineering	Vigil SR	CC	FW	DC		4		40	10
South Africa	Advanced Technologies & Engineering	Vulture	M	FW	IS	(120)	3-4	Up to 200	125	25
South Africa	Denel Aerospace Systems	Bateleur	M, DV	FW	DC	250	18-24	750	1000	200
South Africa	Denel Aerospace Systems	Seeker II	M	FW	IS	220	10	250	275	50
South Africa	Denel Aerospace Systems	Seeker 400	M	FW	DC	(120)	16			100
South Africa	Denel Aerospace Systems	Seraph	M, DV	FW	DC	0.85	1.5	1300	900	80
South Korea	Agency for Defense Development	unnamed	M	FW	DC		24	500	4000	500

(Continued)

Country	Producer(s)/Developer(s)	System Designation	Class	Airframe	Status	Max. Speed (km/h)	Endurance (hrs)	Endurance (km)	MTOW (kg)	Payload Capacity (kg)
South Korea	KAL Aerospace	KUS-7	M	FW	DC	150	2.5	45	70	7
South Korea	KAL Aerospace	KUS-9	M	FW	DC	180	8	80	190	20
South Korea	KAL Aerospace	unnamed prototype 1	M	FW	POC,DC	150	2.5	40	65	
South Korea	KAL Aerospace	unnamed prototype 2	M	FW	DC	178	6	80	130	
South Korea	Korean Aeronautical Research Institute	Durumi	DP	FW	DC	130	30	3300	15	2.5
South Korea	Korean Aeronautical Research Institute	Smart UAV	M, DV	TR	DC	500	5	200	950	40-100
South Korea	Korean Aerospace Industries	Night Intruder 300	M	FW	IS	185	6	120	300	45
South Korea	Korean Aerospace Industries	Night Intruder II	M	FW	IS	185	9	120	270	45
South Korea	Korean Aerospace Industries & Daewoo	Arch-50	CC	RW	MR	150	0.6		300	50
South Korea	Microaerobot	Batwing	M, DV	FW	POC		0.25	1	48	3
South Korea	Microaerobot	Flying Robot	DP, DV	SRW	POC					
South Korea	Microaerobot	FM-07	DP, DV	FW	DC	70	0.6	5	0.5	
South Korea	Microaerobot	M-05	M, DV	FW	POC		0.5	1	80	7
South Korea	Microaerobot	MA1	DP	FW	POC					
South Korea	Microaerobot	MA2	DP	FW	POC					
South Korea	Oneseen Skytech	Cine-Copter	CC	RW	DC		0.6	0.2	37	12
South Korea	Oneseen Skytech	MAVtronix7000	DP	RW	POC		0.25		8	4
South Korea	Oneseen Skytech	X-Copter	DP	RW	MR		2	10	120	30
South Korea	Ucon Systems	RemoEye 002	M	FW	MR	80	1	10	2.4	

(Continued)

Country	Producer(s)/Developer(s)	System Designation	Class	Airframe	Status	Max. Speed (km/h)	Endurance (hrs)	Endurance (km)	MTOW (kg)	Payload Capacity (kg)
South Korea	Ucon Systems	RemoEye 006	M	FW	MR	75	1	10	6.5	
South Korea	Ucon Systems	RemoEye 015	M	FW	MR	170	4	50	15	
South Korea	Ucon Systems	RemoEye H120	M	RW	DC	130	2	50	120	30
South Korea	Ucon Systems	RemoH-C100	CC	RW	MR	60	1	0.3	100	
South Korea	Ucon Systems	RemoH-M100	M	RW	MR	120	4	60	100	
Spain	Aerovision	Fulmar	DP	FW	DC	150	8	50	20	8
Spain	Airview	AV-01	CC	FW	DC	110	2		6	1.5
Spain	Airview	AV-02	CC	FW	DC	120	2		8	2
Spain	Airview	AV-03	CC	FW	DC	130	6		28	8
Spain	Aitem	Dedalo	CC, RV	FW	POC,ND,MR					
Spain	Aitem	Horus	CC, RV	RW	POC,ND,MR					
Spain	Alpha Unmanned Systems	Atlantic	DP	FW	MR	120	6-8	0.8	31	6
Spain	Alpha Unmanned Systems	Atlas	DP	FW	MR	40	2		6.9	1
Spain	Alpha Unmanned Systems	Commando	DP	RW	MR	120	1		8	1.5
Spain	Alpha Unmanned Systems	Sniper	DP	RW	MR	150	2-3	30	14	4
Spain	EADS D&S	Atlante	M	FW	DC			10		60
Spain	INTA	ALO	M	FW	IS	200	2	50	20	6
Spain	INTA	Milano	M	FW	DC	230	20		900	120
Spain	INTA	Siva	M, DV	FW	DS	170	6	150	300	40
Spain	PLATINO Consortium	HADA	M	RW	POC,DC	420	6	200	380	100
Spain	Robotnik Automation	X4	DP	RW	DC					
Spain	Sistemas de Control Remoto (SCR)	Alba 01	M, DV	FW	DC	216	1		18	
Spain	Sistemas de Control Remoto (SCR)	X-Vision	M, DV	FW	DC	150	2		40	

(Continued)

Country	Producer(s)/Developer(s)	System Designation	Class	Airframe	Status	Max. Speed (km/h)	Endurance (hrs)	Endurance (km)	MTOW (kg)	Payload Capacity (kg)
Spain	UAV Navigation	KUAV	DP	FW	MR		5	30	30	
Sweden	CybAero	APID 55	DP	RW	ES	90	3-6	100	160	(55)
Sweden	CybAero & NLR, USA	APID Vantage	M	RW	ES	180	3-5	100	160	15.75
Sweden	Saab	Filur	M, DV	FW	POC,ND	300			55	
Sweden	Saab	Sharc	M, DV	FW	POC,DC	M0.8			60	
Sweden	Saab	Skeldar V-150 prototype	M	RW	POC,DC	(100)	5	100	150	
Sweden	Saab	Skeldar V-150	DP	RW	POC,DC	130	4	100	200	30
Sweden	Saab	Skeldar-M (V-250)	M, DV	RW	POC,DC			180	250	
Sweden	SmartPlanes	Smart-1	DP	FW	MR	(55)	1		1	
Switzerland	Aeromedia	Aerocopter 1	CC	RW	MR		0.16		2	
Switzerland	Aeromedia	Aerocopter 2	CC	RW	MR		0.25		12	
Switzerland	Aeromedia	AeroStar 1	CC	FW	MR		0.5		1	
Switzerland	Aeromedia	AeroStar 2	CC	FW	MR		0.5		2	
Switzerland	Aeroscout	B2-120	CC	RW	DC					20
Switzerland	Aeroscout	T5	CC	RW	DC				25	10
Switzerland	Minizepp	Z10000Pro	CC	LtA	MR	75	2-4			4
Switzerland	Minizepp	Z13000	CC	LtA	MR	60	3-4			13.5
Switzerland	RUAG Aerospace	Ranger	M	FW	IS	240	6	180	280	45
Switzerland	RUAG Aerospace	Super Ranger	M	FW	DC	(234)	20	200	500	150
Switzerland	SenseFly	Swinglet	CC, RV	FW	DC	55	0.5	0.5	0.5	0.15
Switzerland	Skive Aviation	Skive	CC	LtA	MR	50	3-4			8
Switzerland	Skybotix Technologies	Coax	DP	RW	MR		0.3	1	320	
Switzerland	Skybotix Technologies	OS4	CC	RW	IS				<2	

(Continued)

Country	Producer(s)/Developer(s)	System Designation	Class	Airframe	Status	Max. Speed (km/h)	Endurance (hrs)	Endurance (km)	MTOW (kg)	Payload Capacity (kg)
Switzerland	Skybotix Technologies	Planemini	CC	FW	IS					
Switzerland	Skybotix Technologies	Skybox	CC	RW	DC					
Switzerland	Swiss Federal Institute of Technology	Sky-Sailor	CC, RV	FW	MR				2.3	
Switzerland	Swiss UAV	NEO S-300	DP	RW	DC,MR		2.5	60	75	20
Switzerland	Swiss UAV	KOAX X-240	DP	RW	DC	80	1.5			8
Taiwan ROC	Aero Flight Technology Enterprises	Mx-1	DP	FW	DC		5	400		
Taiwan ROC	AIDC-Aerospace Industrial Defence Corp	Fireant	M	FW	DC	180	2			5-10
Taiwan ROC	Chung Shan Inst. of Science & Technology	Blue Magpie	M, DV	FW	?				<5	
Taiwan ROC	Chung Shan Inst. of Science & Technology	Chung Shyang II	M, DV	FW	DC	150	10		450	
Taiwan ROC	National Cheng Kung University	Swan	RV	FW	POC,DC	(80)	2		12	
Taiwan ROC	Chung Shan Inst. of Science & Technology	Kestrel II	M	FW	POC	185	8	100	120	25-30
Taiwan ROC	YoShine Helicopters	Ezycopter Micro	DV	RW	POC,DC					
Taiwan ROC	YoShine Helicopters	Ezycopter UAV	M, DV	RW	POC,DC	120			300	
Tunisia	Tunisia Aero Technologies	Jebelassa	M, DV	FW	POC,DC	140	13		228	
Tunisia	Tunisia Aero Technologies	NasNas	M, DV	FW	POC,DC	130	14		125	
Turkey	Global Teknik	Globiha	M	FW	MR	(83)	1.5-2	5-10	5	
Turkey	Kale & Baykar Technologies	Bayraktar	DP	FW	POC,DC	110	1	20	5	1.5
Turkey	Kale & Baykar Technologies	Malazgirt	M	FW	DC		0.5	20		1

(Continued)

Country	Producer(s)/Developer(s)	System Designation	Class	Airframe	Status	Max. Speed (km/h)	Endurance (hrs)	Endurance (km)	MTOW (kg)	Payload Capacity (kg)
Turkey	Kuzgun High Technology Design	Kuzgun 1	CC	LtA	MR		1.5			0.5
Turkey	METU - Departement of Aerospace Eng	Mini UAV	DP	FW	POC,DC	92	1.5	10	4.5	
Turkey	Tusas Aerospace Industries	Baykus	M, DV	FW	DC					
Turkey	Tusas Aerospace Industries	Marti	M, DV	FW	DC					
Turkey	Tusas Aerospace Industries	Gozcu	M	FW	DC	185		50	85	8
Turkey	Tusas Aerospace Industries	Pelikan	M, DV	FW	DC					
Turkey	Tusas Aerospace Industries	UAV-X1	M, DV	FW	DC	222	7	1200	245	30
Turkey	Tusas Aerospace Industries	Tiha	M	FW	DC	140	24	200	1500	200
Turkey	Vestel	Ari	M	FW	DC	120	0.5	1	1	
Turkey	Vestel	Efe	M	FW	DC	160	2.5	15		
U.A.E.	ATS (ADCOM Group)	Smart Eye	M	FW	DC	220			1300	70-500
U.A.E.	ATS (ADCOM Group)	Yabhon Noble	M	RW	DC				630	
U.A.E.	ATS (ADCOM Group)	Yabhon - H	M	FW	DC	175	8		62.5	5
U.A.E.	ATS (ADCOM Group)	Yabhon - M	M	FW	DC	240	12		280	30
U.A.E.	ATS (ADCOM Group)	Yabhon - R	M	FW	DC	240	30		500	50
U.A.E.	ATS (ADCOM Group)	Yabhon - RX	M	FW	DC	310	42		535	60
U.A.E.	ATS (ADCOM Group)	Yabhon - RX-18	M	FW	POC,DC				1300	
U.A.E.	GAMCO	GRS 200	M, DV	FW	DC					
UK	Aero Sekur	unnamed	M, DV	FW	POC					
UK	Autonomous Vehicles International	Seeker	M	TR	POC,DC		0.25	10	16	4.5
UK	BAE Systems	Corax	M, DV	FW	POC,DC					

(Continued)

Country	Producer(s)/Developer(s)	System Designation	Class	Airframe	Status	Max. Speed (km/h)	Endurance (hrs)	Endurance (km)	MTOW (kg)	Payload Capacity (kg)
UK	BAE Systems	Fury	M	FW	DC					
UK	BAE Systems (airframe Lindstrand Tech.)	GA22	M	LtA	DC	90	24			45
UK	BAE Systems	Havir	M, DV		DC					
UK	BAE Systems	Herti-1A	M, DV	FW	POC,DC	220	30		500	145
UK	BAE Systems	Herti-1B	M	FW	POC,DC	220	20		750	
UK	BAE Systems	Herti-1D	M, DV	FW	POC,DC				350	
UK	BAE Systems	ITAO	M	LtA	DC					
UK	BAE Systems	Kestrel	M, DV	FW	POC,DC					
UK	BAE Systems	Raven	M, DV	FW	POC,DC					
UK	BAE Systems	Reliant Mauser	M, DV	FW	DC					
UK	BAE Systems	Taranis	M	FW	POC,DC				8000	
UK	BAE Systems & Flight Refuelling (airframe)	Phoenix	M	FW	POC,ND	157	4.5	70	180	50
UK	BAE Systems & GE Aviation & Meggit & QinetiQ & Selex Galileo	Mantis	M	FW	POC,DC		24			
UK	Cyberflight	Bushmaster	M	FW	?		1.5	9	2	
UK	Cyberflight	Cybereye	M, DV	FW	DC	(150)	5-6		45	
UK	Cyberflight	CyberOne	M, DV	FW	DC	160	2		12.2	6.8
UK	Cyberflight	Fat Boy	M, DV	FW	DC		5-6		70	
UK	Cyberflight	S.O.D.I	M, DV	FW	DC	128	1	10	3	0.5
UK	Cyberflight	S.O.D.III	M, DV	FW	DC	110	2.5	25	5.4	2
UK	Cyberflight	S.O.D.IV	M, DV	FW	DC	110	0.25	9	0.5	0.25

(Continued)

Country	Producer(s)/Developer(s)	System Designation	Class	Airframe	Status	Max. Speed (km/h)	Endurance (hrs)	Endurance (km)	MTOW (kg)	Payload Capacity (kg)
UK	Cyberflight	Super Swift-Eye	M, DV	FW	POC,DC	140	1			
UK	Cyberflight	Swift-Eye	M, DV	FW	POC,DC	140	0.6	3.2	6.4	
UK	Dragonfly Air Systems	Highland Darter	DV	TW	DC	166				
UK	Dragonfly Air Systems	Skimmer	DV	TR	DC		1-2		20	
UK	Fanwing	STOL UAV	DV	FW	DC	(29)			12	2
UK	Fanwing	Fanwing	DP, DV	FW	POC,DC	72	10		40	15
UK	GFS Projects	Flying Saucer	M, DV	SRW	DC		0.02			
UK	GFS Projects	GFS-7	DP	SRW	POC,DC				5.45	
UK	Kestrel Aerospace	Kestrel UAV	DV	TR	POC,DC	340		1390	295	
UK	Kestrel Aerospace	Lancer	DV	TR	POC,DC	305	12	1667		30
UK	MagSurvey	Prion	CC	FW	MR	90			30	
UK	MBDA & Blue Bear & Cranfield Aerospace	Fire Shadow	M	FW	DC		10		200	
UK	Meggitt Defense Systems	Phantom	M	FW	MR,ND	(180)	4		40	
UK	Meggitt Defense Systems	Spectre	M	FW	POC,ND	240	3-6	160	145	25
UK	Merlin Integrated Solutions Consortium	Optica (BLAC airframe)	DV	FW	DC		8	1046	1315	
UK	Nitrohawk	Nitrohawk	CC	RW	?		0.5	1.7	15	
UK	QinetiQ-Cortex	Tiger	M, DV	TB	POC				3	
UK	QinetiQ-Farnborough	Mercator	M, DV	FW	DS				27	2
UK	QinetiQ-Farnborough & Cranfield Aerospace	MinO	M	FW	POC,DC	144	1		10	
UK	QinetiQ-Farnborough & Cranfield Aerospace	Observer	M, DV	FW	MR	126	2	25	30	4

(Continued)

Country	Producer(s)/Developer(s)	System Designation	Class	Airframe	Status	Max. Speed (km/h)	Endurance (hrs)	Endurance (km)	MTOW (kg)	Payload Capacity (kg)
UK	Roke Manor Research	Unnamed	M, DV	Prf	DC					
UK	Selex Sensors and Airborne Systems	Damsfly	M, DV	FW	DC	(300)				
UK	Selex Sensors & Airborne Systems & U.T.R.I, Italy	Asio	M	SRW	DC	46	0.6	10	4	0.5
UK	Selex Sensors & Airborne Systems & U.T.R.I, Italy	Otus	M	FW	DC	83	1	10		1
UK	Selex Sensors & Airborne Systems & U.T.R.I, Italy	Strix	M	FW	DC	75	1.5	12.5		1
UK	SkyShips	C1000	CC	LtA	POC,DC	55	1		30	7
UK	SkyShips	Cirrus 840	CC	LtA	POC,DC					
UK	Swarn Systems	Owl	DP, DV	RW	DC				1	
UK	Tasuma	CSV-30	M, DV	FW	DC	175	2		22	
UK	Tasuma	Hawkeye	M, DV	FW	DC	85	1		3.2-3.8	
UK	Tasuma	CSV-20	M, DV	FW	DC	140	2		20	
UK	Tasuma	MinO (airframe)	M, DV	FW	DC					
UK	Tasuma	MSV-10	M, DV	FW	DC	55	0.5		4.2	
UK	Tasuma (Airframe) & Flight Refuelling	Raven 1	M	FW	DC	126	2	50	15	
UK	Tasuma (Airframe) & Flight Refuelling	Raven 2	M	FW	DC	180	3	50	84	
UK	Thales UK Tactical UAV Systems	T-TUAV	M	FW	MR	176	20	200	450	150
UK	Universal Target Systems	Spotter	CC	FW	MR					
UK	Universal Target Systems	Vigilant	M	FW	MR		4		25	

(Continued)

Country	Producer(s)/Developer(s)	System Designation	Class	Airframe	Status	Max. Speed (km/h)	Endurance (hrs)	Endurance (km)	MTOW (kg)	Payload Capacity (kg)
UK	VTOL Technologies	Aerial Police Dog	DP	SRW	?	240	1	100	5	1
UK	Warrior (Aero-Marine)	Gull 24	DV	FW	POC,DC	136			18	6
UK	Warrior (Aero-Marine)	Gull 44	DV	FW	POC,DC	170			93	34
UK	Warrior (Aero-Marine)	Gull 68	DV	FW	POC,DC	184			250	94
Ukraine	Scientific Industrial Service	A-10 Phoenix	DP	FW	POC	250	15		750	
Ukraine	Scientific Industrial Service	A-12 Hurricane	DP	FW	POC				18	
Ukraine	Scientific Industrial Service	A-160	DP	FW	POC	208	4-6	150	160	
Ukraine	Scientific Industrial Service	A-5 SeaEagle	DP	FW	POC		5	40	27	7
Ukraine	Scientific Industrial Service	Albatros 4	M	FW	POC?	60-125	2	20	18.3	3
Ukraine	Scientific Industrial Service	Inspektor	M	FW	DC?	160	10	1000	250	50
Ukraine	Scientific Industrial Service	Remez-3	M	FW	POC?	105	1	5	10	3
USA	AAI Corp	Shadow 200 TUAV	M	FW	IS	230	5-6	125	148	25
USA	AAI Corp	Shadow 200 TUAV	M	FW	MR		9			
USA	AAI Corp	Shadow 400	M	FW	IS	185	5	185	201	30
USA	AAI Corp	Shadow 600	M	FW	IS	190	12-14	200	265	41
USA	AAI Corp Aerosonde	Mark 4.4	DP	FW	IS	120	24		16.8	
USA	AAI Corp Aerosonde	Mark 4.7	DP	FW	DC		12			
USA	AAI Corp Aerosonde	Mark V	M	FW	DC		30			
USA	AC Propulsion	So Long	RV	FW	POC,DC	80	?	8	12.5	?
USA	Accurate Automation	LoFLYTE	M	FW	DC	463	0.3		32	
USA	Accurate Automation	X-43A-LS	M	FW	DC	(410)	0.75		81	
USA	Accurate Automation	MLOV	DV	FW	DC		4		20	
USA	Acuity Technologies	AT-3 Owl	M	FW	DC		8		185	
USA	Adaptive Flight	Hornet	DP	RW	DC	55	0.3	1-3	1.1	

(Continued)

Country	Producer(s)/Developer(s)	System Designation	Class	Airframe	Status	Max. Speed (km/h)	Endurance (hrs)	Endurance (km)	MTOW (kg)	Payload Capacity (kg)
USA	Advanced Ceramics Research (BAE Systems)	Coyote	M	FW	MR	(110)	1	37	6.4	1
USA	Advanced Ceramics Research (BAE Systems)	Manta B	M	FW	MR	200	6	37	23.5	6.8
USA	Advanced Ceramics Research (BAE Systems)	SilverFox	M	FW	IS	200	8-10	37	12	2.3
USA	Advanced Hybrid Aircraft	Hornet	DP	LtA	MR?	150		1100	3400	1134
USA	Advanced Hybrid Aircraft	Wasp	DP	LtA	MR?			160	34	5
USA	Advanced Soaring Systems & NASA	Apex	RV	FW	DS		5		272	
USA	AeroCam	23F	DP	RW	MR		0.5	2	10	3
USA	AeroCam	60F	DP	RW	MR		0.3	3	20	10
USA	Aerocross Systems	Echo Hawk	M	FW	DC	260	10		700	
USA	AeroMech Engineering	Fury	M	FW	DC	180	27		40	9
USA	AeroMech Engineering	SuperFly	M	FW	DC	120	1.25		2	0.45
USA	Aerotonomy	Nanos	DP	FW	DC		0.5		0.230	
USA	AeroVironment	Black Widow	M, DV	FW	DC	46	0.5		0.06	0.015
USA	AeroVironment	Centelios	RV	FW	DC					
USA	AeroVironment	DragonEye	M	FW	IS	80	0.9	5	2.6	0.46
USA	AeroVironment	Global Observer	RV, DP	FW	DC		1 Week		4500	450
USA	AeroVironment	Global Observer G0-1	RV, DP	FW	DC				1800	160
USA	AeroVironment	Global Observer G0-2	RV, DP	FW	DC	200	24	2780	4100	450

(Continued)

Country	Producer(s)/Developer(s)	System Designation	Class	Airframe	Status	Max. Speed (km/h)	Endurance (hrs)	Endurance (km)	MTOW (kg)	Payload Capacity (kg)
USA	AeroVironment	GLUAV	M	FW	DC	222	0.5	60	0.7	
USA	AeroVironment	Hawkeye	M	FW	DC					
USA	AeroVironment	Helios	RV	FW	POC,ND	51	14		825	100
USA	AeroVironment	Hiline	RV	FW	DC	55	48		325	
USA	AeroVironment	Hornet	RV	FW	DC					
USA	AeroVironment	MicroBat	RV	FLW	DC					
USA	AeroVironment	NAV	DV	RW	DC		0.3		0.010	0.002
USA	AeroVironment	OAV	M, DV	FW	DC					
USA	AeroVironment	Pathfinder Plus	RV	FW	POC,ND	(57.5)	Weeks		218	11.3
USA	AeroVironment	Pointer (FQM-151A)	M	FW	IS	80	1.5	10	3.8	0.9
USA	AeroVironment	Aqua Puma		FW	IS,ES		2.5	10		1.8
USA	AeroVironment	Land Puma	M	FW	IS,ES		2.5	10		1.8
USA	AeroVironment	Puma	M	FW	MR	95	3-4*	15	4.6	1
USA	AeroVironment	Puma AE		FW	MR	83	2	15	5.9	
USA	AeroVironment	Raven RQ-11A	M	FW	IS	57	1.5	10	2	0.18
USA	AeroVironment	Raven RQ-11B	M	FW	IS,ES	81	1.5	10	1.9	
USA	AeroVironment	SkyTote	M, DV	RW	DC	350	1.3	375	113	23
USA	AeroVironment	SP2S	M	FW	DC					
USA	AeroVironment	Swift		FW	MR	83	1+	10	2.8	0.45
USA	AeroVironment	Switchblade	M	FW	DC	145	0.3-0.6	7		
USA	AeroVironment	Wasp	RV	FW	DC	60	0.6-1.15	2	0.17	0.015
USA	AeroVironment	Wasp II (Land)	M	FW	IS,ES	65	0.8	4	0.275	
USA	AeroVironment	Wasp II (Sea)	M		IS,ES	65	0.8	4	0.290	

(Continued)

Country	Producer(s)/Developer(s)	System Designation	Class	Airframe	Status	Max. Speed (km/h)	Endurance (hrs)	Endurance (km)	MTOW (kg)	Payload Capacity (kg)
USA	AeroVironment	Wasp III	M	FW	IS	65	0.75	5	0.430	
USA	AeroVironment	Wasp VTOL	M	RSW	POC,DC					
USA	Agrarius (& Advanced Ceramics Research)	HawkEye	CC	FW	IS	32-105	6	10	13.6	4.536
USA	Airfoil Aviation	GSR/Pro1 Mount		Prf	MR					
USA	Airscooter	Airscooter E70	CC	RW	MR		0.25		14	2.25
USA	Airscooter	Airscooter G70	CC	RW	MR	92	0.6		14	4.5
USA	Airship Surveillance	L15	DP	LtA	DC	111	50			453
USA	Allied Aerospace (ex MicroCraft)	iStar (OAV)	M, DV	SRW	DS				2.27	
USA	Allied Aerospace (ex MicroCraft) & DARPA	Lift Augmented Ducted Fan	M, DV	SRW	DC	93	1	10	1.8	
USA	Americain Dynamics Flight Systems	AD-150	M	SRW	DC					
USA	Applied Research Associates (ARA)	JTEC	M	FW	DC		18			11.25
USA	Applied Research Associates (ARA)	Nighthawk	M	FW	IS	80	1.5	10	0.72	0.22
USA	Applied Research Associates (ARA) & Irvine Sensors	Tom Hawk	M	FW	DC	55	1	10	0.72	
USA	Arcturus	T-15	DV, DP	FW	DC	166	12+		20	4.54
USA	Arcturus	T-16	DV, DP	FW	DC	147	16		29	8
USA	Arcturus	Tracker	DV	RW	DC					
USA	Atair Aerospace	LEAP I	M	Prf	MR		48-55		1620	112
USA	Atair Aerospace	Insect (LEAP type II)	M	Prf	MR		34		540	90

(Continued)

Country	Producer(s)/Developer(s)	System Designation	Class	Airframe	Status	Max. Speed (km/h)	Endurance (hrs)	Endurance (km)	MTOW (kg)	Payload Capacity (kg)
USA	Atair Aerospace	Micro LEAP (type III)	M	Prf	MR		4		33.7	13.5
USA	AUAV	Boomerang 4	M,DV	FW	DC	157	1	130	9	2.27
USA	AurAayan Aerospace (D-Star Engineering)	AeroLensCraft (ALC)	DV	FW	DC	360	1 Year			
USA	Aurora Flight Sciences	Chiron	RV	FW	DS	344	24		2200	
USA	Aurora Flight Sciences	DA42 MPP	M	FW	MR	360	12		1785	
USA	Aurora Flight Sciences	Excalibur	M	FW	DS	850	3		1180	180
USA	Aurora Flight Sciences	GoldenEye-50	M, DV	SRW	DC	185.3	1		8.1	0.9
USA	Aurora Flight Sciences	GoldenEye-80	M, DV	SRW	DC		2.5	16	74.4	7.2
USA	Aurora Flight Sciences	GoldenEye-100	DP, DV	SRW	DC	296	4		68	9
USA	Aurora Flight Sciences	Hunter II	DV	FW	DC		29		1300	500
USA	Aurora Flight Sciences	Marsflyer	RV	FW	DC	111	2		108	
USA	Aurora Flight Sciences	Odysseus	DV	FW	POC,DC		44000		4500	450
USA	Aurora Flight Sciences	Orion	DV	FW	DC	440	30		5080	1800
USA	Aurora Flight Sciences	Orion HALL	RV	FW	DC	450	100		3174	181
USA	Aurora Flight Sciences	Perseus	RV	FW	MR	85-128	24	3000	1000	80-150
USA	Aurora Flight Sciences	SkyWatch	RV	FW	DC		32		4000	450
USA	Aurora Flight Sciences	Theseus	RV	FW	DS	530	32+		3580	
USA	Aurora Flight Sciences & Athena Technologies	Unnamed	M, DV	FW	DC?	(180)			64.4	
USA	Autonomous Airborne Systems	HOVTOL	DP, DV	FW	DC					
USA	Autonomous UAV	Boomerang	DV	FW	DC		1		10	2.5
USA	BAE Systems	IAV1	M	SRW	DC			4.5	34	

(Continued)

Country	Producer(s)/Developer(s)	System Designation	Class	Airframe	Status	Max. Speed (km/h)	Endurance (hrs)	Endurance (km)	MTOW (kg)	Payload Capacity (kg)
USA	BAE Systems	IAV2	M	SRW	DC			4.4	56.7	11.3
USA	BAE Systems	Skyagent	M	SRW	DC		1	15	56.25	6.75
USA	BAE Systems	SkyEye RAE-Extended	M	FW	POC,ND	189	12	185	567	136
USA	BAE Systems	SkyLynx II	M	FW	DC				150	
USA	BAE Systems	MAV	M, DV	SRW	DC	150	1		7.7	0.68
USA	BAE Systems	MicroStar	M, DV	FW	DC	55	0.3	5	0.14	0.015
USA	BAE Systems	OAV	M, DV	SRW	DC	203	1.5		63	10
USA	Bell Helicopter Textron	Eagle Eye	M, DV	TR	DS,DC	390	4	204	1020	136
USA	Berkut Engineering & Design	Mobius	M, RV	FW	IS	402	20			454
USA	Biorobotics (Case Western Univ. & University of Florida)	MMALV series	CC, DV	FW	DC				0.450	
USA	Blackwater Airships	Polar 400	CC	LtA	IS		60			180
USA	Boeing	Heliwing	M, DV	RW	DC				658	
USA	Boeing	ISIS	M	LtA	POC,DC		months			
USA	Boeing	Little Bird	M, DV	RW	DC					
USA	Boeing	U-MELB	M, DV	RW	DC	233	7	555		1000
USA	Boeing	Unnamed	M	FW	?					
USA	Boeing	VARIOUS	M	RW	DS	300				
USA	Boeing	X-36	M, DV	FW	POC,ND	207-450			565	
USA	Boeing	X-45A	M, DV	FW	DS,DC	M 0.8	2	2400	5529	680
USA	Boeing	X-45B	M, DV	FW	DS,DC	850	2		8618	907
USA	Boeing	X-45C Phantom Ray	M, DV	FW	DS,DC	M 0.85	2	2400	16559	2000

(Continued)

Country	Producer(s)/Developer(s)	System Designation	Class	Airframe	Status	Max. Speed (km/h)	Endurance (hrs)	Endurance (km)	MTOW (kg)	Payload Capacity (kg)
USA	Boeing	X-46	M, DV	FW	DC					
USA	Boeing	X-50 Dragonfly	M, DV	CRW	DS,DC	740	3	278	645	91
USA	Boeing	X-48B	RV	FW	POC,DC	(220)			227	
USA	Boeing & General Dynamics	Unnamed	M	FW	POC,DC					
USA	Boeing (Frontier Systems)	A160 Hummingbird	M, DV	RW	DS,DC	259	30-40	4630	1814	136
USA	Boeing (Frontier Systems)	Maverick	M, DV	RW	DS,DC	217	7	320		180
USA	Bostan	ARCQ			ND					
USA	Brandebury Tool Company	FoxCar	M	FW	IS				160	
USA	Brandebury Tool Company	Scythe	M, DV	FW	IS	230	6		7.7	2.7
USA	Brandebury Tool Company	TigerShark	M	FW	IS		9		135	
USA	Carolina Airships	Guardian 31	DP	LtA	DC	48	3	22	37.5	6.35
USA	Carolina Airships	Guardian 34	DP	LtA	DC	48	3	22	56	11.3
USA	Carolina Unmanned Vehicles	SLURS	M	FW	DC	40	1	10	4.5	1
USA	Charles Stark Draper Laboratory	NAV	RV	RW	DC		0.3		10	2
USA	Charles Stark Draper Laboratory	SARD	RV	RW	DC		0.3		15.7	
USA	Charles Stark Draper Laboratory	Wasp	M,DV	FW	POC,DC		1		3.9	
USA	CIRPAS	Pelican (Cesna-based)	RV	FW	IS	(167)	24		2086	150
USA	Continental Controls and Design	LOCUST MAV	RV	FW	POC,DC	(62)		5	0.57	

(Continued)

Country	Producer(s)/Developer(s)	System Designation	Class	Airframe	Status	Max. Speed (km/h)	Endurance (hrs)	Endurance (km)	MTOW (kg)	Payload Capacity (kg)
USA	Coptervision	CVG 2002	CC	RW	IS	123	0.5		23	
USA	Cyber Defense Systems	Cyberbug	M	FW	IS	(32)	0.75	10	1.2 - 2.7	2.27
USA	Cyber Defense Systems	Cyberscout	DP	FW	IS	185	1	180	31.5	
USA	Cyber Defense Systems	M.A.R.S. HAA	DV	LtA	DC	70	10 days			675
USA	Cyber Defense Systems	M.A.R.S. MAA	DV	LtA	DC	(57)	48			450
USA	Cyber Defense Systems	SA 60 LAA	DV	LtA	DC	(57)	5			
USA	Cyber Defense Systems	SA 90 LAA	DV	LtA	DC	55	48			
USA	Dara Aviation	D-1 Heavy Payload	DP	FW	DC	126	2	200	35	13
USA	Dara Aviation	D-1 Long Mission	DP	FW	DC	146	15	1500	35	4
USA	Dara Aviation	D-1 Short Mission	DP	FW	DC	120	1.5	100	28.5	4
USA	DARPA	NAV	M	FW	DC	18	0.3		8	2
USA	DARPA	Rapid Eye	M	FW	POC,DC					
USA	DARPA	RUGS	M, DV	FW	DC					
USA	DARPA	Vulture	DV	FW	POC,DC		months			
USA	Defense Research Associates	DevilRay	M, DV	FW	DC				3	
USA	DPR Group	SkyForce	DV	FW	DC		20-30		1451	150-294
USA	Dragonfly Pictures	DP-4X Mule	CC, DV	RW	MR	170	5	56	95	13.6
USA	Dragonfly Pictures	DP-4XT	CC, DV	RW	MR		9.4			16
USA	Dragonfly Pictures	DP-5	CC, DV	RW	MR		2.5	170		16
USA	Dragonfly Pictures	DP-5T	CC, DV	RW	MR		1	80		68

(Continued)

Country	Producer(s)/Developer(s)	System Designation	Class	Airframe	Status	Max. Speed (km/h)	Endurance (hrs)	Endurance (km)	MTOW (kg)	Payload Capacity (kg)
USA	Dragonfly Pictures	DP-5X Wasp	CC, DV	RW	DS	219	6	75	322	45
USA	Dragonfly Pictures	DP-5XT Gator	CC, DV	RW	MR	300	6.5	75	718	97
USA	Dragonfly Pictures	DP-6 Whisper	M	RW	DC	122	1	9	36.3	1.1
USA	Dragonfly Pictures	DP-11 Bayonet	M	RW	DC	207	9	90	60	22
USA	Dragonfly Pictures	DP-12 Rhino	M	RW	DC	180	10	90	160	13.7
USA	DRS Unmanned Technologies	Neptune	M	FW	IS	157	4	74	36.3	9
USA	DRS Unmanned Technologies	Sentry HP	1 M	FW	MR	203	6	370	147	34
USA	DRS Unmanned Technologies	Sentry STM-5B	M	FW	IS,ND	(130)	6		148	
USA	Emmen Aerospace	Condor A	M	RW	DC	100	1	50	6.3	0.9
USA	Emmen Aerospace	SuperSwiper	M	FW	DC	115	2	10	4	1
USA	Emmen Aerospace	Swiper	M	FW	DC	75		1	675	
USA	Flight Systems	Tracker	CC	FW	POC,ND	98	1.5	16	6.75	0.9
USA	Flightstar Sportplanes	FlightSpyder II	M, DV	FW	POC,ND					
USA	Freewing Aerial Robotics	Scorpion 100-60		TB	DC	220	6.5	75	215	27
USA	Freewing Aerial Robotics	Scorpion 60-25		TB	DC	185	4	75	45	11
USA	Freewing Flight Technologies	Spirit 100-800		FW	DC	257	2.5-15		340	31-100
USA	Freewing Flight Technologies	Spirit 20-200		FW	DC	192	2.2-5.6		68.2	8.4-16.7
USA	Freewing Flight Technologies	Spirit 400-3000		FW	DC	270	5-24		1380	31-405
USA	Frontline Aerospace	V-star	M, DV	SRW	DC	736	15		1060	180
USA	General Atomics Aeronautical Systems	Altair	RV	FW	IS	400	30	9580	3266	299(i)
USA	General Atomics Aeronautical Systems	Altus	RV	FW	IS	(120)	24	5500	975	150

(Continued)

Country	Producer(s)/Developer(s)	System Designation	Class	Airframe	Status	Max. Speed (km/h)	Endurance (hrs)	Endurance (km)	MTOW (kg)	Payload Capacity (kg)
USA	General Atomics Aeronautical Systems	GNAT 750	M	FW	IS	259	40	2778	511	63.5
USA	General Atomics Aeronautical Systems	I.GNAT	M	FW	IS	230	40	2778	703	91(i)+136(e)
USA	General Atomics Aeronautical Systems	I.GNAT ER	M	FW	IS	220	40	2778	1043	204(i)+136(e)
USA	General Atomics Aeronautical Systems	Ikhana	RV	FW	IS	(400)	30			900
USA	General Atomics Aeronautical Systems	Mariner	M	FW	DC	440	49	15186	5000	522(i)+907(e)
USA	General Atomics Aeronautical Systems	MQ-1 Predator	M	FW	IS	220	40	3704	1040	204(i)+136(e)
USA	General Atomics Aeronautical Systems	Predator B - MQ-9B	M	FW	IS	(400)	32	12264	4536	363(i)+136(e)
USA	General Atomics Aeronautical Systems	Predator C Avenger	M	FW	DC					
USA	General Atomics Aeronautical Systems	Prowler II	M	FW	MR	230	18		340	45
USA	General Atomics Aeronautical Systems	Sky Warrior	M	FW	DC					
USA	Georgia Tech Research Inst.	Entomopter	DV	FLW	DC					
USA	Georgia Tech Research Institute	MarsFlyer	RV	FW	DC					
USA	Georgia Tech Research Institute	UAV	RV	FW	POC,DC					

(Continued)

Country	Producer(s)/Developer(s)	System Designation	Class	Airframe	Status	Max. Speed (km/h)	Endurance (hrs)	Endurance (km)	MTOW (kg)	Payload Capacity (kg)
USA	Griffon Aerospace	Broadsword XL	M	FW	DC	202	4-6		249	45-54
USA	Griffon Aerospace	Outlaw	M	FW	DC	177	4-6		59	6.8-13
USA	Groen Brothers Aviation	Heliplane	DV	RW	POC,DC					
USA	GT Aeronautics	Bandit	M	FW	MR	185	10		80	33.75
USA	GT Aeronautics	Bandito MAV	M	FW	MR				0.84	0.02
USA	GT Aeronautics	Tacamo	M	FW	DC	320				
USA	Guided SystemsTechnologies	Unnamed VTOL	DV	RW	POC,DC		3.3		450	94.5
USA	HEI Group	Blicopter	CC	LtA	POC,DC		10			5
USA	Honeywell	gMAV	DP	SRW	DC					
USA	Honeywell (airframe AAI Corp.)	MAV	M, DV	SRW	IS,ES	90	0.6	11	6.8	0.91
USA	Honeywell (airframe AAI Corp.)	T-Hawk	DP	SRW	IS	75	0.8	10		
USA	Honeywell & AeroVironment	Kestrel (OAV)	M	RW	DC				11.3	
USA	Insitu	GeoRanger	CC	FW	DC	130	15		19	7
USA	Insitu	Integrator	M	FW	DC	165	24		59	23
USA	Insitu	SeaScan	DP	FW	MR	130	15	100	18	(6)
USA	Insitu & Boeing	ScanEagle	DP	FW	IS	126	15	100	18	3.2
USA	IntelliTech MicroSystems	Vector P	RV	FW	IS	185	4	60	34	4.5
USA	Iron Bay	XTM	DP	FW	POC,DC					
USA	Iron Bay	Fatboy	RV	RW	POC,DC					
USA	Iron Bay	Knighthawk	RV	FW	DS					
USA	Iron Bay	Sabre	RV	FW	POC,DC					
USA	ISL Inc. Bosch Aerospace	Lears IV	DP, DV	FW	POC,DC	203	30	150	54.5	27.3

(Continued)

Country	Producer(s)/Developer(s)	System Designation	Class	Airframe	Status	Max. Speed (km/h)	Endurance (hrs)	Endurance (km)	MTOW (kg)	Payload Capacity (kg)
USA	ISL Inc. Bosch Aerospace	SASS LITE	DP	LtA	POC,DC	70	12-24	100	770	225
USA	ISL Inc. Bosch Aerospace	WASP	DP	Prf	POC,DC	73	1		10	
USA	Kaman Aerospace & Lockheed Martin	K-Max Burro	DP	RW	DC	185		150	5443	2721
USA	Kuchera Defence	Falcon	DP	RW	DS					
USA	L3 - BAI Aerosystems	Evolution	M	FW	IS	80	0.75	10	2.95	0.45
USA	L3 - BAI Aerosystems	Evolution-XTS	M	FW	DC	75	0.75	10	3.7	0.7
USA	L3 - BAI Aerosystems	Exdrone	M	FW	IS	362	2.5	90	43	9
USA	L3 - BAI Aerosystems	Isis	M	FW	DC	158	12	90	193.2	34
USA	L3 - BAI Aerosystems	Javelin	DP	FW	IS	105	2	8	6.8	1.45
USA	L3 - BAI Aerosystems	Neptune	M	FW	POC,DC					
USA	L3 - BAI Aerosystems	Porter	M	FW	MR		4		100	25
USA	L3 - BAI Aerosystems	Scimitar	M	FW	?					
USA	L3 - BAI Aerosystems	Tern	M	FW	IS	125	2	50	59	11.3
USA	L3 - BAI Aerosystems	Viking 100	M	FW	DC	120	6-8	50-75	68	9
USA	L3 - BAI Aerosystems	Viking 105	M	FW	MR	120	10-14		68	
USA	L3 - BAI Aerosystems	Viking 300	M	FW	DC	126	8-10	50-75	143	
USA	L3 - BAI Aerosystems	Viking 400	M	FW	DC	145	10-12	50-75	221	27
USA	L3 - Geneva Aerospace	Dakota	RV	FW	IS	185	4.5	315	108	36
USA	L3 - Geneva Aerospace	Mobius	M	FW	MR	400	24-30	276	1325	
USA	Lew Aerospace Inc	E-CLASS	M	FW	POC,DC	192	16	1184	13.6	9
USA	Lew Aerospace Inc	Inventus E	DP	FW	MR	157	2	166	2.7	
USA	Lew Aerospace Inc	Inventus S-1	DP	FW	DC	220	30	3700	22.7	22-34
USA	Lew Aerospace Inc	S-CLASS	M	FW	POC,DC	233	20	3535	158	67

(Continued)

Country	Producer(s)/Developer(s)	System Designation	Class	Airframe	Status	Max. Speed (km/h)	Endurance (hrs)	Endurance (km)	MTOW (kg)	Payload Capacity (kg)
USA	Lew Aerospace Inc	SSS-CLASS	M, DV	FW	POC,DC	233	30	3200	824	360
USA	Lite Machines	Voyeur	M	RW	DC	111	0.5	32	1.8	
USA	Lockheed Martin	Cormorant	M, DV	FW	POC,DC		3		4500	450
USA	Lockheed Martin	Desert Hawk	M	FW	IS	92	1	10	3.18	0.5
USA	Lockheed Martin	Desert Hawk III	M	FW	IS	80	1.5	10+	3	0.8
USA	Lockheed Martin	HALE-D	M	LtA	POC,DC	36	2 weeks			22.5
USA	Lockheed Martin	High Altitude Airship	M	LtA	POC,DC		month			
USA	Lockheed Martin	High Altitude Airship	M	LtA	DC	130	30			230
USA	Lockheed Martin	LOCAAS	M	W	DC	(370)	0.5	185	45.4	
USA	Lockheed Martin	Morphing UAV	M, DV	TW	POC,DC					
USA	Lockheed Martin	NAV	DV		DS				10	
USA	Lockheed Martin	P-175 Polecat	M, DV	FW	POC,DC		4		4000	450
USA	Lockheed Martin	Sky Spirit	M	FW	MR?	185	10	50-90	81.5	11.34
USA	Lockheed Martin	Sky Spirit ER	M	FW	DC?	185	22	50-90	80	27.2
USA	Lockheed Martin	Stalker	M	FW	MR		2		6.4	
USA	Lockheed Martin & Bell Helicopter Textron	UCAR	DV	RW	POC,ND					
USA	Lockheed Martin & Boeing	Darkstar	M, DV	FW	DS,DC	550	12		3901	
USA	Lockheed Martin & Boeing	Darkstar B	M, DV	FW	DS,DC					
USA	Lockheed Martin & Boeing & Alliant Techsystems	HTV-3	M, DV	FW	DC		M6			
USA	Lutronix & DARPA	Kolibri	M, DV	RW	POC,DC?					

(Continued)

Country	Producer(s)/Developer(s)	System Designation	Class	Airframe	Status	Max. Speed (km/h)	Endurance (hrs)	Endurance (km)	MTOW (kg)	Payload Capacity (kg)
USA	Marcus UAV	Marcus 100	CC	FW	MR	50	3		2.7	
USA	Micropropulsion	NAV	DV	FLW	DS		0.3		10	2
USA	Miraterre Flight Systems	Dragon Slayer	M, DV		POC,DC	150	0.6	5	0.3	
USA	Mission Technologies (MI-Tex)	Backpack	M	FW	DS,DC		2		4.5	
USA	Mission Technologies (MI-Tex)	Buster	M	FW	IS	120	2-4	10	5.9	1.32
USA	Mission Technologies (MI-Tex)	Hellfox	M	FW	MR	220	8		159	
USA	Mission Technologies (Mi-Tex)	Mini-Vanguard	M	FW	IS	200	4		48	
USA	Mission Technologies (Mi-Tex)	Vixen	M	FW	MR				91	
USA	MLB	Bat	DV	FW	POC?	120	4	370	11.3	1.8
USA	MLB	Micro Dot	DV	FW	DC					
USA	MLB	Trochold	DV	FW	DC	98	0.3			
USA	MTC Technologies (BAE Systems)	SpyHawk	M	FW	MR					
USA	NASA Dryden	I-2000	DV	FW	DC				6.8	
USA	Nascent Technology	AHMMH-1 XS	RV	RW	MR		3		8	1
USA	Naval Research Lab.	Alice	M	FW	POC,DC	(120)	2			
USA	Naval Research Lab.	AME	RV	FW	POC,DC					
USA	Naval Research Lab.	BITE-Wing	M, DV	FLW	POC,DC				0.0195	
USA	Naval Research Lab.	Crystal Sun	M, DV	RW	POC,DC					

(Continued)

Country	Producer(s)/Developer(s)	System Designation	Class	Airframe	Status	Max. Speed (km/h)	Endurance (hrs)	Endurance (km)	MTOW (kg)	Payload Capacity (kg)
USA	Naval Research Lab.	Dragon Eye	M	FW	POC,DC	(65)	0.5-1	5-10	2.95	0.225
USA	Naval Research Lab.	Dragon Warrior	M	RW	POC,DC	185	3-5	92	154	16
USA	Naval Research Lab.	Duster	M	FW	POC,DC		22		136	
USA	Naval Research Lab.	Eager	M, DV	RW	POC,DC				50	
USA	Naval Research Lab.	Extender	M	FW	POC,DC	(73)	2.2		14	
USA	Naval Research Lab.	Finder	M	FW	POC,DC	185	10	620	27	5.1
USA	Naval Research Lab.	Flyrt	M	FW	POC,DC		0.5		32.7	11.3
USA	Naval Research Lab.	Ghost/Dakota	RV	FW	POC,DC	185	2		81	
USA	Naval Research Lab.	Hawkeye	M, DV	FW	POC,DC				22.5	
USA	Naval Research Lab.	Ion Tiger	RV, DV	FW	POC,DC	‘	24			2.25
USA	Naval Research Lab.	Laura	M, DV	FW	POC,DC					
USA	Naval Research Lab.	MAC-1	M, DV	FW	POC,DC					
USA	Naval Research Lab.	Mares	RV	FW	POC,DC					
USA	Naval Research Lab.	Mite	M, DV	FW	POC,DC				0.0085	
USA	Naval Research Lab.	NDM-1/2/3	RV	FW	POC,DC					
USA	Naval Research Lab.	Pendopter	M, DV	FLW	POC,DC					
USA	Naval Research Lab.	Samara	M, DV	FW	POC,DC				0.3	
USA	Naval Research Lab.	Sea ALL	M, DV	FW	POC,DC	(65)	0.5-1	5-10	2.04	0.225
USA	Naval Research Lab.	Sender	M, DV	FW	POC,DC	167	2	93	4.54	
USA	Naval Research Lab.	SIERRA	M, DV	FW	POC,DC	101			141	
USA	Naval Research Lab.	Spider-Lion	DV	FW	POC,DC				2.6	
USA	Naval Research Lab.	Sr Telemaster	M, DV	FW	POC,DC				11.36	
USA	Naval Research Lab.	Swallow	M, DV	FW	POC,DC	110	2		28	4.54
USA	Naval Research Lab.	Telemaster	M, DV	FW	POC,DC				29.5	

(Continued)

Country	Producer(s)/Developer(s)	System Designation	Class	Airframe	Status	Max. Speed (km/h)	Endurance (hrs)	Endurance (km)	MTOW (kg)	Payload Capacity (kg)
USA	Naval Research Lab.	VLIIRDT	M, DV	FW	POC,DC					
USA	Naval Surface Warfare Center	Guardian Griffin	M	Prf	DC	(48)	14-16	92	230	136
USA	Navmar Applied Sciences	Tiger Shark	M	FW	IS	120	10	32.8	128	13.5
USA	Navmar Applied Sciences	Mako 2	M	FW	DC	130	10	80	65	11
USA	Navmar Applied Scn. & L3 - BAI Aerosyst.	XPV-2 Mako	M	FW	IS	140	8.5	75	130	13.61
USA	Neany (Titan Aircraft airframe)	Arrow	M	FW	MR	245	5		680	90
USA	Neural Robotic Industries	AutoCopter	CC	RW	IS		0.8	37	22.7	6.8
USA	Nextgen Aeronautics	MFX-1	DV	FW	POC					
USA	Nextgen Aeronautics	MFX-2	DV	FW	POC					
USA	Northrop Grumman (airframe Swift Engineering)	KillerBee KB-2	M	FW	DC	200	12-24	90	19.5	6.8
USA	Northrop Grumman (airframe Swift Engineering)	KillerBee KB-3	M	FW	DC	190	12-24	90	39	13.6
USA	Northrop Grumman (airframe Swift Engineering)	KillerBee KB-4	M	FW	DC	(108)	15		73	13.5
USA	Northrop Grumman (airframe Swift Engineering)	KillerBee KB-X	M	FW	MR	195	12-24	90	163	54.4
USA	Northrop Grumman	MALD	M	FW	DC	850	0.75		48	
USA	Northrop Grumman	RQ-4A Global Hawk	M	FW	IS	(630)	35	22230	12110	900
USA	Northrop Grumman	RQ-4B Global Hawk	M	FW	MR	(570)	36	22780	14630	1360
USA	Northrop Grumman	RQ-4N	M	FW	DC					
USA	Northrop Grumman	Global Hawk Block 30	M	FW	DC					

(Continued)

Country	Producer(s)/Developer(s)	System Designation	Class	Airframe	Status	Max. Speed (km/h)	Endurance (hrs)	Endurance (km)	MTOW (kg)	Payload Capacity (kg)
USA	Northrop Grumman	RQ-8A FireScout	M	RW	DC	230	6	204	1202	(91)
USA	Northrop Grumman	RQ-8B FireScout	M	RW	DC	230	8+		1429	272
USA	Northrop Grumman	Switchblade	M, DV	FW	DC					
USA	Northrop Grumman	X-47A	M, DV	FW	DS,DC			2800	2678	
USA	Northrop Grumman	X-47B	M, DV	FW	DS,DC	320	12	2407	2500	2000
USA	Northrop Grumman & MD Helicopter & CarterCopter	UCAR	M, DV	RW	POC,ND	300	10	700		
USA	Octatron	SkySeer	DP	FW	DS	37	0.75		1.5	
USA	Optimum Solutions	Condor 300	CC	FW	MR	125	4		18	
USA	Oregon Iron Works	Sea Scout	DP, DV	FW	POC,DC		4		135	11
USA	Orion Aviation	Model 706 Seabat	DP	RW	DC		4		91	
USA	Piasecki Aircraft	Air Guard	M	RW	DS					
USA	Piasecki Aircraft	Air Scout	M	SRW	DS					
USA	Piasecki Aircraft	Turais	M	FW	POC,DC	370	6	277	350	90
USA	Pioneer UAV Inc. (ceased trading)	Pioneer	M	FW	IS,ND	210	5.5	185	200	45
USA	Prioria	Maveric	M, DV	FW	MR	60	1	4	1.15	0.150
USA	Procerus Technology	Unicorne 1	DP	FW	MR	98	2		3.1	
USA	Procerus Technology	Unicorne 2	DP	FW	MR	70	1		1.8	
USA	Procerus Technology	Unicorne 3	DP	FW	MR	70	1		1.35	
USA	Proxy Aviation Systems	SkyWatcher	M	FW	DC	(320)	15		1450	290
USA	Proxy Aviation Systems	SkyRaider	M	FW	POC,DC	(320)	30		1632	225
USA	Pusher Development Corp.D	Unnamed	DP	FW	ND					
USA	Raspet Flight Research Laboratory	Owl	M, DV	FW	DC		24+			

(Continued)

Country	Producer(s)/Developer(s)	System Designation	Class	Airframe	Status	Max. Speed (km/h)	Endurance (hrs)	Endurance (km)	MTOW (kg)	Payload Capacity (kg)
USA	Raytheon Missile Systems	Cobra	RV	FW	POC,DC	150	3		45	11
USA	Raytheon Missile Systems	MALD	M	FW	DC	800	0.3-0.6		113	
USA	Raytheon Missile Systems	MALD-J	M	FW	DC					
USA	Raytheon Missile Systems	SilentEyes	M	FW	DC	150-185			4.5	
USA	Rotomotion	SR 20	CC	RW	IS	50	0.2-0.4	0.8		7.5
USA	Rotomotion	SR 30	CC	RW	MR	45	1.5			4.5
USA	Rotomotion	SR 100	CC	RW	IS	50	0.45	0.8		8
USA	Rotomotion	SR 200	CC	RW	MR	80	4	0.8		22.7
USA	RP Flight Systems	Shaker AP	CC, RV	FW	DC					
USA	RP Flight Systems	Slipstream II	CC, RV	FW	DC					
USA	RP Flight Systems	Slipstream III	CC, RV	FW	DC					
USA	SAIC	LEWK	M, DV	FW	DC	278	8		363	91
USA	SAIC	Skybus 30K	M	LtA	IS	65	12			150
USA	SAIC	Skybus 80K	M	LtA	MR		24			227
USA	SAIC	Vigilante 496	DP	RW	DC	139	5	278	500	136
USA	SAIC	Vigilante 502	M	RW	DC	217	9	416	500	68
USA	Sanswire-TAO	SAS-51	CC, RV	LtA	DC					
USA	Sanswire-TAO	Skysat	DP	LtA	DC					
USA	Sanswire-TAO	Stratellite	DP	LtA	DC	140	72		180	900
USA	Scaled Composites	Proteus	RV	FW	DC	500	14		5670	900
USA	Sikorsky Aircraft	Cypher II = Dragon Warrior	M	RW	DC	130	2.5		115	
USA	Sikorsky Aircraft	X-2 UAV	M	RW	POC,DC	460	5	200		

(Continued)

Country	Producer(s)/Developer(s)	System Designation	Class	Airframe	Status	Max. Speed (km/h)	Endurance (hrs)	Endurance (km)	MTOW (kg)	Payload Capacity (kg)
USA	Sikorsky Aircraft & Raytheon	UCAR	M, DV	RW	POC,ND					
USA	Southwest Research Institute (SwRI)	HiSentinel50	M	LtA	DC	36.8				22.5 94
USA	Systems Research & Development	Archangel	M, DV	FW	POC,DC	120	30		42.6	
USA	Systems Research & Development	Super Archangel	M, DV	FW	POC,DC	110	16		64	
USA	Systems Research & Development	Wraith	M, DV	FW	DC					
USA	Tactronix-Tactical Airspace Group (TAG)	TAG-M2600	DP, DV	RW	DC	200	2		654	250
USA	Theiss Aviation	P.L.A.N.C	M	FW	DC	(40)			2.4	
USA	Theiss Aviation	Super Ferret	M	FW	DC	(54)			3.6	
USA	Thorpe Seeop Corp.	P10	DP	FW	MR					22.68
USA	Thorpe Seeop Corp.	P10A	DP	FW	MR					22.68
USA	Thorpe Seeop Corp.	P10B	DP	FW	MR					22.68-45
USA	Thorpe Seeop Corp.	P40	DP	FW	MR					22.68
USA	Thorpe Seeop Corp.	P4000	DV	FW	DC					
USA	Thorpe Seeop Corp.	P7108	DP	FW	MR					4.5
USA	Thorpe Seeop Corp.	RMI Spinwing	DV	RW	DC					
USA	Thorpe Seeop Corp.	TS1000	DP	FW	MR					31.75
USA	Thorpe Seeop Corp.	TS2000	DP	FW	MR				550	181
USA	Trek Aerospace	DragonFly	DP	TR	DC	380	3		485	204
USA	Trek Aerospace	OVIWUN	DV	TR	DC	75	0.3	1.6	5.4	0.5

(Continued)

Country	Producer(s)/Developer(s)	System Designation	Class	Airframe	Status	Max. Speed (km/h)	Endurance (hrs)	Endurance (km)	MTOW (kg)	Payload Capacity (kg)
USA	UAV Solutions	Talon 45	DV	FW	DC		0.75		1.1	
USA	UAV Solutions	Talon 180	DV	FW	DC		18		16.2	
USA	University of Florida	SUAV	RV	FW	IS	(80)	2	5		
USA	University of Kansas	Meridian	RV	FW	DC			13		489
USA	University of Michigan	Bat	DV	FLW	DC				0.113	
USA	USAF Research Lab.	SensorCraft	M, DV	FW	DC					
USA	US Army & SAIC (Caravan airframe)	COPA	M	FW	DC					
USA	Veratech	Phantom Sentinel	DV	RW	POC,DC				1.8	
USA	Veratech	X-Pro	DV	RW	POC,DC					
USA	Victory Systems	Mini-UAV	DP	FW	DC	55	1	9		0.45
USA	Victory Systems	VTOL UAV	M	RW	DC		6			450
USA	Vought Aircraft	Kingfisher II	M	FW	DC	565			4308	1133

Index

K. Dalamagkidis et al., *On Integrating Unmanned Aircraft Systems into the National Airspace System*, Intelligent Systems, Control and Automation: Science and Engineering 54, DOI 10.1007/978-94-007-2479-2,

CPSIA information can be obtained
at www.ICGtesting.com
Printed in the USA
LVHW082232141218
600501LV00001B/9/P

9 789400 724785